# The Green Bottom Line

**Environmental Accounting for Management**
Current Practice and Future Trends

Contributing Editors
Martin Bennett and Peter James

# The Green Bottom Line

ENVIRONMENTAL ACCOUNTING FOR MANAGEMENT
Current Practice and Future Trends

Contributing Editors
MARTIN BENNETT AND PETER JAMES

Greenleaf **Publishing** 1998

© 1998 Greenleaf Publishing unless otherwise stated.

Published by *Greenleaf* Publishing

*Greenleaf* Publishing is an imprint of
Interleaf Productions Limited
Broom Hall
Sheffield S10 2DR
England

Typeset by Interleaf Productions Limited and printed on environmentally
friendly, acid-free paper from managed forests by The Cromwell Press,
Trowbridge, Wiltshire.

British Library Cataloguing in Publication Data:

The green bottom line : environmental accounting for management
   : current practice and future trends
   1. Environmental management   2. Environmental auditing
   I. Bennett, Martin   II. James, Peter
   658.4'08

ISBN 1874719071

# Contents

# Acknowledgements

THE IMMEDIATE origins of this book lie in a report on the 'state of the art' in environmental accounting which we prepared for British Telecom during 1995 and 1996. We are very grateful to Chris Tuppen both for commissioning that report and for helping us refine our ideas through his incisive comments and practical experience. We are also grateful to our colleague on that project, Geoff Lane of Coopers & Lybrand (now PriceWaterhouseCoopers), for his valuable inputs.

The British Telecom report was undertaken partly under the auspices of Ashridge Management College, where Peter James then worked full-time and Martin Bennett was an associate. We are therefore most appreciative of the assistance and inputs of staff from the College and its research group, particularly the Director of Research, Dr Laurence Handy, and Margaret Dawson, Terrylynn Knott and Sarah Pratt.

We are also grateful for the support and advice of colleagues at our current home of the University of Wolverhampton Business School, in particular to Professor Les Worrall, Associate Dean of Research and Consultancy, and Mark Price, Divisional Manager of the Business School's Financial, Information and Operations Management Division; and to Chris Thompson and Hilary Price for all their administrative support.

During the three years that the book has been considered, assembled and edited, we have benefited from the insights of many individuals, especially Roger Adams of the Association of Chartered Certified Accountants, Bill Blackburn of Baxter International, Peter Hopkinson and Andy Hughes of the University of Bradford, Jonathan Selwyn of the UK Centre for Economic and Environmental Development, and many of the contributors to this volume. We have also benefited from the support and insights of our colleagues on the European Eco-Management Accounting project: Matteo Bartolomeo, Jaap Bouma, Peter Heydkamp, Wim Hafkamp and Teun Wolters. We are grateful for their permission to include two of the case studies that we generated for that project, on Xerox and Zeneca respectively.

The book would also have been impossible without the editorial support and patience of John Stuart, Sue Pearson and Dean Bargh at Greenleaf. Last but not least, we thank our wives, Rosalie and Sue, and our children, Lorna, Geoff, Helen and Nicholas. Without their co-operation and support, and their readiness to take a liberal approach to accounting for domestic time, this book could not have been produced.

# Forewords

MONEY is the language of business. As the World Business Council for Sustainable Development has always recognised—and has recently argued in Greenleaf's *The Sustainable Business Challenge*—translating environmental, and indeed social issues, into financial terms is therefore a vital element in motivating business to take action.

As this book shows, there is much that companies can do right now to reduce some of the environment-related costs they are currently incurring. They can also gain new benefits—for example, by systematically considering opportunities for new products created by environmental regulation and changing public attitudes. And, as a major WBCSD co-ordinated scenario study has shown, the scale of these costs and benefits will greatly increase in the longer run. One reason for this will be the more precise costing of external environmental and social impacts and their internalisation to company accounts through taxes and other measures.

Even where we cannot precisely quantify the financial benefits of environmental action we can at least recognise the risks of inaction. We know in the insurance sector that these risks can be substantial. We know too that the best protection against their occurrence comes from alert and proactive managers who understand the environmental risks that their organisation faces and take the precautionary measures that can prevent them.

Environment-related management accounting is therefore an essential tool for tomorrow's business. As managers of one of the world's largest green funds, the Storebrand Scudder Environmental Value Fund, we will certainly expect that the companies we invest in have understood its precepts and are implementing some of the practical accounting measures that are outlined in the book.

Jan-Olaf Willums
*Senior Vice President*
*Storebrand, Norway*

BAXTER has always aspired to be one of the leaders in corporate environmental management. One of the most important tools in achieving this is our annual environmental financial statement, which calculates the costs and benefits of our environmental actions. It shows that our environmentally beneficial activities bring positive financial results—and provide net benefits year after year. These figures have persuaded Baxter managers and stakeholders that our green bottom line is highly positive and that, simply stated, good environmental programmes make good business sense. The result is strong support for our environmental initiatives.

Readers wanting more information on the development of our statement can find it in the case study within this volume. The case is the outcome of many years of collaboration between Baxter and the authors, who have not only studied our actions but also helped develop our reports and procedures through their feedback.

Baxter is an international company that takes a global perspective on environmental issues. One of our internal priorities is to highlight the green bottom line in all of our facilities around the world. Externally, we are members of the CERES initiative to achieve uniform reporting standards in all countries. Hence, I welcome a book like this which collects together the best work from Europe and North America and provides an international perspective on the current position and future opportunities of environment-related management accounting.

The following chapters demonstrate that many companies can benefit by getting a better understanding of the financial costs and benefits of environmental action. It also provides information and insight on what they can do in practice. I wish it had been around when we started our own environmental accounting journey in the early 1990s!

William Blackburn

*Vice President, Environment, Health and Safety,*
*Baxter International*

SINCE 1992 we have published an environmental performance report giving details of BT's environmental impacts and targets for future improvement.

While we have won a number of awards for our reports, we have always been conscious that our readership would like more financial information. The concept is good but we could find no one who was applying such a concept. 'Simple,' people would say, 'just list your environmental costs and benefits.' 'But what', we would ask, 'do you exactly understand by environmental costs and benefits? After all, they are not covered by any accounting standards.'

To help answer this question, BT brought together in 1995 the editors of the present volume together with the company's financial auditors, Coopers & Lybrand, to prepare a report on the state of the art in environmental accounting in industry. I am very pleased that the work we initiated then has led to the even more in-depth study presented here.

In BT we are very keen to progress towards full sustainable development accounting and reporting—the so-called 'triple bottom line' incorporating financial, social and environmental considerations. This book covers one dimension of this complex and inter-related triad and will, I am sure, make a significant contribution to our better understanding of environmental accounting.

Ian Ash
*Director, BT Corporate Communications*

FOR THE LAST SIX YEARS, the US Environmental Protection Agency has worked collaboratively with industry, academia and other sectors to advance eco-efficiency in business practices by improving managerial accounting for environmental costs and benefits. Through research, case studies, training and other activities we have worked together to move these concepts into the classroom and boardroom. Today, industry and government alike are realising the benefits of such techniques as environmental life-cycle costing, materials tracking and environmental activity-based costing. Who could not be intrigued by a set of concepts that promises opportunities for meeting both productivity and environmental performance objectives?

We are pleased that the editors chose to include five EPA-sponsored documents in this wide-ranging, international compendium of environmental accounting resources. This text creates a convenient reference for those who are new to environmental accounting issues, as well as for those companies who wish to benchmark their environmental accounting efforts against companies on both sides of the Atlantic. There has been a recent explosion of interest in environmental accounting worldwide, as evidenced by the growth in international membership in EPA's Environmental Accounting Network and by daily informational requests from such countries as India, Iraq, Vietnam, Australia, Chile, Indonesia and Japan. The book promises the start of a more concerted international effort to raise awareness of the benefits of, and possibilities for, advancing environmental accounting.

Our next challenge will be to facilitate the integration of these techniques into companies' mainstream business processes. We have seen environmental management systems begin to integrate environmental issues throughout a company's operations. We have watched as corporations decentralise environmental, health and safety (EH&S) departments and spread that staff across their facilities. Similarly, environmental accounting should be applied by personnel throughout a firm. A heightened understanding and awareness of a firm's environmental costs and benefits informs decision-making. It is not the purview of EH&S to make more informed decisions—it is everyone's. Nor should we expect EH&S to assume full responsibility for protecting the community and environment. That responsibility belongs to everyone who makes decisions that have an impact on the environment, who designs products and processes, and who selects and handles materials. Indeed, very often those who are in the best position to identify 'greener' options—e.g. engineers, chemists and procurement professionals—possess the will to do so, but lack the tools to ensure that their alternatives are financially justifiable. Readers of this book are asked to consider how the concepts described herein can be incorporated into the assessment methodologies, information systems and other decision-making tools used by all the functional areas of a company. By better leveraging materials and cost data, all parties have a chance to do well while doing good.

Susan McLaughlin
*Program Manager, Environmental Accounting Project,*
*US Environmental Protection Agency*

DUE TO increasing regulatory pressure from governments, and public demand for a better environment, industry has developed tools to monitor and manage the environmental impact of its operations successfully.

Tools such as environmental impact assessment, environmental audits, life-cycle assessment and environmental accounting allow companies to identify problems as well as areas for improvement. Others, such as ecodesign, help identify solutions. Environmental reporting enables companies to monitor progress and to communicate with the public.

It is clear that industry's adoption of more systematic approaches to environmental management has led to the improvement of environmental performance and to Cleaner Production, often linked with economic benefits related to the reduced cost of pollution treatment, savings on raw materials, and a good public image.

Thousands of firms are already benefiting from such an approach, and many examples have been documented, as reported in particular in the UNEP International Cleaner Production Information Clearinghouse (ICPIC). More broadly, better environmental management is being seen as a key source of competitive advantage for industrial companies. A whole range of benefits can accrue from improved environmental performance, ranging from reduced effluent charges to better community relations. As environmental protection plays a more central role in companies' operations, it is also being recognised that better environmental management and better management are the same thing: a company that manages its impacts on the environment well is a well-managed company.

This volume is therefore of considerable importance as it provides concrete evidence from the USA, UK, Germany and other countries that there can be financial benefit from environmental action. Furthermore, these examples demonstrate that, as the costs of not acting are rising, proactive environmental management is the key strategic issue for business. They demonstrate that treating pollution is an expense; preventing it is an investment.

The suggestions and practical guidelines offered in this volume should assist companies in initiating or further developing their own 'green bottom line'.

Jacqueline Aloisi de Larderel
*Director, UNEP IE*

# Introduction

Martin Bennett and Peter James

THERE IS AN increasing interest in identifying and analysing the financial costs and benefits of environmental actions by business (Cairncross 1991, 1995; DeSimone and Popoff 1997; Gitjenbeek, Piet and White 1995). For example:

- Many managers and others see it as an increasingly important bottom line issue that merits greater attention (DeSimone and Popoff 1997; Fussler with James 1996). The reasons for this include identifying opportunities to reduce or gain better value from environmental expenditures, and/or to provide means of overcoming internal barriers to environmental improvement and to highlight business opportunities for eco-efficient products and processes.

- Environmentalists and green consultancies regard it as one of the most effective means of persuading companies to give greater salience to environmental issues (Elkington 1997).

- Policy-makers likewise view it as the main driver of environmental improvement and are consequently changing the emphasis of environmental public policy from the traditional command-and-control philosophy to one based on incentives such as eco-taxes (EEA 1996; OECD 1994, 1997; President's Council 1996).

- Investors see it as a sign that environmental issues are both well managed and making a contribution to their long-term returns (Bennett and James 1998a; Schmidheiny and Zorraquin 1996).

These developments require action by companies and their staff, particularly those located within the functions of environmental management, accounting and financial management. Indeed, many authors and publications have called for and helped to develop a new quasi-discipline of 'environmental accounting' in response to these pressures. Unfortunately, there has been little consistency in the way that the words have been used, with consequent confusion as to what is meant by them and their practical implications for business.

The aims of this book are to:

- Dispel some of this confusion by distinguishing between different forms of environmental accounting
- In particular, seek to define and elaborate a sub-category of environment-related management accounting that is primarily focused on providing management with financial or neo-financial (e.g. productivity) data to support internal decision-making (see Chapter 1 for a detailed definition and discussion)
- Collect together existing international contributions to the topic of environment-related management accounting (for example, by the US Environment Protection Agency or the European Union's Eco-Management Accounting [ECOMAC] project)[1] and supplement these with additional contributions written especially for this volume or its precursor, the Spring 1997 special issue of the journal *Greener Management International*
- Examine future trends within the field
- Provide applied insight and guidance for practitioners

In order to provide focus to the book, and to collect together as many significant contributions as possible within the topic area, four areas of potential relevance have been excluded. One of these areas is the communication of 'green bottom line' data about the financial costs and benefits of environmental action to external stakeholders. This is important but already well covered in other publications (for example, SustainAbility/UNEP 1996). We also feel that, until some of the internal accounting issues surrounding the generation and analysis of environment-related financial data are better characterised and resolved, there is a danger of reporting erroneous or misleading information (see Bennett and James 1998a for similar arguments with regard to environment-related performance data generally).

As we discuss in Chapter 1, we believe that one important objective of environment-related management accounting is to help to achieve sustainable development.

---

1. The ECOMAC (eco-management accounting) project is a European-wide empirical research study being conducted in Germany (IBM Deutschland), Italy (Fondazione ENI Enrico Mattei), Netherlands (EIM, Erasmus University and TNO) and the UK (UK Centre for Environment and Economic Development and University of Wolverhampton). It is largely financed by the European Union. The study has involved interviews with 84 companies and detailed case studies of 15 companies in the four countries. Two of the UK case studies—Xerox and Zeneca—have been reproduced in this volume. Further details of the project can be obtained from Bartolomeo, Bennett and James (1998), Bouma and Wolters (1998) or from the project co-ordinator, Teun Wolters of EIM at fax +31 79 341 5024 or e-mail two@eim.nl. We are most grateful to all these individuals and other project team members for the friendship and stimulating discussion they have provided during the gestation of this book and the prior special issue.

However, a second area that this volume does not address is that of the social dimensions of sustainability. While this topic is important, we feel that there is more than enough to cover with regard to the environmental facets of sustainability. It is also unclear whether this will require the same conceptual approach or means of implementation within business as social sustainability. Finally, there is already a developing literature on the subject (Elkington 1997; Zadek, Pruzan and Evans 1997).

The third associated topic that we do not cover in this volume is that of the relationship between green bottom line activities or outcomes, and the financial or stock market performance of an enterprise. The reasons for this are similar to those for reporting, namely a combination of preferring to focus in detail on the means of generating financial and non-financial data within companies, a belief that the topic is well covered elsewhere (see Chapter 1 for references) and a feeling that discussion of the issue will be improved by a more sophisticated understanding of green bottom line concepts and techniques.

Finally, we have also excluded any detailed papers on the generation, analysis and use of non-financial information. Although, as we discuss in Chapter 1, this is an important element of environment-related management accounting, it is a topic that is covered elsewhere (for example, Bennett and James 1998a; Bartolomeo, Bennett and James 1998). In contrast perhaps to some other areas of management accounting, we feel the more urgent need is the development of financial data and techniques.

The ultimate justification for these exclusions, we believe, is the range and quality of the material of the papers that are included. These are divided into four sections:

- **Part 1** deals with the general concepts of environment-related management accounting.
- **Part 2** contains empirical research on the topic.
- **Part 3** provides case studies from individual companies.
- **Part 4** provides practical suggestions on how organisations can implement environment-related management accounting.

The following paragraphs provide short summaries of all the papers contained within parts 1–3, while a final section identifies some common themes.

## ▮ The Context of Environment-Related Management Accounting

Chapter 1, by Bennett and James, first defines what environment-related management accounting is and considers why it should be undertaken. The reasons include: demonstrating impact on the income statement (profit and loss account) and/or balance sheet of environment-related activities; identifying cost reduction and other improvement opportunities; prioritising environmental actions; guiding product pricing, mix and development decisions; enhancing customer value; future-proofing investment and other decisions with long-term consequences; and helping to develop sustainable business. Achieving many of these benefits requires a modification of existing activities rather than establishing completely new ones, but it is important to create new cross-functional processes and to establish an environment-related management accounting 'champion'. The chapter then describes relevant forms of financial and non-financial data, and considers the main items in the financial report to assess their relevance for environmental management. It reviews the main current management accounting techniques, and considers how these can be adapted to reflect environmental factors and support environmental management. It goes on to propose environmental value analysis as a means of evaluating the relationship between an organisation's economic value added and its environmental impacts, and thereby helping to achieve sustainable business. It also develops a conceptual model to position these areas.

The chapter concludes that there are inevitably problems in measuring some of the broader (and probably most significant) effects, but that, as the information available is expanded and improved, and as environment becomes a significant cost issue for business, there is an increasing potential for environment-related management accounting to make a substantial contribution both to business success and to sustainable development, although its development is likely to be discontinuous in nature. An appendix also contains a detailed discussion of the precise nature of environment-related costs and benefits in business.

Chapter 2, which reproduces a US Environmental Protection Agency report, also discusses and defines the key concepts and terms within the field of environmental accounting generally and environment-related management accounting in particular. At its heart is the well-known 'total cost assessment' (TCA) classification of environmental costs into: conventional; potentially hidden; contingent; and image and relationship costs. Conventional costs include costs of raw materials and energy, which are usually addressed in costing and investment appraisal and which have environmental relevance. Potentially hidden costs are environment-related costs that are captured by accounting systems but are then lost in overheads or other ways. Contingent costs are those that may be incurred at some point in the future. Image and relationship costs are intangible, but may be an important determinant of future cash flows and profitability. The paper also discusses the differences between internal and external costs, and notes the possibility that the latter may be progressively internalised through taxation and other means. The final sections discuss the ways in which environment-related management accounting can be applied within companies to the areas of cost allocation, capital budgeting and process and product design.

Schaltegger and Müller continue this discussion of the practice of environment-related management accounting in Chapter 3. Through detailed examples, they demonstrate how environmental costs can be lost in overheads and/or be wrongly allocated to products and processes. They suggest an alternative approach, based on activity-based costing (ABC). They also discuss four possible bases for allocating environmental costs in production: volume of emissions or wastes treated; toxicity of emissions and waste treated; total environmental impact of emissions and waste treated; and relative costs of treating different kinds of emissions. A final section notes the difficulties of properly considering environmental issues within conventional investment appraisal techniques. It particularly emphasises the way in which investments can create—or close down—future options and the consequent need to introduce some form of option pricing into calculations. The authors conclude that there is much that companies can do to deal with these issues, but that any changes will face considerable internal opposition.

In Chapter 4, Epstein and Roy examine the capital investment decision processes conventionally used by business, and conclude that there is scope to improve these by the integration of environmental considerations. Their paper relates environmental considerations to five competitive priorities for business, and identifies the likely barriers to change in many companies. One of these is that environmental projects are still usually perceived as compliance-driven and therefore go through an evaluation process different from other, non-environmental, proposed investments. Although this will enhance the chances of approval of purely compliance-driven projects (since they do not have to meet the same financial hurdles as conventional projects), it makes it more difficult to evaluate the benefits of discretionary, non-compliance projects that offer environmental benefits. The paper proposes a single common framework for appraising all types of project, with three 'screens' through which all should be required to pass before approval: technical, environmental and financial. The environmental screen would include life-cycle analyses and monetising external environmental costs. For the financial screen, the paper proposes innovative techniques such as option assessment, option screening, scenario forecasting, decision trees and Monte Carlo simulations. This will require a cross-functional approach involving staff from operations, engineering, accounting, design, legal, marketing and environmental management.

Chapter 5 summarises a report of the US Environmental Protection Agency on the topic of environmental liabilities. Its survey of research finds that most work has focused on providing information for financial reporting, and therefore does not adequately meet the needs of internal decision-makers and planners for information on actual or potential liabilities. Even though these are hard to quantify, it argues that it is preferable to use even uncertain estimates rather than to ignore them completely. This is particularly true when investment and other decisions have significant environmental consequences. The paper discusses a number of techniques for valuing liabilities and discusses them in terms of the main liability categories of: compliance liabilities; remediation costs; compensation liabilities; fines and penalties; punitive damages; and natural resource damages.

The issue of liabilities is an important one in Italy, according to Bartolomeo's survey of environment-related management accounting there in Chapter 6. He offers a detailed discussion of Agip's approach to calculating provisions for the decommissioning of oil-fields and Italia Petroli's calculation of the risks and consequences of fuel spillage from petrol stations. He also discusses the *bilancio ambientale* (environmental balance sheet) technique developed by the Fondazione ENI Enrico Mattei and applied in a number of Italian companies. This integrates data on material and energy flows and emissions with cost data for treatment and other environment-related costs. He concludes that, compared to North America or northern Europe, Italian initiatives have been driven more by the need to comply with legislation and provide data to external stakeholders than by cost minimisation or the generation of environmental benefits.

Bouma also provides an overview of national activities, in this case in the Netherlands, in Chapter 7. That country has ambitious environmental goals and is using innovative policy means such as eco-taxes and binding covenants with industry associations to achieve them. Both these developments require companies to develop greater knowledge of their current and potential environmental costs. Bouma draws attention to the substantial Dutch literature on environmental accounting which has developed in response but has not yet been translated into English. He then describes three specific areas of research that are of interest to an international audience. One is work for the Central Bureau of Statistics on definitions of environmental costs, which demonstrates the discrepancies in the field and concludes that the management accounting axiom of 'different costs for different purposes' applies to the environmental field as much as any other. A second area is research undertaken for Dutch Railways on the internal and external environmental costs of rail transport and how these can be allocated between individual operating units. The third is work for the Dutch Ministry of Environment on identifying and quantifying the externalities created by building construction, operation and disposal of buildings, and use of this information in decisions about new construction or refurbishment. His conclusion is that there are major differences in the relevance and practice of environment-related management accounting between nations and regions.

In Chapter 8, Burritt develops further the points made by Bouma about the fluidity of environmental costs and benefits. The central theme of all the previously (and subsequently) discussed papers—and indeed almost all the work in the field of environment-related management accounting—is that of providing decision-makers and others with a better representation of reality through more accurate knowledge of environmental costs. However, Burritt provides a more iconoclastic view. He challenges the traditional rationalist assumption that the principal purpose of management accounting is to provide accurate and relevant decision-support information. He argues that no information is objective, since it is always more or less subtly shaped by the context and purpose of its creation. He therefore believes that the purpose of management accounting information, including that used in an environmental context, should not be accuracy for its own sake but rather to influence the behaviour of managers and others. This has been the philosophy of Japanese management accounting, which has (for example) tended to inflate apparent staff costs to encourage increased labour productivity. An environmental equivalent would be creation of internal taxes to penalise poor environmental practices. Such taxes could be set as much or more on the basis of sustainable development needs than of actual costs.

## ∎ Empirical Studies

Chapter 9, by Ditz, Ranganathan and Banks, summarises what is probably the most cited work in the field, the *Green Ledgers* project of the US World Resources Institute (Ditz, Ranganathan and Banks 1995). This involved detailed pilot projects on identifying environmental costs in five large companies: Amoco Oil, Ciba-Geigy, Dow Chemical, Du Pont and S.C. Johnson Wax. The most striking outcome was the finding that these comprised almost 22% of non-feedstock operating costs at Amoco's Yorktown refinery, compared to initial estimates of 3%. Almost as striking was the discovery that they accounted for over 19% of the manufacturing costs of a Du Pont agricultural pesticide and a similar percentage of non-raw-material costs for a Ciba-Geigy product. The chapter also discusses the project's examination of pollution prevention initiatives by small-to-medium sized initiatives in Washington state, and found similar evidence that environmental costs can often be substantial. The authors conclude with five recommendations for business. One is to inform decision-makers of the environmental costs they create. The second is to increase the accountability of managers for environmental costs and benefits. The third is to develop indicators that anticipate future costs and other measures of performance. The fourth is to create incentives to address the causes of current and future costs. The last is to incorporate environment-related management accounting into ongoing business processes.

Chapter 10, by Shields, Beloff and Heller, reports on a benchmarking study at the US and Mexican operations of five companies: Celanese Mexicana, Ciba-Geigy, Grupo Primex and two anonymous companies, designated International Refineries and Specialty Refiners. The study took place in 1995–96 and was financed and first published by the US Environmental Protection Agency's Environmental Accounting Project. It found a wide variety of approaches, although none of the companies was using ABC or tracking intangible or less tangible costs. All were bringing environment into investment appraisal processes but were prioritising the environment-related projects that were required for compliance, possibly creating a bias towards acceptance of end-of-pipe projects at the expense of those focused on pollution prevention. Environmental operating and maintenance costs were being tracked in most cases with the information used in the same way as non-environmental costs data, i.e. to identify problem areas in cost performance or to demonstrate the cost impact of process improvement. In general, the information generated was less useful than non-financial information for day-to-day management but more effective in justifying the need for intervention to senior management.

The study's overall conclusion was that the success of initiatives was dependent on: favourable attitudes among top management; the presence of an internal champion; and adequate funding and follow-through that includes integration into everyday decision-making. It also found that companies already placing strong emphasis on integrating environment into all business considered it less important to differentiate between environmental and non-environmental costs—emphasising rather the allocation of all costs to products and processes—than those that were not.

Chapter 11 by Haveman and Foecke presents the findings of a US Environmental Protection Agency study of environmental accounting at 24 American electroplating facilities. It found that most 'conventional' environmental costs, such as wastewater treatment and hazardous waste disposal, were recognised and captured in new project or process evaluations. However, there were substantial hidden costs associated with (1) the loss of plating chemicals and process solutions, which became fully apparent only when materials balances were prepared; and (2) the labour costs of environmental management. Electroplaters were more willing to factor the former into economic evaluations than the latter, on the grounds that reducing chemical and solution losses could have a direct effect on cash flows whereas the latter might not. One of the most valuable uses of environment-related management accounting in electroplating was found to be the generation of greater understanding of and interest in facility processes, whose cost importance was generally undervalued. However, gathering and tracking information on environmental costs was often expensive compared to potential benefits and may often be practical only when there are other, non-environmental, uses for the information. Moreover, the study found that the largest elements in facility environmental overheads were not necessarily the best focus for action, since tackling others could produce additional, non-environmental, benefits.

Earl, Moilanen and Clift also deal with the issue of investment appraisal in Chapter 12. They discuss the difficulties of incorporating uncertainties about the future into conventional appraisal techniques such as net present value (NPV) or payback. They then describe a tool for addressing this uncertainty, particularly when it is related to the actions of external stakeholders such as environmental groups. This tool is termed 'stakeholder value analysis' and has two main components. The first is a means of defining, valuing and processing the views of different stakeholder groups on projects proposed by companies. The second is a financial appraisal model, similar to the TCA approach, known as the Paras model (after the consultancy that developed it). The authors believe that this is superior to TCA because it explicitly takes risk into account—particularly the risks of 'stakeholder derailment' revealed and quantified from the stakeholder value analysis—and produces risk-weighted 'expected monetary values' for a project. They provide a detailed case study of the use of the technique at British American Tobacco and the benefits that resulted.

In Chapter 13, Bierma, Waterstraat and Ostrosky address the issue of life-cycle costing, with particular emphasis on the supply and use of chemicals. They note that there are substantial environment-related costs associated with this, e.g. wastage in processes and costs of disposal. However, these are often hidden by poor materials tracking data and inaccurate overhead allocations, and/or are not allocated to the budgets of those responsible for causing them. One means of reducing costs is to replace a conventional, 'hands-off', supplier–customer relationship with one in which the supplier provides a chemical management service. This service can be attractive to both parties if it is based on a 'shared-savings' approach, i.e. the benefits arising from reductions in environment-related and other costs of chemicals are shared between suppliers and customers. The paper examines the implementation of such schemes at General Motors, Ford, Chrysler and Navistar. One plant alone—Chrysler's Belvidere assembly facility—saved over $1 million from reducing its chemical wastage, as well as improving product quality. It concludes that an important part of the success of the schemes was changes in accounting systems to provide better data on chemical usage and wastage within the facilities.

Chapter 14 describes and utilises a total cost assessment (TCA) model. This aims to improve the consideration of environmental factors in capital investment decisions by—in particular—expanding the 'cost inventory' of costs and benefits brought into the analysis beyond what is usual in most companies' appraisal procedures. Reiskin, Savage and Miller of the Tellus Institute report the results of a study of 22 cases where both a conventional and a TCA analysis had been carried out, and analyse in depth the evaluation in one of those companies. The effect of the TCA analyses was in most (though not all) cases to enhance the attractiveness, in conventional financial terms, of potential investments in environment-related projects such as clean technologies and process changes to eliminate the use of toxic materials. The in-depth study examines a TCA exercise at a small printing company that was considering investment in a computerised pre-press system. This would be a fundamental change in their production process, which offered both improved efficiencies and customer service in a number of ways, with effects on several costs. The environment-related benefits included reductions in the usage of darkroom chemicals, and reduced transport since previously out-sourced activities could now be brought in-house. The result was that, even though the total initial cost of the investment was found to be slightly higher than had been previously estimated, due to the identification of additional costs, its financial attractiveness was substantially enhanced through a near-tripling of net present value and rate of return, and corresponding reduction in the payback period.

# ▐ Case Studies

In Chapter 15, Bennett and James examine Baxter's well-known annual corporate 'Environmental Financial Statement', which attempts to quantify the financial costs and benefits of its environmental activities. The 1996 corporate statement concluded that, even when calculated on a relatively conservative basis, the total benefits were $105 million, over five times higher than the costs of $20 million, and that environment was therefore making a positive contribution to the company's financial well-being. The costs involved are primarily for staff and production equipment. The main benefits have been savings achieved in environment-related costs that would otherwise have been incurred: for example, on the disposal of wastes and the purchase of resources such as ozone-depleting substances and packaging. These savings were largely prompted by programmes initiated by Baxter's environmental management function and implemented by operational management. The effect has been to demonstrate that, over time, good environmental management has been not merely a 'hygiene factor' cost incurred only in order to comply with the law, but a positive benefit for the company that can be measured financially. Another benefit of the exercise is the creation of better relationships between the environmental and other functions. This is a consequence of both the cross-functional interactions involved and the enhanced credibility of environmental management that is generated by the figures.

Chapter 16 summarises a case study on Ontario Hydro developed by the US Environmental Protection Agency. This focuses on the topic of full-cost accounting. Although recent managerial changes at the utility have outdated some of the details of the case, its account of the objectives and methods of the exercise still holds considerable interest for other companies venturing down this route. The case begins by describing the drivers of the exercise and the definitions of full-cost accounting that were developed. It then describes the composition, *modus operandi* and outcomes of the Full-Cost Accounting Team which was assembled within the utility. These outcomes included the development of corporate guidelines, the introduction of full-cost accounting into planning activities, the identification of internal environmental costs and the introduction of communication and education programmes. Ontario Hydro's own learning from the exercise included: the importance of positioning full-cost accounting as something that makes business as well as environmental sense; the need for a senior executive to champion its value and use for business decisions; the need for cross-functional approaches, especially those that link environmental and financial staff; and the problems caused by confusing and conflicting terminology within environmental accounting. However, it also reported that it had found it impossible to monetise all environmental externalities.

In Chapter 17, Schroeder and Winter describe and analyse a smaller-scale initiative to reduce costs through waste minimisation at the Swiss engineering company, Sulzer Hydro. The exercise revealed that environmental costs accounted for around 5% of turnover and that it would be cost-effective to reduce them by a third. The work began with the creation of a detailed 'eco-balance', i.e. the identification of all flows of materials and energy into, through and out of the plant. This is an increasingly common approach in Germanic countries, but has hitherto been less used in the UK and USA (see Chapter 1). The authors conclude that, as with most companies, Sulzer's existing materials accounting systems do not reflect environmental considerations and therefore need to be amended if environment-related management accounting is to be successful.

In the Xerox case, which forms Chapter 18, Bennett and James examine packaging initiatives in that company's UK subsidiary. In particular, the introduction of reusable 'totes', which can accommodate a wide variety of different products, has reduced costs by several million pounds annually as well as delivering environmental benefits. The change was devised and developed by a cross-functional Quality Improvement Team (QIT). This found that much of the data needed for a cost analysis of the overall packaging chain was not available from accounting systems but had to be collected directly from operational sources. However, the initial identification of logistics, including packaging, as an area worth review was enabled due only to the systems previously set up to collect and report logistics-related costs across the chain as a whole. Accurate measurement and tracking of this meant that the scale of the challenge, and the potential benefits of improvements, could be assessed, and quantitative targets defined. One implication is that companies that have developed more sophisticated internal financial reporting systems (such as Xerox's corporate logistics system) are more likely also to have developed support systems whose data can then be used for further purposes. The totes project also raised the endemic management issue of the appropriate balance between central planning and control on the one hand—which was needed in this case to spot the chain-wide opportunities and to develop a plan to realise them—but on the other hand the need to implement changes with care and tact, in order to avoid appearing as corporate intrusion which might disempower and demotivate local managements. The latter may be best achieved if any changes can first be trialed with a limited number of co-operative operating companies before being implemented more widely, as in this project.

In Chapter 19, Bennett and James examine 'Cost of Waste' initiatives at Zeneca's Huddersfield site. Its products are high-performance chemicals produced in multi-stage processes which can generate large amounts of waste in aggregate. Process and product data was collated, and calculations made of the purchase costs of materials discarded in the process and the overall costs of disposing of wastes—with the former turning out to be the major element. The exercise revealed that Huddersfield's 'costs of waste' amounted to tens of million of pounds annually and that there was a potential to save £10 million per annum—much of which has now been achieved. One lesson from the case is that organisational factors are important. A crucial aspect of Zeneca's success was the location of the Process Technology Department at the Huddersfield site, which gave it day-to-day involvement with site processes and good personal relations with site production staff. The research also found that the success of environment-related management accounting initiatives does not necessarily rely on the involvement of accounting specialists, as that function played little direct part in the Zeneca Cost of Waste project. This was partly because the company found that much of the data it needed for the exercise was already in existence in process records and product specifications. Its main activity was in pulling these together and in understanding areas where action was likely to be fruitful—a process in which works chemists were especially important.

Finally, Greene considers the issue of costing and internalising externalities in Chapter 20. With the notable exception of Ontario Hydro (see Chapter 16), most empirical work to date in environment-related management accounting has addressed only internal costs, on the basis that this offers the opportunity for substantial improvements in practice in a relatively uncontentious area. Greene proposes that companies wishing to go beyond mere regulatory compliance in their environmental management should redraw the boundaries of their definitions of 'environmental costs' to include external as well as internal costs, in assessing the relative costs and benefits of alternative courses of action. Their motive may be proactive 'beyond-compliance' environmental management as a matter of corporate principle, or, more pragmatically, as a way of anticipating the future, in the expectation that continuing environmental legislation will mean that in time today's external costs will become tomorrow's internal costs. The chapter describes three alternative approaches to calculating external costs, and provides a detailed exposition of how each could, with hindsight, have been applied to a common situation—a US tyre manufacturer in the 1970s, considering four alternative methods to deal with its toxic wastes, with varying degrees of environmental responsibility. The analysis computes the lowest-cost options for each of a range of assumptions, and concludes that the result is highly sensitive to several variables, such as the assumed discount rate, and the speed with which anticipated environmental legislation is assumed to come into force. It finally recognises and addresses four objections to external cost accounting, and concludes that this offers a means to integrate environmental and financial management in business.

## ▌ Common Themes

A number of common themes can be found in these various papers. They include:

- Findings that environmental costs are considerable for many types of organisation (although care has to be taken that these are not unduly exaggerated), but are often not recognised

- A consequent bias against pollution prevention projects that provide long-term cost savings as opposed to either no action at all, or investment in end-of-pipe approaches

- Problems with availability and reliability of data

- A view that the easiest initiatives to take are those involving point decisions such as the appraisal of proposed capital investments

- A recognition that, in practice, much of the activity of environment-related management accounting is an organisational change process, involving the alteration of perceptions of what constitutes appropriate responses to environmental issues, and the creation of new tools to do so

- A recognition that identification of costs is not sufficient on its own, but that it needs to be accompanied by proper cost allocation and/or the use of appropriate decision techniques

- An emerging view among some observers that, in the long term, the financial returns from new products and services may equal or even exceed those arising from increased resource productivity, so that greater attention needs to be paid to accounting for this area

- Findings that the form and content of environment-related management accounting are greatly influenced by contingent variables such as the nationality, strategic objectives and organisational form of an enterprise
- Warnings about the potential dangers of environment-related management accounting legitimising unsustainable ways of doing business: for example, by suggesting that no action should be taken unless there are strong financial drivers for doing so
- A feeling that, at present, and for the foreseeable future, environment-related management accounting is not a distinctive activity but a 'virtual' one that is focused on changes in the existing activities of management and financial accounting, environmental management and general management
- A recognition that effective tracking of energy and material flows and other forms of non-financial data are an important aspect of environment-related management accounting and that, for this and other reasons, accountants in business and as a profession will be only one of several groups of players who are involved
- A perception that the costs of changing ongoing accounting systems to reflect environmental issues have to be commensurate with the benefits—which may often be insufficient to justify major change. However, major opportunities can arise when systems are being changed for other reasons, e.g. with the introduction of activity-based costing systems.

For these and other reasons, current work has generally been defined by traditional boundaries and activities and has focused on developing models or guidelines that modify traditional costing and investment appraisal processes to take better account of environmental factors. In the long run, however, new and more visionary approaches may need to be developed, such as life-cycle costing and the internalisation of external costs, if sustainable business is to be achieved.

Chapter 1 examines some of these points in greater detail and attempts to elucidate them through the development of a simple conceptual framework. Before doing this, however, it is of interest to consider how the contents and themes of the book relate to the broader 'parent' discipline of management accounting.

## ▌ The Relationship between Environment-Related Management Accounting and Generic Management Accounting

In many ways, environment-related management accounting is merely the application of many of the ideas of advanced management accounting to a particular area. For example:

- It forms one element of the 'balanced scorecard' approach advocated by Kaplan and Norton (1992, 1993, 1996a, 1996b).
- It can be seen as a specific application of ABC (see Chapter 1) which focuses on environment as a key cost driver.
- Its emphasis on end-of-life and other costs that are downstream and upstream from the organisation itself, relates to broader debates on the topic of product-life costing.

- The environmental critique, that conventional methods of investment appraisal have difficulties in dealing with uncertainty and long-term strategic benefits, links into more general discussions on these topics.
- Its emphasis on future threats and opportunities reflects the argument that management accounting in general must become more strategic and less focused on short-term controlling and reporting (Johnson and Kaplan 1987).

However, the traffic between environment-related management accounting and broader management accounting can also be two-way because:

- Environment-related management accounting provides interesting practical examples of some of the generic themes within the discipline of management accounting as a whole.
- Environment-related management accounting also has distinctive issues or examples which may spread to other areas of performance measurement in future.

Some elements of environment-related management accounting that illuminate generic challenges include:

- How far management accounting of any kind is essentially about the collection and manipulation of quantitative data, or also has a qualitative, 'process', dimension
- The trade-offs between simple measures that can easily be understood and used but may not capture a complete picture of performance, and more complex ones with the opposite characteristics
- The primary objectives of management accounting, in any particular organisation or situation: for example, how far the emphasis is on control, on developing awareness or on motivating continuous improvement

One unusual characteristic of environment-related management accounting is the need to evaluate different kinds of data—the 'apples and pears' problem. Another is the high degree of external interest in the area—sometimes resulting in a requirement to disclose certain data—and the consequent emphasis on effective communication and verification. As similar requirements for disclosure are being, or could be, adopted in other areas, such as the performance of regulated utilities, experience in the environmental area of 'management accounting in a goldfish bowl' has a much wider significance.

In the light of these points, we hope that this volume will be of interest to management accountants generally as well as to all who have a particular interest in the topic of environment-related management accounting. Indeed, its ultimate aim is to make the distinction irrelevant by creating an awareness that environment is already a 'bottom line' issue and will be even more so in future.

# Part One

## General Concepts

# 1 The Green Bottom Line

Martin Bennett and Peter James

THIS CHAPTER begins by addressing three questions:

- What is environment-related management accounting?
- Why should it be undertaken?
- Who should do it?

It then identifies relevant sources of financial and non-financial information and discusses the ways in which existing management accounting techniques can be modified to take account of environmental issues. A final section draws conclusions and is followed by an appendix on definitions of environmental costs and benefits.

## ▮ What is Environment-Related Management Accounting?

The term 'environmental accounting' has been used to cover both national and firm-level accounting activities, the processing of both financial and non-financial information, and the calculation and use of monetised external damage costs as well as those that are internal to the firm (see Chapter 2). For clarity, Figure 1 distinguishes six different domains of environmental accounting that are relevant to the firm level, based on their boundaries of attention—an individual organisation, the supply chain of which it forms part and the whole of society—and the extent to which they focus on financial and/or non-financial information. The six domains that emerge can be defined in this way (the two life-cycle definitions are based on the US Environmental Protection Agency discussion in Chapter 2):

| | Organisation | Supply chain | Society |
|---|---|---|---|
| **Financial focus** | Environment-related financial management | Life-cycle cost assessment | Environmental externalities costing |
| **Non-financial focus** | Energy and materials accounting | Life-cycle assessment | Environmental impact assessment |

**Figure 1:** Domains of Firm-Level Environmental Accounting

1. **Energy and materials accounting:** the tracking and analysis of all flows of energy and substances into, through and out of an organisation

2. **Environment-related financial management:** the generation, analysis and use of monetised information in order to improve corporate environmental and economic performance

3. **Life-cycle assessment:** a holistic approach to identifying the environmental consequences of a product or service through its entire life-cycle and identifying opportunities for achieving environmental improvements

4. **Life-cycle cost assessment:** a systematic process for evaluating the life-cycle costs of a product or service by identifying environmental consequences and assigning measures of monetary value to those consequences

5. **Environmental impact assessment:** a systematic process for identifying all the environmental consequences of the activities of an organisation, site or project

6. **Environmental externalities costing:** the generation, analysis and use of monetised estimates of environmental damage (and benefits) created by the activities of an organisation, site or project

Firm-level environmental accounting can potentially encompass all of the six domains but, in practice, is centred in the first two as the areas where accountants' experience and accounting techniques (as opposed to those of, say, environmental managers and environmental management techniques) have the most to contribute.

The literature on firm-level environmental accounting initially focused—and, to a considerable extent, still does—on external accountability to stakeholders outside the company, rather than on serving the needs of management. There are two distinct aspects to this:

• A broad concept of accountability to all of a company's stakeholders

• The traditional financial accounting focus of providing accurate and reliable information on the financial position of companies to their shareholders

In both cases, the emphasis is on collecting, verifying and reporting information to audiences outside the organisation, as opposed to the internal audience of the organisation's own management.

The broad accountability approach is founded on the premise that the responsibility of companies should not be seen—as in the traditional micro-economic theory that still largely shapes company law—as limited to maximising profits or value for the benefit of their owners (shareholders) alone. On the contrary, the activities of companies have wider impacts on society and the environment, and an enlightened company will recognise this and ensure that it maintains good relationships with all its stakeholder groups in order to preserve its implicit 'licence to operate' (RSA 1994). This was the main theme in much of the early literature on environmental accounting (Bebbington and Thompson 1996; CICA 1992; Grayson, Woolston and Tanega 1993; Müller *et al*. 1994; Gray, Bebbington and Walters 1993; Gray, Owen and Adams 1996; Owen 1992; Zadek, Pruzan and Evans 1997) and has been largely responsible for prompting many companies to publish corporate environmental reports (KPMG 1997; Lober *et al*. 1997; Owen, Gray and Adams 1997; SustainAbility/UNEP 1997). Even some authors who have seen themselves as following a management accounting approach—i.e. one that focuses on provision of information for internal decision-making—have, in practice, placed considerable emphasis on its role in generating information for external stakeholders (Birkin and Woodward 1997a–f).

There has also been a narrower concern, particularly within the accountancy profession and among financial regulators such as the US Securities and Exchange Commission (SEC), that regular financial reports by companies to their shareholders may be significantly inaccurate. It is said that these do not adequately reflect the effect on the business of environmental issues, particularly in the US where 'Superfund' liabilities can be substantial (Ethridge and Rogers 1997; Schoemaker and Schoemaker 1995). The accounting profession in Europe and internationally has also considered this and has provided guidance to its members (ASB 1997; FEE 1996; IASC 1997; ICAEW 1996), although the prevailing consensus seems to be that existing financial accounting practices, so long as they are properly applied, are adequate to deal with environmental effects on business and do not require change.

Both these bodies of work can be seen as adopting a 'financial accounting' approach, i.e. with a focus on reporting to external stakeholders. However, there is now a growing literature that adopts a genuine 'management accounting' approach that does focus on providing information to support internal decision-making (although, of course, much of this data may be of value to external stakeholders also). The starting point for this was probably the well-known '3P' (Pollution Prevention Pays) initiative introduced by 3M during the 1970s. This was expanded during the 1980s and early 1990s by further pollution prevention initiatives introduced by companies and/or government-sponsored programmes in the Netherlands, USA and other countries. These required more precise data on the costs and benefits of environmental action and therefore spawned new methodologies such as the 'total cost assessment' technique developed by the Tellus Institute for the US Environmental Protection Agency (White, Becker and Goldstein 1991; see also Chapter 14). The EPA has since sponsored a number of studies and publications on the topic—many of which are summarised in this volume—and the Tellus Institute has continued with its applied research and application. Other important US contributions have been made by Bailey and Soyka (1996), Ditz, Ranganathan and Banks (1995), Epstein (1996b, 1996c), IMA (1995) and Rubenstein (1994). In Europe, the topic has been addressed by, *inter alia*, IIIEE and VTT (1997), Schaltegger, Müller and Hindrichsen (1996), Tuppen (1996) and Wolters and Bouman (1995).

This book is positioned within this management accounting approach and contains contributions from most of the authors and organisations cited. We see this focus as being complementary—rather than an alternative—to a financial accounting approach. It addresses different needs and is also necessary in order to provide many of the data that are of interest to external stakeholders.

Our working definition of environment-related management accounting is therefore:

> **The generation, analysis and use of financial and non-financial information in order to optimise corporate environmental and economic performance and achieve sustainable business.**

The term 'environmental' precedes 'economic' in order to indicate an environmental bias. As we discuss below, the main aim at present must be to overcome the barriers to environmental action that can be created by current management accounting practices. However, there will be occasions when even modified practices reveal trade-offs between environmental and economic parameters which will result in the latter being given priority over the former. For this reason, we use the term 'environment-related management accounting' in our following discussions to signal that the activity is focused on meeting corporate as well as societal objectives.[1]

We include the term 'sustainable business' to indicate that, although much of the practical action generated by environment-related management accounting involves adaptation of existing activities, such as management accounting and environmental management, part of its objective is to support the goals of sustainable development (see below).

A final point is that environment-related management accounting relies heavily on non-financial information, particularly regarding inputs, outputs and flows of energy, materials and water (see below). Some would see the development of this information as a primary objective (for example, Birkin and Woodward, 1997a–f). However, we would argue that, at present, such information is a means rather than an end for environment-related management accounting. Its ultimate objective is to provide information to support environment-related decision-making by mainstream business managers. While this may sometimes require 'raw' physical data, we believe that the need is more often for either productivity measures (e.g. materials consumption or waste generation per unit of production) or information expressed in financial units. This is because:

- For profit-seeking firms, the ultimate objective (maximising shareholder value, or profitability) is expressible in monetary form, and information that can be expressed in the same or related (e.g. productivity) terms is always likely to attract more immediate attention.

- The financial side of management is relevant to all functions, including environmental management. Not only do environmental budgets need to be managed, but proposals for action that can be justified in terms of conventional methods of financial investment appraisal and product costing, for example, are more likely to be successful.

A supporting point is that environmental and operational managers are fully capable of developing and using such data and are often doing so in practice. Hence, there is no need to invent a new discipline or activity to accomplish this. Indeed, to do so

---

1. Note that these are the authors' opinions and terminology, and would not necessarily be accepted and used by all the other contributors to the book.

could be counter-productive because it may foster resentment and defensiveness among line staff about territorial aggrandisement by accountants.

More pragmatically, there is little evidence that the accountancy and finance functions are greatly involved in energy and materials accounting activities in most companies or have the interest and expertise to do so in the near future. The Zeneca case study in Chapter 19, for example, found that the substantial savings that followed such an exercise at the company's Huddersfield site were almost entirely driven by operational staff and had only a marginal accounting involvement.

At first sight, this argument may appear to be in conflict with advocates such as Kaplan and Norton (1992, 1993, 1996a, 1996b), Simmonds (1991) and Wilson (1997), who have argued for the development of strategic management accounting and, as part of this, greater use of non-financial data and indicators. However, we would argue that their views are less relevant to an area that usually has a relative abundance of non-financial data and a shortage of financial data. Moreover, their arguments have had—at least as yet—only limited impact on management accounting practice. While there is certainly more attention being paid to the strategic use of non-financial data and indicators through 'balanced scorecards', anecdotal evidence suggests that it is more often strategic planning, business excellence and other functions that are implementing it, rather than accountancy and finance. Research in other areas of management accounting has also found that practice can be slow to adapt (Drury *et al.* 1993) and that initiatives in new or developing areas such as non-financial performance measurement are often taken by functions other than accounting.

For all of these reasons, we would suggest that the immediate priorities for environment-related management accounting are the generation, analysis and use of financial or neo-financial (e.g. indicators of resource productivity) information, and modifying and adapting the established techniques of management accounting and financial management to take account of environmental issues.

## ▌ Why Undertake Environment-Related Management Accounting?

The primary aim of environment-related management accounting is to better inform and otherwise support decision-making processes that are influenced by environmental factors—which are primarily those of accounting and financial management, environmental management and operational management.[2] Some of the specific objectives that this creates can be summarised as:

- Demonstrating the impact on the income statement (profit and loss account) and/or balance sheet of environment-related activities

- Identifying cost reduction and other improvement opportunities

- Prioritising environmental actions

- Guiding product pricing, mix and development decisions

- Enhancing customer value

2. See Bartolomeo, Bennett and James (1998) for a more detailed discussion of objectives, based on research conducted for the ECOMAC project.

- Future-proofing investment and other decisions with long-term consequences
- Supporting sustainable business

*Income Statement and Balance Sheet Impact.* As many of the chapters in this book show, there is growing evidence that environment can have significant impacts on expenses, revenues, assets and liabilities and that these impacts are often underestimated. Making such financial impacts apparent can make it easier to take, and win support for, further environmental initiatives.

In the US, most attention has focused on the balance sheet issue of environment-related liabilities. This is a consequence of the high levels of damage claims and fines, and of specific legislation such as that requiring the clean-up of contaminated land. It has been estimated that American industry may be under-provided for 'Superfund'-related clean-up liabilities by up to a trillion dollars (Schoemaker and Schoemaker 1995). Liabilities are less in the UK and other European countries, but still significant for some companies. They may become more significant if proposed legislation on the topic comes into force.

Investment in environment-related assets can also be significant: the chemical industry has estimated that up to 20% of its new capital investment in recent years has been to deal with environmental problems. This is financially significant because these assets have to be financed but, to the extent that the need for them is driven by compliance rather than by commercial business criteria, they do not generate any direct return.

European attention has been focused more on opportunities to reduce or avoid expenses than on liabilities, and this topic is increasing in importance in the USA too. Initiatives are usually taken on a one-off basis (see below), but an aggregate measure of savings can be a useful means of demonstrating that environmental management can be a profit contributor rather than merely an additional cost burden on business, and of building bridges between environmental staff and mainstream management. 3M calculates the accumulated first year's savings from initiatives carried out under its Pollution Prevention Pays (3P) programme, while Baxter, as we discuss in Chapter 15, produces an annual environmental financial statement with details of expenses and savings. So far, less attention has been paid to the revenue opportunities arising from environmental action, but these too may be significant in future.

*Cost Reduction and Improvement.* A number of corporate programmes, practical demonstration projects and research studies have shown that waste minimisation and similar initiatives can create savings and cost avoidance. In the first phase of the Aire and Calder Valley study (Johnston 1994), for example, potential improvements worth £2 million per annum were identified across the eleven industrial sites studied, with more longer-term possibilities in prospect when the project had run longer. Of the proposals stimulated by the project, 72% had payback periods of either zero or less than twelve months. Similarly impressive results have been reported by other waste minimisation and energy efficiency projects—including those at Sulzer, Xerox and Zeneca, as Chapters 17–19 describe. In addition, these and other initiatives, such as product redesign, can sometimes increase product quality and therefore sales revenues.

Of course, once the 'low-hanging fruit' has been gathered, there may be a point at which further cost reductions are not available (Walley and Whitehead 1994). However, if regulatory and social demands continue to increase and to create new potential costs for business, this point may be delayed for some time. Even after many years of waste minimisation initiatives, Dow, for example, continues to expect to find a large number of waste minimisation and similar projects that can provide annual returns on capital of at least 30%–40% over the coming decade (McLean and Shopley 1996).

*Prioritise Environmental Actions.* If they are fortunate, companies will need to prioritise between a number of win–win improvement opportunities. If not, they may need to prioritise between environmental improvements that do not create any net economic benefit but which may nonetheless have differing rates of (negative) return. Du Pont, for example, calculates the costs of different means of meeting given emission reduction targets as a means of achieving this.

*Guide Product Pricing/Mix/Development Decisions.* To maximise product profitability, it is vital that accurate product cost information is available and is taken into account when setting prices. This information also allows poorly performing products to be changed or dropped from the product range. As Chapter 9 shows, a study by the World Resources Institute found, at several of the companies that they examined, that, although environmental costs were significant, they were not being fully identified and allocated to products, so that pricing was not reflective of real costs (Ditz, Ranganathan and Banks 1995). As previously noted, environment can also influence the lifetime costs of products: for example, by requiring end-of-life disposal routes. Gaining a better understanding of these costs—as with the Philips model for considering end-of-life disposal costs (Brouwers and Stevels 1997)—allows timely action to be taken to minimise or avoid them through redesign and/or to put more cost-effective disposal routes in place.

In the long run, too, many markets are likely to be shaped by environmental factors—including the changing cost structures resulting from eco-taxes and other developments. Although this threatens some existing products and services, it also creates opportunities for others (Fussler with James 1996; Porter and van der Linde 1995b). Gaining a better understanding of medium to long-term environmental costs and benefits can help to neutralise threats and ensure that opportunities are taken.

*Enhance Customer Value.* Environmental actions taken within discrete portions of product chains can sometimes be economically and/or environmentally suboptimal, so co-ordinated action can provide higher returns for all of the chain members involved. One example from our research was a company providing a chemical in a small disposable container. The containers were expensive to buy and incurred waste disposal costs for the customer. Changing to re-usable containers reduced procurement costs for the supplier and eliminated the customer's waste costs. Demonstrating a detailed business case for such actions can spur improvement and also provide opportunities to develop closer relationships with customers.

*Future-Proofing Decisions.* Many investment and product development decisions are determined by levels of costs and benefits arising some years in the future. Unanticipated environmental factors can often affect these costs and benefits, sometimes to the point where returns become negative. Many of the chapters in this book discuss how the risks of this can be reduced through better analysis of environment-related costs and benefits (see also below).

*Supporting Sustainable Business.* There is increasing discussion of the implications of sustainable development for business, which are clearly considerable (DeSimone and Popoff 1997). They include:

- Radical improvements in environmental performance: a minimum 'factor-four' reduction in environmental impact is needed for the delivery of final goods and services to consumers, according to some estimates (von Weizsäcker, Lovins and Lovins 1997)

- 'Eco-innovation', i.e. development of new products and processes that are capable of meeting these objectives (Fussler with James 1996)

- A long-term perspective in decision-making, with greater emphasis on the impacts of decisions on future generations

- A greater degree of internalisation of external environmental costs to business

This implies the need for environmental and management accounting systems to collect new types of data, such as those relating to environmental effects throughout the entire product chain. It also suggests that more attention needs to be paid by accountants and others to identifying and raising internal awareness of long-term cost trends.

The section on environmental value analysis towards the end of the chapter discusses one way in which environment-related management accounting can operationalise these ideas.

## ▮ Who are the Environment-Related Management Accountants?

Our answer to this is: anyone who is involved in generating financial and neo-financial information about the business impacts of environmental issues. Hence, many environmental and operational managers and some management accountants are already practising environment-related management accounting. In this respect, environment-related management accounting is largely a 'virtual' activity enhancing what already exists rather than creating something completely new. This is primarily:

1. Making better use of, or modifying, existing sources of data and generating new ones

2. Making better use of, or modifying, existing management accounting techniques

One additional task is to foster the longer-term perspective that allows the challenge of sustainable development to be addressed by the business (see above). A second is the need to create processes that bring together the accounting, environmental and other functions to achieve both specific objectives and a more general awareness of each other's concerns and activities. An obvious example is to include environmental managers in both regular investment appraisal and other business case procedures, and also irregular accounting change activities such as activity-based costing (ABC) or business process re-engineering. Conversely, accountants and accounting data need to be included within environmental management systems (McLaughlin and Elwood 1996).

Researchers have found that the benefits from such processes can be as important as any specific outcomes. That is also the conclusion of many practical initiatives such as the environmental financial statement developed by Baxter (see Chapter 15). AT&T also believed this to be so important that it created a 'green accounting team' to develop good relationships and a common understanding of worthwhile initiatives that might be undertaken (see Fig. 2).

This combined requirement to change the conduct of existing tasks, to establish a longer-term view of the business implications of sustainable development, and to

IN THE MID-1990s, AT&T established an ambitious design for environment (DFE) initiative, aimed at identifying whole-life environmental impacts and costs of computing and telecommunications equipment in order to make environmental considerations a priority during the design stage (examined in detail in US EPA 1995b). It created six cross-functional teams examining areas such as 'green accounting', 'life-cycle analysis', 'supply-line management' and 'product takeback'. The green accounting team combined representatives from accounting, operations, environmental management and other business functions, from both the corporate centre and business units. Members were invited to join both for their functional and specialist expertise and for their ability to use their influence to support the implementation of the team's proposals. The team was co-chaired by two executives with backgrounds in management accounting and environmental engineering respectively.

Several of the team had previous experience with activity-based costing (ABC) and activity-based management (ABM) initiatives at AT&T and saw their environmental tasks as extensions of this. Their aim was to identify those costs for which environmental factors are the main cost drivers, and track these through to products and processes. Hence, their working definition of environment-related costs was those costs where environmental professionals are the best placed to identify both the cost drivers and the means to affect them.

The team's main objective was to integrate environmental considerations into existing management accounting systems and to support long-term strategy development and decisions—in particular on product and process design.

One early example was the use of lead in soldering processes, which has adverse environmental impacts. The conventional costing system simply spread soldering costs across all products by general apportionment on the basis of the costs of materials used. However, on investigation, it was found that the true driver of both the environmental impact and the associated business costs was the number of soldering operations performed, which varied widely between different products. One product would go through the process only once, another ten times—but the existing system would apportion the same cost to both. Making this visible by changing the costing system provided the incentive to look for ways of reducing costs and environmental impacts through product redesign and the use of different types of flux. Similarly, the main driver of quantities and therefore costs of chemicals wastes in batch production systems was found to be not the volume of production but the number of set-ups required, and therefore the size of batches.

The team produced several outputs, including a glossary of environmental accounting—to reduce internal and external misunderstandings—and a checklist for sites to identify areas of costing weakness. One particularly important tool that was developed by them was the 'Green Activity Matrix'. This lists the various costs that have an environmental element, in two dimensions:

- The first dimension is categorised by the type of cost incurred: people, materials and supplies, services and consulting fees, depreciation on equipment, energy and utilities, etc. These correspond with the general ledger codes used in financial accounting systems.
- The second dimension lists some 30–40 types of environment-related activity such as obtaining permits, treating on-site waste, handling/storing/disposing of hazardous wastes, and environment-related training.

The cells in the matrix are completed with the respective amounts of cost. This provides a link between the general ledger system, which collects costs by the types of resource, and the business activities that drive the amounts of costs incurred. This can be used to identify which activities are the most significant in driving the largest proportions of costs.

Unfortunately, the break-up of AT&T in 1996 ended the team's activities (although many of its specific activities have continued in its successor companies). However, its experience shows what can be achieved through the creation of cross-functional processes and a practical model of how to do so.

**Figure 2:** The Green Accounting Team at AT&T

initiate and maintain new processes, suggests that there is a need for a concrete manifestation of environment-related management accounting in the form of an organisational 'champion'. To be effective, he or she will require a combination of personal dynamism and vision and accounting, business and environmental know-how. In principle, this could be found and therefore located in any business function, but the critical need at present to change the attitudes and actions of accountants and the accounting function suggests that the champion will have maximum impact if located there.

The seeds of such a development have been sown by professional management accountancy institutes who have called for more involvement in environmental management by their members. The UK Chartered Institute of Management Accountants (1997) has commented that

> the forward-thinking management accountant should be taking an active role in environmental management…as he or she has key skills to apply to the process, including the provision of advice relating to strategy formation and the effective use of resources.

Parker, commenting for the US Institute of Management Accountants' Foundation for Applied Research on the Tellus Institute's study of environmental cost accounting for capital budgeting (White *et al.* 1995; White and Savage 1995), has also observed that

> corporate accountants and financial managers are not necessarily in the best position to recognize and understand the trend toward transforming internal costs. In many cases, recognition of what is at stake comes from non-financial professionals. But management accountants, aware of the strategic value of environmental accounting and aided by decision support tools, can wake up senior management to the necessity for analyzing environmental costs. This leadership can change senior management's perception that management accountants are simply corporate scorekeepers (Parker 1995: 53).

There is less evidence to date of individual companies taking the initiative, but this is likely to change in future.

## ▌ The Practice of Environment-Related Management Accounting

Figure 3 provides a graphical representation of what this involves in practice. It is divided into three vertical levels to indicate a progression from the foundation of non-financial and financial data, through the techniques that process this into information (i.e. outputs that are useful for managers and stakeholders), to the highest level—objectives (see discussion above). The next section discusses the base of the triangle, i.e. relevant data, in more detail, while the subsequent section considers the techniques that convert this into information.

## ▌ Key Data for Environment-Related Management Accounting

This section discusses the existing data sources and systems that can provide inputs to environment-related management accounting, the extent to which they take environment into account (or conceal its importance) at present, and possible ways in which they can be modified or supplemented to reflect environmental considerations in future.

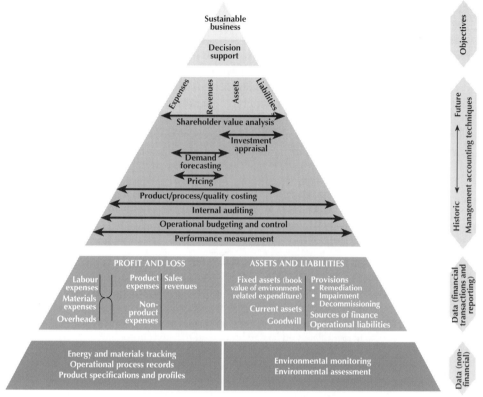

**Figure 3:** The Environment-Related Management Accounting Pyramid

### Non-Financial Data

As Kaplan and Norton (1992, 1993, 1996a, 1996b) and others have stressed, non-financial data is an important element in all areas of management accounting. This is particularly true of environment-related management accounting, whose ultimate 'raw material' is data on physical and energy flows and stocks and their impacts upon the environment. This is collected in operational process records, material resource planning systems, resource planning, emissions monitoring and other systems, which are managed by production, environmental and other non-accounting functions. Moreover, Shields, Beloff and Heller's study of five oil and chemical companies in the USA and Mexico (see Chapter 10) found that such non-financial information was more useful for day-to-day environmental decision-making than was financial data. We have argued in the Introduction that the generation of non-financial data should not be seen as a central objective of environment-related management accounting, on the grounds that it is either already being provided—or could more easily be provided in future—by functions such as environmental management or production. However, this is not to say that accounting techniques or accounting professionals cannot play a role in this area. The potential for this has been explored in detail by the previously mentioned ECOMAC project (Bartolomeo, Bennett and James 1998). One of the key opportunities that has emerged from this and other studies is to obtain a better understanding of the extent and financial implications of energy, materials and water flows through organisations (see also Chapters 6, 11, 13 and 17 in this volume). This is important for many reasons, but especially because the full costs of wasted materials (i.e. including their purchase price and the costs of processing them to the point where they become waste) are often the single most significant environmental cost (see the appendix below and the Zeneca case in Chapter 19). Except for specific industry sectors, such as chemicals and pharmaceuticals, which have always been concerned with the yields and detailed characteristics of their processes, few organisations in the UK and USA appear to have a full picture of their energy and material flows; indeed, obtaining such a picture is usually the first step in successful waste minimisation programmes and often creates non-environmental business benefits.

As yet, there are no American or British equivalents (at least in the public domain) of the Germanic 'eco-balancing' approach practised by Kunert and other companies, which builds a picture of all energy and material flows on a periodic basis (Bennett and James 1998a; Birkin and Woodward 1997d; James, Prehn and Steger 1997). This is then used as the basis for day-to-day 'eco-controlling' in order to reduce environmental impacts (Hallay and Pfriem 1992; Hopfenbeck and Jasch 1993; Schulz and Schulz 1994). One potential task for environment-related management accounting in the US and UK is therefore to apply these ideas and associated techniques there.

### Financial Data

The accounting function in business (and non-business organisations) has, potentially, three distinct objectives:

- The day-to-day **operational** needs of initiating and recording transactions, and of managing assets and liabilities such as, respectively, working capital and bank loans
- Supplying regular external **financial reports** to shareholders, to provide them with reassurance that their assets are being safeguarded and their interests are being met. This is a legal obligation for companies in all advanced economies
- Providing **management** within the organisation with **information** that is relevant to its function of making decisions and ensuring that the organisation's activities and outputs are kept under proper control.

The first objective is essential for day-to-day operations. The second is a legal obligation. The third objective is optional, and the extent to which financial information and techniques are used in management is discretionary and will reflect the particular management style adopted. However, some internal financial processes are near-universal in organisations of any significant size, such as budget-setting and budgetary control, and tracking costs through the organisation to the responsible department and manager.

To support these three objectives in large organisations, an integrated system of accounting data capture and collection is needed—ledgers, registers, cash books, etc. Although the principal factors in the design of these systems are likely to be determined by the needs of the first two objectives, they also represent a financial database that is available for use by management.

This subsection reviews the breadth of the financial data that are usually available, based for convenience on the classification of the legally required published financial reports. These are centred on two core financial statements. The first is the income statement, which aggregates expenses and revenues throughout a given financial period (usually a year). The second is the balance sheet, which summarises a company's assets and liabilities at a particular point in time (usually at the end of the same financial period). These four basic categories of expenses, revenues, assets and liabilities are reflected in Figure 3.

***Expenses and Revenues.*** The basic accounting systems within organisations (the book-keeping systems and ledgers) will generally capture and collate expenses in terms of a combination of two parameters:

- The type of resource being acquired and consumed: materials, labour, services, depreciation, etc.
- The functional area of the business in which the expenses are incurred: production, selling and distribution, general and administrative, etc.

These classifications reflect the sources of the data in the various subsystems of a normal business accounting system. Labour costs will be available from payroll systems; materials costs from materials management systems which draw their data from invoices and bills of materials; the depreciation charge from a register of fixed assets; etc. The appendix discusses in detail which of the general pool of expenses and income can be classified as environment-related.

Traditionally in management accounting (including here its junior sibling, cost accounting), there has been more emphasis on dealing with costs than with revenues. This reflects the origins of cost accounting in the production function, due in large part to the financial requirement to report in published financial reports the historic cost of stocks (inventories). This requires detailed analyses of production costs.

There is no equivalent external compulsion to carry out detailed analyses of revenues. Only relatively recently in most organisations, in terms of the historic development of accounting, has the marketing function achieved the importance that it has in modern business. Hence, the traditional emphasis has been on production and costs rather than on marketing and revenues.

The accounting systems will need to capture and collect data on revenues as they arise through transactions, as evidenced by shipping documentation and invoices, and will need to analyse these in sufficient detail to support management—as a minimum, by type of product, customer, market sector and distribution channel.

Reclassifying accounting data after its initial entry is sometimes impossible and almost always time-consuming and costly. Hence, the secret of success in all areas of management accounting—including that related to environmental issues—is to capture any data necessary for analysis (such as the purpose of expenses)—when the data are entered. However, modifying existing systems can also be costly. The lowest-cost option is to build in environmental considerations when systems are being changed for other reasons: for example, because of the introduction of an activity-based costing system (see below). A key task for environment-related management accounting is therefore to ensure that the needs of environmental management are considered when changes are being made. The opportunity is to obtain better-quality data, but almost as important is to avoid a deterioration in the quality of existing data. In one company that we researched, a re-engineering exercise resulted in previously separate categories for energy purchases—electricity, gas, etc.—being collapsed into a single energy category. As a result, it lost the ability to calculate easily its energy-related carbon dioxide emissions.

**Assets.** Accountants identify three broad categories of asset: fixed (or long-term) assets, current assets and goodwill (a particular type of long-term asset). Fixed assets are those with a useful life beyond a single accounting year and are (with some exceptions) stated in the balance sheet at their original historic cost, reduced by depreciation provided to date in respect of the portion of their useful life that, to date, has expired. The high cost and long life of assets such as pollution control equipment and landfill sites means that these can be significant fixed assets, and the depreciation on them can also be significant, although this is often excluded in calculations of environmental costs.

An alternative method of valuing fixed assets is at their replacement cost. This is of potential environmental significance because rising environmental standards often mean that the cost of building new environment-related facilities is much higher than those that are being replaced (see below). This is the case with landfill at Zeneca's Huddersfield site (see Chapter 19). The company has a long-standing site which is fully depreciated. Hence, only operating expenses are charged back to product and process cost centres. As the site has many years' life, this can be practically justified but, at current rates of waste generation, a replacement landfill facility will have to be built at some point. This is likely to be very expensive and will therefore result in an immediate increase in recharged costs as depreciation is included in the figures. However, the conventional method of basing these recharges on historic costs means that managers who take decisions that affect the volume of wastes generated, through process control and product design, are not encouraged by the system to take into account also the opportunity cost that is indirectly incurred as landfill capacity is consumed.

The main current assets for most companies are cash balances, debtors (accounts receivable), and stocks and work in progress. Although environment has some tangential relevance to these—changes in environmental legislation could result in stocks becoming more difficult or impossible to sell—this is not for most a major area of concern.

Goodwill is an asset with whose treatment the accounting profession has been struggling—with only limited success—for some time. Conventional accounting practice recognises and includes goodwill in company balance sheets only when money is directly outlaid to acquire it, when one company is purchased by another. The goodwill then arising is the amount by which the purchase consideration exceeds the value of the tangible net assets acquired, and represents what the acquirer is prepared to pay for the present value of the amount by which the acquirer's future profits are expected to exceed a normal rate of return. However, the true value of goodwill in any company should also include what it has built up within the business as a result of operating over time and building up a reputation among customers, even though this is not represented by any specific outlays and is therefore not captured by the accounting system. Several authors have suggested that environment is an important determinant of company reputation, although the precise extent of this is difficult to quantify (DeSimone and Popoff 1997; Charter 1992).

**Liabilities.** Liabilities can be distinguished by type into three broad categories: sources of finance; liabilities arising from normal operations; and provisions.

In most major corporations, the raising of finance and the balancing of debt and equity is handled within the finance function by a treasury management function which is separate from the financial controlling activities of then allocating, managing and accounting for this finance within the business. Environmental performance is increasingly significant for treasury management, since the extent of risks being borne by a company, including those that are environment-related, can affect access to and the cost of raising new capital. Several studies (for example, Butler 1997) have shown that funds that invest in companies with good environmental records have matched and sometimes out-performed the market average—thereby lowering the cost of raising equity—and, conversely, that those investing in companies that have experienced major environmental incidents have been depressed, making new equity more expensive (Blumberg, Korsvold and Blum 1997). Kvaerner, for example, has paid slightly lower interest rates on loans because of its good environmental record and consequently greater credit-worthiness. Some analysts and insurers are coming to see evidence of good environmental management by a company as indicative of the quality of its management generally.

Liabilities arising from normal operations include trade and most sundry creditors, corporation tax due, and tax collected but not yet paid over in connection with PAYE and VAT, etc. These may be affected by any events occurring within the business, including environment-related events, but are unlikely to be particularly significantly affected by environmental management.

Provisions are amounts allocated to cover any likely future liabilities or losses that have arisen but have not yet been settled (provisions included under 'liabilities' will exclude any provisions made in connection with the impairment in value of assets, such as arising from depreciation of fixed assets or from the obsolescence of inventories, which will be reflected in the balance sheet as a reduction in the value of the related assets). There is considerable concern that significant liabilities could exist, in respect of (for example) remediation or future decommissioning costs, which are frequently not fully provided for in company financial reports. Chapter 5 provides an overview of how potential remediation liabilities can be assessed as an aid to both internal decision-making and external reporting. Barth and McNicholls (1994) and Schoemaker and Schoemaker (1995) also provide good overviews.

Until now, the emphasis has been on quantifying liabilities arising from past events but, while this remains important, Brent Spar and other developments have focused attention on potential future liabilities. In order to be proactive and to ensure that environmental liabilities are not under-provided, Chevron has introduced a systematic cross-functional process to identify, evaluate, measure and disclose its environmental liabilities, and then monitor their remediation (Lawrence and Cerf 1995).

### Techniques

The second tier of the pyramid in Figure 3 identifies the main management accounting techniques that are used to process the data arising from the financial and non-financial systems into information for management's benefit: performance measurement, operational budgeting and control, internal auditing, costing, pricing, demand forecasting, investment appraisal and shareholder value analysis. All of these are actually or potentially relevant to environment. One important distinction is between those that are concerned with current data and those (broadly, demand forecasting, investment appraisal and shareholder value analysis) that make projections into the future. Subsequent sections discuss the environmental relevance of each of these techniques.

*Performance Measurement.* Performance measurement is a growing field in all areas of business. Traditional performance measurement in the UK and USA, particularly at higher levels of management, has focused on meeting financial targets. However, over the last decade, the quality movement and other drivers have focused attention on the importance of non-financial performance measures, and schemes such as the European Quality Award and Kaplan and Norton's 'balanced scorecard' now provide templates for this (Kaplan and Norton 1992, 1993, 1996a, 1996b). To date, environment-related performance management has largely been developed by environmental managers and has limited interaction with management or others involved in strategic performance measurement (Bennett and James 1998). However, as the Introduction and other authors (for example, Epstein 1996c) have noted, environment should be an important part of a balanced scorecard for many companies and more interaction is needed in future. Many of the practical challenges of developing and implementing environment-related performance measurement are also generic to all areas of performance measurement, and greater interaction would facilitate mutual sharing of experience and learning. Finally, the dependence of environment-related management accounting on the non-financial data that is generated by environmental, operational and other functions makes it important that its needs are considered during each of their own performance measurement processes.

*Operational Budgeting and Control.* The setting of budgets is an important means of implementing strategic objectives, while tracking budgetary outcomes can be a valuable means of monitoring how well objectives are being achieved. Budgeting is relevant to environmental management for three reasons. First, environmental actions will require resources that need to be specified within budgets. Second, budgetary outcomes can be a useful means of checking whether environmental goals are being achieved: for example, over-budget expenditures on energy provide an early warning that energy efficiency targets are unlikely to be achieved. Finally, as has been previously noted, identifying and allocating environmental costs to specific budgets provides a powerful incentive for action to be taken.

*Internal Auditing.* An external audit to verify the published financial report is required of companies by law. In addition, most companies of a significant size also operate an internal audit function. This is optional, and its responsibilities, activities and position in the organisation are at the discretion of each company, though professional guidance on internal auditing generally is available from bodies such as the Institute of Internal Auditors.

Traditionally, internal audit has been a part of the finance function, reporting to the finance director and primarily concerned with internal checks and controls on financial data and activities. One motive in many organisations was to minimise the costs of external auditors, by using internal audit as a more cost-effective means of providing assurance on the quality of internal controls and thereby to reduce the quantity of detailed checking work that the external auditors needed to do directly themselves. More recently, the role of internal audit has expanded to include the audits of (in particular) computer systems, operations and the quality of management. Internal audit has become central to risk assessment and management in many companies, and its importance has been enhanced by recent concerns over corporate governance. One effect of this has been that internal audit now reports within the organisation, increasingly frequently, not within the finance function but at chief executive level, often with a further reporting line to an audit committee at board level.

Internal audit is potentially relevant to environmental management in several ways, especially for companies for whom environmental issues may represent potentially significant risk factors to their businesses. One area of apparent potential overlap or co-operation is the environmental audit, though this will depend on the objectives of the particular environmental audit and how far the competences that it requires may be outside the scope of the other activities of internal audit. In most organisations, most environmental audits are handled by a specialist corporate environmental management function, though some environment-related checks may be included in regular internal audit programmes.

Internal audit can also be relevant through its original purpose of checking and confirming the accuracy and integrity of information—the information that is used internally by management, as well as what is published externally. This function includes both the integrity of the data that are captured and collected through the organisation, and how these are then converted into usable information and disseminated in reports to management and/or published externally. For example, the Environmental Issues Unit of British Telecom enlists the support of their colleagues in BT's internal audit function to help to assure the integrity of the information that they plan to publish in their annual environmental report, prior to its further verification by an external party.

**Costing.** Costing is perhaps the area of greatest activity within environment-related management accounting (Ditz, Ranganathan and Banks 1995; Epstein 1996b, 1996c; Russell, Skalak and Miller 1995). It is also addressed to a greater or lesser degree by most of the chapters in this volume. This section discusses six main issues associated with costing:

- Activity-based costing
- Quality costing
- Product costing
- Life-cycle costing
- Cost projection
- Strategic costing

*Activity-based costing (ABC).* Traditional costing techniques have been based on specific categories of direct cost such as labour and materials, plus a residual overhead. The latter is then frequently either allocated to products or processes on a more or less arbitrary basis—for example, the EPA's study of the electroplating industry described in Chapter 11 found that square footage of product was the easiest way to do this for environment-related costs—or written off as a period cost and therefore not tracked through to products or processes at all. Indeed, in many companies, the main part of environment-related costs such as energy, water, waste disposal and the salaries of environmental staff are likely to be included in overheads (White *et al.* 1995; White and Savage 1995). These practices mean that, where products or processes have high environmental costs, the figures can be hidden from decision-makers. This decreases the motivation to reduce the costs and can also create a bias against pollution prevention projects (Hamner and Stinson 1993).

One potential solution to this problem—which is common to other areas of management accounting—is cost system redesign (Drury and Tayles 1998). As Schaltegger and Müller discuss in Chapter 3, an approach that is of particular relevance to environment is activity-based costing (ABC). This tries to create more meaningful cost information by tracking costs to products and processes on the basis of the underlying 'drivers' that cause those costs to be created in the first place. The amount of cost lost in overheads is thereby greatly reduced. As a result, product prices can be set more accurately, and significant cost drivers can be targeted for cost reduction measures. Where environment is a significant cost driver, it will be highlighted naturally by ABC activities. However, there is usually considerable scope for more proactive environmental concern, either by building a more detailed picture of environmental cost drivers and categories where these have already emerged as important or by highlighting them when this is not the case (Kreuze and Newell 1994). Schaltegger and Müller explore this issue at a conceptual level in Chapter 3, while Bierma, Waterstraat and Ostrosky (Chapter 13) show the close relationship between ABC and environment-related management accounting at Chrysler's Belvidere plant. However, Chapter 10 by Shields, Beloff and Heller found that ABC was not widely used by the North American companies that they studied. The previously mentioned ECOMAC survey (Bouma and Wolters 1998) reached similar conclusions for Europe. Hence, supplementary routes will be needed to introduce environment-related management accounting for the foreseeable future.

*Quality costing.* Several authorities have identified the links between total quality management programmes and good environmental management (Roth and Keller 1997; Davies 1997). Quality costing aims to measure in financial terms the benefits of good quality management, and is complementary to ABC. The rationale of quality costing is to highlight the costs of non-quality in order to stimulate motivation to reduce these and to prioritise possible actions. Conventional quality costing distinguishes three types of cost:

- **Failure:** the costs of putting right or otherwise dealing with defects, arising through either internal failure or external failure (i.e. those defects that occur in use by customers)
- **Monitoring:** inspection and other costs to ensure that defects are eliminated or detected
- **Prevention:** costs of avoiding defects

The finding from cost-of-quality studies is frequently that, in the long run, total costs are minimised when the emphasis is placed on prevention rather than on either monitoring or the toleration of failures. However, without these studies, this might not be apparent, since failure costs include several that are intangible and/or at some distance in the organisation from the point in the operational process at which the loss in quality occurred.

Quality costing techniques can easily be applied to the environmental area (Hughes and Willis 1995). A Dutch study that used this model to calculate the 'costs of non-environment' found that, on a narrow definition of environment as the costs of dealing with pollution and wastes, they amounted to around 2% of total operating costs (Diependaal and de Walle 1994). To be valuable in the environmental field, 'failure costs' probably need to be defined more broadly so that they include what might be called 'indirect failure costs' or the 'costs of inefficiency', i.e. the costs of purchasing and processing materials and energy that end up as waste (see appendix).

*Product costing.* Producers need accurate information about the cost make-up of their products in order to determine price and identify cost reduction opportunities. Users need data about the total costs of products that they are buying in order to compare alternatives that have different proportions of acquisition and operating costs. Designers need both types of information in order to create products that have reasonable purchase and running costs. Environmental costs are important in all these cases and there can be detrimental consequences if they are not properly identified and allocated.

As several case studies demonstrate (Ditz, Ranganathan and Banks 1995), it is not uncommon for a small number of products to generate a disproportionately large share of total emissions and wastes. If these costs are not allocated to individual products but instead are treated as a general overhead, then 'clean' products will appear to have higher costs than is actually the case, while 'dirty' products will appear to be cheaper to produce than they really are.

*Life-cycle costing.* Environmental costs are increasing in every stage of the product life-cycle. Green taxes are being introduced on many types of raw materials, emissions, wastes and products at the end of their lives. Certain disposal routes for materials and products are being banned or are subject to stringent regulation which makes them very expensive. As the nuclear and oil industries have discovered, it can also be much more costly to decommission equipment than was originally anticipated. Producers can potentially incur liabilities as a result of environmental problems related to their products.

These changes can, to some extent, be incorporated within the emerging management accounting concept of life-cycle costing (Bailey 1991). This means extending horizons beyond the purchase costs of products to consider all the costs that will be incurred over their operating lifetime—including, in principle, the environmental costs involved in buying, using and disposing of the product. However, the dispersal of responsibility between and within suppliers and their customers can often obscure this. Bierma, Waterstraat and Ostrosky (Chapter 13), for example, demonstrate that this is a serious problem with regard to the whole-life costs of chemicals. It can therefore be sensible for the various parties to work together to identify and calculate these at the time of purchase. Two particular areas that a number of organisations have already started to examine are the costs of dealing with emissions or wastes from the operation of equipment, or of disposing of products at the end of their lives. Chapter 6 by Bartolomeo provides examples of this from Italy.

Of course, this interpretation of life-cycle costing is a narrow one which ignores all environmental costs incurred before equipment is acquired, or downstream costs for which an organisation has no responsibility. A broader definition is therefore required if environment-related management accounting is to be useful in the area of product development, which must consider these areas.

Some authors have seen the ultimate goal of life-cycle costing as being the monetisation of all impacts identified by life-cycle assessment (LCA). However, the difficulties of achieving consensus about even the relatively simple issue of the most appropriate means of undertaking LCAs, quite apart from the contentious issue of reaching agreement on appropriate conversions from physical to monetary units, make this unrealistic for the foreseeable future.

More limited work has taken place on calculating financial costs and benefits as an input to design for environment (DFE) initiatives. These aim to reduce life-cycle impacts by taking action in the design stage, e.g. making recycling easier by making equipment easier to disassemble. Brouwers and Stevels (1997) have described an end-of-life costing model developed for this purpose at Philips. Kainz, Prokopyshen and Yester (1996) also describe an exercise at Chrysler to calculate the whole-life costs of two designs. This found that, although a design that contained mercury was cheaper to purchase, its whole-life costs were greater as the wastes generated in its production then had to be treated as special wastes.

Wood, the former leader of AT&T's green accounting team, has noted the opportunities to extend these initiatives into a more strategic approach which she terms 'environmental life-cycle costing' (Wood 1998). Monsanto provides one example of this by giving its salespeople a checklist to identify opportunities to reduce its own and/or customers' environment-related costs, to their mutual benefit (Tuppen 1996). And the Xerox case in Chapter 18 demonstrates the potential for cross-chain initiatives to change completely perceptions of key business activities and consequently to reveal major environment-related savings opportunities.

*Cost projection.* Projecting future costs is an important part of investment appraisal and is also valuable for other purposes. Environment can be an important determinant of these future costs. This is highly visible with new legislative or regulatory demands. However, forward-looking companies will also be considering the potential costs of possible future legislation or other environmental action. One indication that this may happen is when costs in one country are much lower than in others. Another is when there are large external damage costs created by environmental impacts that are not yet reflected in the company's internal financial calculations, but could be in future as a result of governmental or social action. There is a growing amount of research that suggests that these externalities are considerable for energy production (Cookson 1997; EC 1995; Oak Ridge and Resources for the Future 1992–96; Office of Technology Assessment 1994), transport and other economic activities. However, there is considerable controversy about the results of such research and the methodology employed so that, from a business perspective, the figures are best regarded as indicative rather than exact (Hongisto 1997).

Even so, Epstein (see Chapter 4) and other researchers (CICA 1997; Tuppen 1996) have argued that companies making capital investment and other decisions with long-term financial consequences might be wise at least to consider the implications of these. Several business leaders and companies, notably Ontario Hydro (see Chapter 16) and two senior Dow executives (Popoff and Buzzelli 1993), have also advocated the use of 'full-cost accounting' (including external costs) by companies, although little has happened in practice as a result.

*Strategic costing.* As Burritt notes in Chapter 8, costing is not always about creating an accurate reflection of real costs. It can also be a strategic tool to encourage or discourage certain inputs, activities and outputs by influencing relative prices: for example, by putting a high overhead on labour to encourage moves towards automated production (Bromwich and Bhimani 1994). In principle, it is easy to apply this approach to environment: for example, by introducing internal taxes on energy consumption or on waste disposal. The level of these, and their trajectory over time, could be based on long-term cost projections. In practice, it is difficult to do this because of fears about competitive disadvantage if other companies do not follow suit. Nonetheless, more use is likely to be made of this approach in future.

**Pricing.** Pricing requires consideration of customers and competitors as well as costs, so accounting techniques are only one aspect. However, adequate cost analysis is an essential part of pricing decisions, which may be distorted by any inaccuracies in costing systems.

Life-cycle costing provides the framework to consider costs not only within the organisation itself, but also along the product chain, by including as well as internal costs also costs incurred upstream (by suppliers) and downstream (by customers and consumers). This can help to identify opportunities where modest extra spending by the company may increase value for the customer disproportionately, which can be reflected in an increased selling price and/or increased sales volume. As Bennett and James's study (Chapter 15) demonstrates, Baxter International has generated substantial savings in materials costs for itself through packaging redesign. As well as this benefit, reducing the quantity of packaging that the final user has to dispose of is becoming an increasingly significant selling point in countries such as Germany which have strict legislative controls.

**Demand Forecasting.** Environmental factors are already shaping many markets and will influence more in future. This influence takes two forms: the volume of a product or a service that can be sold, and the price at which it is sold. Sales volumes of a number of products—for example, CFCs—have already been largely or completely curtailed by law as a result of environmental considerations, and the likelihood is that more will be withdrawn from the market (or 'sunsetted') in future. Customers may also discriminate against products with poor environmental performance, especially if better-performing ones offer similar value. Sunsetting and other environmental developments also create opportunities for new products. Indeed, it may be that the revenue streams from future eco-efficient products—i.e. those that offer greater customer value and better environmental performance—will have far greater impact than any of the other areas discussed in this chapter. Of course, it is not usually possible to do more than guess at the amounts of potential future revenues from hypothetical new products, and consequently less attention has been paid to this area in the environment-related management accounting literature than to methods of cost analysis. However, it is important that it should receive more attention in future.

**Investment Appraisal.** Environmental factors can be significant in determining the ultimate returns from new investments. It is therefore important that they are identified and considered during the early stages of investment decision-making (Kite 1995; Rückle 1989). This not only allows major problems to be avoided, but also provides an opportunity for remedial action at a stage when the costs of doing so can be relatively low.

Many companies are currently bringing environment into capital budgeting by requiring qualitative assessments of impacts arising from major investments. This can be done in two main ways:

- By widening the range of costs and benefits that are taken into account
- By adapting appraisal techniques

A 1995 Tellus Institute survey of US companies, for example, found that over 60% of respondents are now considering the costs of emissions and waste monitoring, treatment and disposal in project evaluations (White *et al*. 1995; White and Savage 1995). However, there are still many costs that are excluded from most evaluations, as Chapter 11 confirms for the US electroplating sector.

Research suggests that most investments in US and UK companies—including those related to environment—are appraised on the basis of relatively high discount rates (Bouma and Wolters 1998). This means that the long-term benefits that often result from environmental action frequently have a low, or even zero, net present value. Many observers also believe that conventional techniques do not properly consider the issue of risk (Busby and Pitts 1997). Hence, new or modified appraisal techniques might be required. Appraisal techniques can be adapted to take account of the long-term benefits of environmental actions and/or the potential risks of investments with serious environmental impacts. This can be done, for example, by applying lower or higher discount rates to environmentally significant investments, or for long-life projects by extending the period for which future benefits are considered beyond the usual truncation point. Chapters 4, 10, 12 and 14 discuss these issues in greater detail.

*Shareholder Value Analysis.* In recent years, there has been an increasing interest in measuring shareholder value (Rappaport 1986; Stewart 1991, 1994). This has in part been in recognition of the principle that, at least in the UK and USA, in law (and so far as capital markets are concerned), corporations exist primarily for the benefit of equity investors. It is also a correction for generally perceived deficiencies of conventional measures of accounting profitability as the main indicator of business performance.

The term is often used only loosely but, when used more precisely, defines shareholder value as the present value of the company's future cash flows, discounted at an appropriate rate that reflects the risks involved. As environment can affect all of the main parameters in this equation—future expenses, revenues and cost of capital—it is therefore an important element to be considered in any calculations.

The ultimate aim of shareholder value analysis is to influence equity valuations, both directly by influencing capital market perceptions and indirectly by making it a priority issue for internal managers. Several studies have suggested that financial analysts and fund managers are either ignorant of or unconcerned about the environmental performance of the companies in whom they invest (Business in the Environment 1994; UNDP 1997). However, other more recent studies have established a strong case for trying to establish a link between environment and shareholder value, and/or have provided evidence that such a link already exists (Barber, Daley and Sherwood 1997; Blumberg, Korsvold and Blum 1997; Cohen, Fenn and Naimon 1995; *EAAR* 1997; Müller *et al*. 1994; Müller *et al*. 1996; Schaltegger and Figge 1997; Schmidheiny and

Zorraquin 1996; Verschoor 1997). Schaltegger and Figge (1997) have also developed a detailed framework, with examples, to analyse the linkages. An important element in their analysis is the importance of environment-related financial risk, which has also been addressed by the Centre for the Study of Financial Innovation. It has developed an environmental equivalent to the well-established financial risk-rating processes of Moodys and Standard & Poor (CSFI 1995).

Establishing the existence of links between environment and shareholder value is not sufficient to influence market perceptions. Companies also need to communicate effectively, in terms relevant to the financial markets, the significance of environmental issues to their long-term business success and the adequacy of the efforts that they are taking to manage them (ACBE 1996; Kreuze, Newell and Newell 1996).

***Environmental Value Analysis.*** This is the relationship between an organisation's economic value added and its environmental impacts. Although there have been few attempts to measure this to date, it is an important issue, which in principle can be evaluated in two ways. The first is by developing relational measures. The output measure can take a variety of forms—for example, turnover or profits—but, as value added is a more direct measure of the net economic contribution made by a company, it is widely considered to be the most appropriate. Calculations can then be made of value added per tonne of emission or per unit of environmental impact or, alternatively, tonnes emitted or units of environmental impact per £ of value added. These give a crude measure of how efficiently organisations—or, in aggregate, industries—are using environmental resources.

However, knowing that an organisation is using resources efficiently says little about whether its use is sustainable. Sustainability implies limited 'eco-capacity', i.e. a finite availability of physical resources such as fossil fuels and biological materials, and of environmental 'sinks' such as the atmosphere. The costs of exceeding this eco-capacity can, in principle, be calculated and then disaggregated to the level of an individual business via taxes—for example, a carbon tax—or other means. The relationship between these 'costs of unsustainability' and value added can therefore serve as a crude measure of an enterprise's sustainability.

Of course, in a world where all such costs are internalised through taxes and other measures, sustainable value added would be equal to economic value added, but this is far from being the case at present. Hence, approximations to sustainable value added can be produced by taking estimates of damage costs. In the case of environmental damage costs, figures are available for many impacts, although there is limited consensus about the best basis of calculation or their accuracy. In the case of social damage costs, few figures are available and this situation seems unlikely to change for the foreseeable future.

Only one organisation has so far made even a crude attempt to calculate its sustainable value added. This is the Dutch computer services company BSO Origin, who, in its 1992 environmental report, calculated its main environmental impacts and then converted these into financial amounts to represent the imputed costs of those impacts. The data for this were based on calculations of long-term costs of control in the Dutch National Environmental Protection Plan. This gives a net cost of each environmental impact individually, and of all its environmental impacts in aggregate, which can be compared with the value added as calculated through its conventional business accounting processes.

The methodology of this can easily be criticised, both for the bases on which costs per unit of impact are calculated, and on how far upstream and downstream costs should legitimately be included. At the present level of understanding of business (and other) impacts on the environment, it is difficult to assess what meaning if any can be attributed to the values generated; BSO recognises this, and claims only that its system indicates orders of magnitude rather than precise values. However, the BSO exercise is best seen as a first experiment in devising a comprehensive system that recognises and quantifies all of the environmental impacts of a business, irrespective of the quality of current legislation and regulation in the country of operation.

## ▮ Conclusions and Future Trends

As the previous discussion—and the following chapters—demonstrate, there is now a growing and rich theoretical and practical body of work on the topic of environment-related management accounting. There is also a trend towards integration of the work and practices being carried out within individual countries. US practice and research is becoming well known in Europe, and practitioners and researchers in individual European countries are also interacting to a greater degree.

This interaction has demonstrated that the relevance and form of environment-related management accounting is influenced by many contingent variables, such as organisational structure and strategic objectives. These are often related to national circumstances, which also have a direct influence: for example, by requiring collection of detailed environmental expenditure statistics.

One broad difference is between the two sides of the Atlantic Ocean. In general, the US has a tougher environmental liabilities regime and higher regulatory penalties, but somewhat lower resource costs, than Europe. Hence, much of the focus of environment-related management accounting in the former has been on recognising and avoiding liabilities and penalties. In Europe, there has been relatively greater attention paid to the systematic analysis of energy, materials and waste flows to identify opportunities for reduction, and also to consideration of further internalisation of the externalities created by resource consumption and transport use.

Within Europe, there is also a divide between, on the one hand, the UK and, on the other, Germany and several other countries in continental Europe. British companies—like those of the USA—are generally strongly influenced by capital markets, tend to have shorter decision-making horizons and are more likely to consider the creation of shareholder value as their principal corporate objective. By contrast, German and many other European companies are less dependent for finance on capital markets, usually have longer decision-making horizons and place greater relative weight on the interests of other stakeholders such as employees and communities. All of these have a considerable influence on environment-related management accounting in practice: for example, with regard to the introduction of systematic eco-balancing approaches (see above and James, Prehn and Steger 1997). However, there is currently a great deal of convergence both within Europe and between Europe and America, and it may be that some of these differences will become less important in future.

Despite these variations, the clear finding of most of the chapters in this book—

and other work in the field—is that there can be considerable business benefits from the application of environment-related management accounting and the development of 'green-bottom-line' frameworks. In most cases at present, these applications will be relatively simple ones, such as adjustments to investment appraisal procedures or ad hoc 'costs of waste' initiatives. However, as the Baxter, Xerox and other case studies in this volume indicate, there is also the opportunity for more advanced initiatives in organisations with substantial environmental costs and/or potential environment-related financial benefits. Although the numbers of these may be relatively low at present, they are likely to increase in future.

Environment-related management accounting can also reach beyond these utilitarian financial goals by helping to implement the goals of sustainable development within the business community. Of course, sustainability is about far more than economics, just as business encompasses many elements other than income statements and balance sheets. Nonetheless, environment-related management accounting can be a significant driver of action through demonstrating the long-term financial implications of sustainability, and creating a vision of the most appropriate responses. In this respect, it has what McAuley, Russell and Sims (1997) have termed a 'narrative' role of making sense of a complex world as well as a 'logico-scientific' one of developing an accurate representation of reality. One practical implication of this is the introduction of internal taxes, as advocated by Burritt (Chapter 8). Even when this is impractical, such a role can at least reduce any danger that environment-related management accounting could introduce a systematic bias towards environmental inaction. This could occur if immediate financial drivers are limited and net financial benefit is seen as the only justification for action.

However, it is important to avoid exaggerating the speed at which environment-related management accounting is likely to be adopted by business. There remain many internal barriers, of which the most significant is the difficulty of considering issues of risk and long-term benefit within high-discount-rate investment appraisal models. The strength and durability of these barriers will be determined primarily by the extent to which regulation and other political and social drivers increase the costs to business of poor environmental performance and enhance the financial incentives for environmental improvement. To date, these are much slower in developing than we and most environmentally concerned observers would like. As a result, the development of environment-related management accounting is likely to be discontinuous in nature. Most organisations will probably experience occasional bursts of activity as, on the one hand, external events such as the introduction of new taxes raise actual costs and/or business's awareness of them, or, on the other hand, internal changes, such as the introduction of ABC, provide opportunities.

This is consistent with the findings of a recent study to which the present authors made a contribution (Tuppen 1996). This concluded that most actions being taken at present are mainly relatively simple ones such as the identification and allocation of energy and waste disposal costs that either did arise, or could well have arisen, for non-environmental reasons.[3] However, it noted eight practical environment-related management accounting options that could be introduced by the report's sponsors, BT, and—by extension—other European companies.

One interpretation of these conclusions could be that environment-related management accounting is merely an instance of empire-building by academics, and that the simple tasks identified can easily be handled by existing accounting and environmental management accounting activities and staff. A similar criticism might be that the concept is merely a form of aggrandisement by the accounting profession, and that the tasks identified are already—and can in future be—accomplished successfully by environmental and operational managers. In either case, environment-related management accounting would be merely creating a new bottle for old wine which is already maturing nicely.

We have some sympathy for these points, and generally believe that the long-term aim of those interested in the field should be to make environment a part of everyday management accounting. However, we also note the evidence from many of the following chapters that a number of environment-related costs are often not identified by normal procedures, and that it therefore requires a systematic environment-related management accounting exercise to identify them and to drive action to reduce them. Equally, while we would certainly argue that the role of environment-related management accounting is, like management accounting generally, to support decision-makers in other functions, our experience also suggests that environmental and other operational managers often lack sufficient understanding of accounting concepts and techniques to utilise fully the information that these can provide. A final point is that some of the longer-term issues—such as the progressive internalisation of externalities or the broader requirements of sustainable development—can be difficult to integrate into management activities that normally focus on day-to-day operational issues. Hence, for the foreseeable future, the adoption of environment-related management accounting is likely to be an extraordinary rather than an ordinary activity, and one that will require unusual champions—such as financially astute environmental managers or environmentally aware accountants—to drive it. We hope that this volume will encourage the development of more such individuals within business.

## ▮ Appendix: Defining Costs and Benefits

There are no standardised definitions of environmental costs and benefits, despite their centrality to almost every discussion of environment-related management accounting (Department of the Environment 1996). In some countries, organisations are required to submit data on costs for tax and/or statistical purposes, but the definitions used vary between countries. They also tend to be biased towards defensive expenditures such as expenditure on pollution control equipment rather than more proactive expenditures such as, for example, expenditures to prevent pollution at source.

The management accounting maxim of 'different costs for different purposes'—which is based on a recognition that cost and benefit data are context-dependent and that universal consistency of definition is therefore impractical—is valid for the field of environment-related management accounting also (see Chapter 2). However, it is clear that the scale of imprecision in the field creates a need for some clarification of the alternative definitions that are available.

In broad terms, environmental costs and benefits from a management accounting

---

3. The report, entitled *Environmental Accounting in Industry: A Practical Review*, can be obtained, free of charge, from BT Environment Unit, PP1A57, Angel Centre, 403 St Johns St, London EC1V 4PL; tel: +44 (0)171 843 5266, or freephone: 0800 731 2403; fax: +44 (0)171 843 7881.

perspective are those for which environment-related factors such as current or likely future environmental legislation are a significant (though not necessarily the only) driver. A pragmatic definition (and one adopted by the AT&T 'green accounting team' described in Fig. 2) is that they are types of costs and benefits where the expertise of environmental professionals is important to their identification and management.

In the case of costs, this results in three generic categories:

- Internal environmental costs, i.e. expenses that are wholly or partially driven by environmental considerations
- External environmental costs, i.e. financial outgoings or other quantifiable disbenefits that are incurred outside the organisation but are not internalised within its accounts (see below and Chapter 20)
- Environmental opportunity costs, i.e. foregone benefits such as higher-than-necessary energy or waste costs

Clearly, the first of these categories is the easiest to quantify, although, as we discuss below, not necessarily the most important.

The potential financial benefits of environmental initiatives can be summarised as:

- Revenues arising from environmental action, e.g. the sale of materials recovered as a result of recycling, or additional or maintained sales of products and services whose markets are strongly influenced by environmental considerations
- Savings and avoided costs as a result of environmental action, e.g. through better use of energy and materials, or by introducing waste minimisation schemes that avoid the need to incur pollution control expenditures
- Intangible benefits arising from environmental actions, such as enhancing the value of a brand or reducing environment-related risks

### Internal Environment-Related Costs

From an environmental management perspective, the internal environment-related costs of a business or other organisation consist of two primary elements:

- Direct environmental expenses, i.e. those that are primarily related to environmental purposes
- Business-integrated environmental expenses, i.e. those that are partially related to environmental purposes but are also influenced by other business objectives

There is also a third, more ambiguous, category of resource expense, i.e. the costs of purchasing, handling and processing energy, materials, water and other resources to the point at which they become either products or wastes.

*Direct Environmental Expenses.* Following the quality costing model, these can be classified into three categories:

- **The costs of failure:** the costs of putting right, but only *after* they have arisen, environmental impacts or their potential causes. Examples include the capital and operating expenses of remediating past environmental impacts such as cleaning up contaminated land, and the use of end-of-pipe technologies to capture environmentally hazardous wastes, effluents and emissions.

- **Monitoring:** inspection and other costs to ensure that impacts are eliminated or detected, e.g. costs of air- and water-sampling equipment
- **Prevention:** costs of preventing environmental impacts, e.g. additional expenses of using water-based rather than solvent-based cleaners

Many environmental expenses are also incurred in the expectation that they will prove to be an investment that will justify itself through future benefits, through either cost savings or enhanced revenues. Baxter's Environmental Financial Statement (see Chapter 15) reports two separate categories of environmental expense: 'Remediation and waste disposal costs' and 'Costs of proactive programme'. The costs of the proactive programme, in contrast to those of remediation and waste disposal, are incurred in the expectation that they will generate future benefits: for example, the time spent by staff in redesign of products and their packaging. The total of these expenses can then be compared with the resulting benefits to provide an approximate cost–benefit evaluation.

There is no clear boundary between these proactive environmental expenses and business-integrated expenses: in both cases, there are usually both environmental and business benefits. One distinction that can be made is on the basis of the primary purpose of the expenditure. Proactive environmental expenses are likely to be driven primarily by environmental management staff, and are strongly influenced by environmental considerations. Business-integrated expenses, on the other hand, are those that are driven by non-environmental staff, for commercial reasons, and would probably be undertaken even if there were no benefit to the environment. However, this is a difficult demarcation to attempt to apply, and some would argue it can be distracting as it may encourage a competitive focus on the sources of initiative, which may discourage the cross-functional co-operation between environmental and operational managers which is necessary to achieve improvements.

*Business-Integrated Environmental Expenses.* Improvements in environmental performance are increasingly achieved by incorporating environmental concerns into normal operations, investment decisions, etc.: for example, into a decision to invest in new technology that reduces wastes and therefore has both environmental and business benefits. Often this can be done without incurring any additional expense. When this is not the case, it can be extremely difficult—and frequently somewhat arbitrary—to determine the precise proportion of the expenses that has been incurred for environmental rather than for business purposes. Essentially, this requires that the actual cost of the investment, etc., is compared against the amount that *would* have been spent, hypothetically, if environment had not been a consideration in the decision. The environmental portion of the cost is then the incremental amount in excess of this comparator, which—since by definition it will not actually be incurred—is a hypothetical figure that must be estimated. However, some companies have felt that this is worth doing when environmental considerations have driven a significant proportion of any expenditures.

A special case of business-integrated expenses is the time that mainstream business staff spend on environmental management. A significant proportion of these costs may be hidden in cost codes that do not make this apparent. For example, a large chemicals manufacturer found through an activity analysis of its staff that nearly 3% of the time of operational staff was being spent on the non-value adding task of capturing and collecting environment-related production data (e.g. reading meters and taking samples) in order to meet its legal reporting requirements. This identified the potential to invest in new computerised environmental information systems, to reduce this burden.

*Resource Expenses.* Resource expenses differ from direct and business-integrated environmental expenses in being an integral part of doing business rather than an unwanted overhead. On the other hand, the winning, processing, distribution and use of resources such as energy and materials is a major source of environmental impact. The possibility that resources are not being used in a sustainable manner is also considered by many to be an environmental issue.

Even more importantly, resource expenses are a major example of opportunity costs. Research suggests that the 'costs of resource inefficiency', i.e. the costs of purchasing and processing materials and energy that end up as waste, can be very significant in many companies and usually outweigh direct and business-integrated environmental expenses combined. This is especially true in Europe where the liability costs associated with accidents or contamination are less onerous than in the USA. The German textile producer Kunert, for example, has calculated that its costs of resource inefficiency amount to around 10% of turnover (Kunert 1995).

In practice, of course, the laws of thermodynamics mean that no organisation can completely eliminate waste. Hence, the practical use of the concept of the costs of resource inefficiency requires the identification of some feasible level of wastage with which the current situation can be compared. Nonetheless, the case of Zeneca in Chapter 19 is just one that demonstrates how wide this gap can sometimes be.

One further point to note with regard to the costs of resource inefficiency is that conventional thinking sees the identification and elimination of these as the objective and practice of day-to-day operational management. Hence, by the criterion identified above—areas of cost and benefit where the expertise of environmental professionals is important—this is not, or should not be, environment-related in any sense. In practice, however, there is much evidence that many organisations do not routinely identify their total costs of inefficiency and that waste minimisation and similar environment-related initiatives can often provide the first effective stimulus to do so. This is true of several of our case studies, notably Xerox (Chapter 18) and Zeneca (Chapter 19).

## ▇ Internal Environment-Related Benefits

*Revenues.* The most obvious source of environment-related revenues is those arising from recycling and similar schemes, which can sometimes be significant: for example, Baxter identifies over $5 million in recycling income in 1996. However, for most companies, greater benefits are available by making improvements that have effects upstream in the '3 Rs' sequence (reduce–re-use–recycle), e.g. by changing processes in order to avoid wastes arising in the first place rather than merely by maximising the income from recycling them. It may also be risky, particularly for an environmental management function that is still establishing its position in the organisation, to encourage a perception in the rest of the company that good environmental management is represented by managing bottle banks and waste paper collection systems. This could have the effect of positioning environmental management as peripheral to the main business and make it more difficult to achieve a positive involvement in mainstream business processes such as capital investment decisions.

In the longer term, a much more significant revenue stream is likely to be from products or services that are sold at least partially on the basis of environmental considerations. This is difficult to quantify, for reasons that are similar to those that make calculation of business-integrated environmental expenses difficult: for example, the difficulty of apportioning when environment is only one of several drivers. This is even more true when environmental attributes have the role of maintaining existing markets that might otherwise have been lost. Nonetheless, both are likely to be of considerable importance in future; indeed, some would see them as the most significant 'green bottom line' element in coming decades (Fussler with James 1996).

*Savings and Cost Avoidance.* 'Savings' are any direct cost savings resulting from environment-related actions, as indicated by reductions in the absolute amounts of spending on a cost item—for example, hazardous waste disposal costs—from one year to the next. However, this measure alone may be misleading. Even if real improvements in environmental performance have been achieved, the absolute amount spent could still increase rather than decrease if either the volume of output has increased more than proportionately, or the prices charged for the product or service that is being purchased have increased.

The latter is particularly likely with environment-related costs, and is one of the main stimuli for companies to take action. For example, the landfill tax recently introduced in the UK means that waste disposal costs have increased by more than would otherwise have been justified by either general inflation or market-generated price changes. Similarly, the costs per kilo of ozone-depleting substances such as CFCs have increased several-fold over recent years since the Montreal Protocol. The real value of the improvements in performance is therefore indicated best by a comparison of actual current spending, not against previous years' spending in absolute terms, but against an estimate of what the cost *would* have been if that improvement had not taken place but market prices and business volumes had continued to increase. This is a hypothetical figure and therefore less easy to calculate and justify than an actual figure taken from an accounting ledger, but a more realistic measure of real benefit. It is the basis adopted by Baxter International in its calculations of 'cost avoidance' in preparing its Environmental Financial Statement (see Chapter 15).

*Intangible Benefits.* Benefits such as those arising from enhanced customer perceptions of the company, or improved staff morale, are also real but even more difficult to attempt to measure. In practice, it is usually unlikely to be worthwhile to attempt to quantify the benefits, at least not in monetary form, though this should not mean that their existence is then overlooked. To avoid this, Baxter note at the foot of their Environmental Financial Statement, without quantification, 'Examples of Undetermined Savings' (and also, for transparency, 'Examples of Undetermined Costs').

*Reduction of Risks.* One important kind of intangible (and sometimes tangible) benefit is a reduction of environment-related financial risks. Well-publicised business failures such as Barings Bank have helped to encourage an increasing awareness of the financial benefits of risk management in business generally. The environment can pose several risks for many businesses, and proactive environmental management aims positively to anticipate and reduce possible risks as well as to deal with the consequences of those risks that unfortunately have actually been realised. A company's competence in managing its risks, including environmental risks, is also relevant to its financial stakeholders, such as investors, bankers and insurers, and can affect the costs of insurance and of raising new capital (*Business and the Environment* 1998; Lascelles 1993; Leggett 1995; Mansley 1995).

There are several types of possible environment-related risk for a business:

- The risk of a major incident or catastrophe such as Bhopal or the Exxon Valdez

- Adverse environmental impacts over a period of time, such as leakage of toxic wastes into land and underground water-courses. These may be more difficult for a business to cope with than a major incident, since these gradual risks are becoming more difficult, if not impossible, to insure against.

- Damage to reputation and public image, which can have an effect on the business not only through the marketplace but also through the ease with which it can obtain new permits or changes to existing permits, which may depend on the perceptions of local communities

- Where a business's operational process and cost structure is based on environmentally unsustainable assumptions—for example, that private road transport will continue indefinitely to be readily available at (relatively) low cost—there is a risk that changes through (for example) environmental taxes could mean that a previously profitable business becomes no longer viable.

# 2 An Introduction to Environmental Accounting as a Business Management Tool

Key Concepts and Terms[1]

US Environmental Protection Agency

THE CENTRAL PURPOSE of this chapter is twofold: (1) to orient readers to key concepts often referred to as 'environmental accounting'; and (2) to explain how the terms that refer to environmental accounting are currently being used, so that confusion about the terms does not impede progress in understanding and applying the core concepts.

EPA's Environmental Accounting Project produced the primer on which this chapter is based at the behest of stakeholders who suggested that an important step in promoting environmental accounting is to clarify key concepts and terms to facilitate more widespread adoption of environmental accounting practices.[2]

---

1. This is an edited version of a longer report of the same title published by the Environmental Protection Agency in 1995 (US EPA 1995a). It was prepared by ICF, Inc. for EPA's Environmental Accounting Project. The full report can be downloaded from the Project's website at *http://www.epa.gov/opptintr/acctg.*
   This chapter refers to environmental accounting activities at several companies in North America. These examples are by no means exhaustive of the many laudable efforts under way to implement environmental accounting at firms in many different industries. Moreover, by mentioning these examples, EPA is not necessarily endorsing their approaches or terminology.

2. In December 1993, a national workshop of experts drawn from business, professional groups, government, non-profits and academia produced an 'Action Agenda' 'to encourage and motivate businesses to understand the full spectrum of environmental costs and integrate these costs in decision-making' (US EPA 1994). The Agenda identifies four overarching issue areas that require attention in order to advance environmental accounting: (1) better understanding of terms and concepts; (2) creation of internal and external management incentives; (3) education, guidance and outreach; and (4) development and dissemination of analytical tools, methods and systems. The purpose of this document is to help address the first recommendation. The US Chamber of Commerce, the Business Roundtable, the American Institute of Certified Public Accountants, the Institute of Management Accountants, AACE International (the Society of Total Cost Management) and the US EPA co-sponsored the workshop.

It was prepared as a starting-point for readers who have questions about environmental accounting. The intended audience includes business managers and other professionals who wish to understand environmental accounting. In addition, people involved with activity-based costing, total quality management, business re-engineering, or design for the environment should find environmental accounting to be compatible with and potentially helpful to their programmes.

This chapter focuses on environmental accounting as a management tool for a variety of purposes, such as improving environmental performance, controlling costs, investing in 'cleaner' technologies, developing 'greener' processes and products, and informing decisions related to product mix, product retention and product pricing. The chapter does not cover all of these potential applications but does summarise how environmental accounting can be applied to cost allocation, capital budgeting and process/product design. Specific applications of environmental accounting are illustrated in case studies that EPA has prepared which document companies' programmes to implement environmental accounting.[3]

## ▌ Introduction

The term 'environmental accounting' has many meanings and uses. Environmental accounting can support national income accounting, financial accounting or internal business managerial accounting. This chapter focuses on the application of environmental accounting as a managerial accounting tool for internal business decisions. Moreover, the term 'environmental cost' has at least two major dimensions: (1) it can refer solely to costs that directly impact a company's bottom line (here termed 'private costs'); or (2) it can also encompass the costs to individuals, society and the environment for which a company is not accountable (here termed 'societal costs'). The discussion in this chapter concentrates on private costs because that is where companies starting to implement environmental accounting typically begin. However, much of the material is applicable to societal costs as well.

## ▌ Why Undertake Environmental Accounting?

Environmental costs represent one of the many different types of cost that businesses incur as they provide goods and services to their customers. Environmental performance is one of the many important measures of business success. Environmental costs and performance deserve management attention for the following reasons:

1. Many environmental costs can be **significantly reduced or eliminated** as a result of business decisions, ranging from operational and housekeeping changes, to investment in 'greener' process technology, to redesign of processes/products. Many environmental costs (e.g. wasted raw materials) may provide no added value to a process, system or product.

2. Environmental costs (and, thus, potential cost savings) **may be obscured in overhead accounts or otherwise overlooked.**

---

3. For more information on EPA's activities in this area or for copies of the case studies, please contact the EPA's Pollution Prevention Information Clearinghouse by telephoning +1 202 260 1023.

3. Many companies have discovered that **environmental costs can be offset by generating revenues** through sale of waste by-products or transferable pollution allowances, or licensing of clean technologies, for example.

4. Better management of environmental costs can result in **improved environmental performance and significant benefits to human health** as well as business success.

5. Understanding the environmental costs and performance of processes and products can promote **more accurate costing and pricing** of products and can aid companies in the **design of more environmentally preferable** processes, products and services for the future.

6. **Competitive advantage** with customers can result from processes, products and services that can be demonstrated to be environmentally preferable.

7. Accounting for environmental costs and performance can support a company's development and operation of an overall **environmental management system**.

EPA's work with key stakeholders leads it to believe that, as businesses more fully account for environmental costs and benefits, they will clearly see the financial advantages of pollution prevention (P2) practices. Environmental costs often can be reduced or avoided through P2 practices such as product design changes, input materials substitution, process redesign and improved operation and maintenance (O&M) practices. For example, increased environmental costs may result from use of chemical A (e.g. a chlorinated solvent), but not from chemical B (e.g. an aqueous-based solvent). This is true even though chemical A and chemical B can be substituted for one another. Another example: some environmental compliance costs are required only when use of a substance or generation of a waste exceeds a defined threshold. A company that can reduce chemical use below such thresholds or employ substitutes for regulated chemicals can realise substantial cost savings from design, engineering and operational decisions.

In two of the most thorough reports on the subject of pollution prevention in the industrial community, the not-for-profit group INFORM (1985, 1992) studied 29 companies in the organic chemical industry in 1985 and again in 1992. This research found that chemical 'plants with some type of environmental cost accounting programme' had 'an average of three times as many' P2 projects 'as plants with no cost accounting system' (1992: 31). The study also showed that the average annual savings per P2 project in production facilities, where data were available, were just over $351,000, which equalled an average savings of $3.49 for every dollar spent. Not only were substantial savings and returns on investment documented for P2 projects, but an average of 1.6 million pounds of waste were reduced for each project.

Results such as these have highlighted the potential benefits of environmental accounting to the business community. For example, responses to a questionnaire administered by George Nagle of the Bristol-Myers Squibb Company at the Spring 1994 Global Environmental Management Initiative (GEMI) Conference showed that corporate professionals are placing a high priority on environmental accounting (GEMI 1994b). Of the 25 respondents to the informal survey, half stated that their company had some form of a tracking system for environmental costs. All but two reported that they believed environmental accounting issues would be more important to their companies in the near future.

## ▌ What is Environmental Accounting?

Different uses of the umbrella term 'environmental accounting' arise from three distinct contexts (see Fig. 1).

- **National income accounting** is a macro-economic measure. Gross Domestic Product (GDP) is an example. The GDP is a measure of the flow of goods and services through the economy. It is often cited as a key measure of our society's economic well-being. The term 'environmental accounting' may refer to this national economic context. For example, environmental accounting can use physical or monetary units to refer to the consumption of the nation's natural resources, both renewable and non-renewable. In this context, environmental accounting has been termed 'natural resources accounting'.

- **Financial accounting** enables companies to prepare financial reports for use by investors, lenders and others. Publicly held corporations report information on their financial condition and performance through quarterly and annual reports, governed by rules set by the US Securities and Exchange Commission (SEC) with input from industry's self-regulatory body, the Financial Accounting Standards Board (FASB). Generally accepted accounting principles (GAAP) are the basis for this reporting. Environmental accounting in this context refers to the estimation and public reporting of environmental liabilities and financially material environmental costs.

- **Management accounting** is the process of identifying, collecting and analysing information principally for internal purposes.[4] Because a key purpose of management accounting is to support a business's forward-looking management decisions, it is the focus of the remainder of this chapter. Management accounting can involve data on costs, production levels, inventory and backlog, and other vital aspects of a business. The information collected under a business's management accounting system is used to plan, evaluate and control in a variety of ways:

  1. Planning and directing management attention

  2. Informing decisions such as purchasing (e.g. make versus buy), capital investments, product costing and pricing, risk management, process/product design and compliance strategies

  3. Controlling and motivating behaviour to improve business results

| Type of environmental accounting | Focus | Audience |
|---|---|---|
| 1. National income accounting | Nation | External |
| 2. Financial accounting | Firm | External |
| 3. Managerial or management accounting | Firm, division, facility, product line or system | Internal |

**Figure 1:** Contexts of Environmental Accounting

4. 'Management accounting is the process of identification, measurement, accumulation, analysis, preparation, interpretation, and communication of financial information used by management to plan, evaluate, and control within an organisation and to assure appropriate use of and accountability for its resources' (Institute of Management Accountants 1981).

Unlike financial accounting, which is governed by generally accepted accounting principles (GAAP), management accounting practices and systems differ according to the needs of the businesses they serve. Some businesses have simple systems, others have elaborate ones. Just as management accounting refers to the use of a broad set of cost and performance data by a company's managers in making a myriad of business decisions, environmental accounting refers to the use of data about environmental costs and performance in business decisions and operations. Figure 2 lists many types of internal management decisions that can benefit from the consideration of environmental costs and benefits. This chapter later summarises how environmental accounting can be integrated into cost allocation, capital budgeting and process/product design.

## ◼ What is an Environmental Cost?

Uncovering and recognising **environmental costs** associated with a product, process, system or facility is important for good management decisions. Attaining goals such as reducing environmental expenses, increasing revenues and improving environmental performance requires paying attention to current, future and potential **environmental costs**. How a company defines an environmental cost depends on how it intends to use the information (e.g. cost allocation, capital budgeting, process/product design,

| ☑ Product design | ☑ Capital investments |
| --- | --- |
| ☑ Process design | ☑ Cost control |
| ☑ Facility siting | ☑ Waste management |
| ☑ Purchasing | ☑ Cost allocation |
| ☑ Operational | ☑ Product retention and mix |
| ☑ Risk management | ☑ Product pricing |
| ☑ Environmental compliance strategies | ☑ Performance evaluations |

**Figure 2:** Types of Management Decision Benefiting from Environmental Cost Information

other management decisions) and the scale and scope of the exercise. Moreover, it may not always be clear whether a cost is 'environmental' or not; some costs fall into a grey area or may be classified as partly environmental and partly not. Whether or not a cost is 'environmental' is not critical; the goal is to ensure that relevant costs receive appropriate attention.

## Identifying Environmental Costs

Environmental accounting terminology uses such words as 'full', 'total', 'true' and 'life-cycle' to emphasise that traditional approaches were incomplete in scope because they overlooked important environmental costs (and potential cost savings and revenues) (see, e.g., Bailey 1991). In looking for and uncovering relevant environmental costs, managers may want to use one or more organising frameworks as tools. This section presents examples of environmental costs as well as a framework that has been used to identify and classify environmental costs.

There are many different ways to categorise costs. Accounting systems typically classify costs as:

1. Direct materials and labour
2. Manufacturing or factory overhead (i.e. operating costs other than direct materials and labour)[5]
3. Sales
4. General and administrative (G&A) overhead[6]
5. Research and development (R&D)

Environmental expenses may be classified in any or all of these categories in different companies. To better focus attention on environmental costs for management decisions, the *EPA Pollution Prevention Benefits Manual* (US EPA 1989) and the Global Environmental Management Initiative (GEMI) environmental cost primer (GEMI 1994a) use similar organising frameworks to distinguish costs that generally receive management attention (termed the 'usual' costs or 'direct' costs) from costs that may be obscured through treatment as overhead or R&D, distorted through improper allocation to cost centres or simply overlooked (termed 'hidden', 'contingent', 'liability' or 'less tangible' costs).[7] Figure 3 lists examples of these costs under the labels 'conventional', 'potentially hidden', 'contingent' and 'image/relationship' costs.

*Conventional Costs.* The costs of using raw materials, utilities, capital goods and supplies are usually addressed in cost accounting and capital budgeting, but are not usually considered environmental costs. However, decreased use and less waste of raw materials, utilities, capital goods and supplies is environmentally preferable, reducing

---

5. Manufacturing or factory overhead typically includes indirect materials and labour, capital depreciation, rent, property taxes, insurance, supplies, utilities, repairs and maintenance, and other costs of operating a factory.

6. General and administrative costs may be pooled with sales costs (i.e. SG&A) or as part of 'technical, sales and general administrative' costs (i.e. TSGA).

7. The EPA's *Pollution Prevention Benefits Manual* (US EPA 1989) introduced the terminology that distinguishes between 'usual', 'hidden', 'liability' and 'less tangible' costs. This framework was largely adopted in *Finding Cost-Effective Pollution Prevention Initiatives: Incorporating Environmental Costs into Business Decision Making* (GEMI 1994a), which uses the terms 'direct', 'hidden', 'contingent liability' and 'less tangible' costs.

| Potentially hidden costs | | |
|---|---|---|
| **Regulatory** | **Upfront** | **Voluntary (beyond compliance)** |
| Notification | Site studies | Community relations/outreach |
| Reporting | Site preparation | Monitoring/testing |
| Monitoring/testing | Permitting | Training |
| Studies/modelling | R&D | Audits |
| Remediation | Engineering and procurement | Qualifying suppliers |
| Record-keeping | Installation | Reports |
| Plans | |    (e.g. annual environmental reports) |
| Training | **Conventional costs** | Insurance |
| Inspections | | Planning |
| Manifesting | Capital equipment | Feasibility studies |
| Labelling | Materials | Remediation |
| Preparedness | Labour | Recycling |
| Protective equipment | Supplies | Environmental studies |
| Medical surveillance | Utilities | R&D |
| Environmental | Structures | Habitat and wetland protection |
| Financial assurance | Salvage value | Landscaping |
| Pollution control | | Other environmental projects |
| Spill response | **Back-end** | Financial support to environmental |
| Storm-water management | |    groups and/or researchers |
| Waste management | Closure/decommissioning | |
| Taxes/fees | Disposal of inventory | |
| | Post-closure care | |
| | Site survey | |
| Contingent costs | | |
| Future compliance costs | Remediation | Legal expenses |
| Penalties/fines | Property damage | Natural resource damages |
| Response to future releases | Personal injury damage | Economic loss damages |
| Image and relationship costs | | |
| Corporate image | Relationship with | Relationship with lenders |
| Relationship with customers |    professional staff | Relationship with host communities |
| Relationship with investors | Relationship with workers | Relationship with regulators |
| Relationship with insurers | Relationship with suppliers | |

**Figure 3:** Examples of Environmental Costs Incurred by Firms

both environmental degradation and consumption of non-renewable resources. It is important to factor these costs into business decisions, whether or not they are viewed as 'environmental' costs. The dashed line around these **conventional costs** in Figure 3 indicates that even these costs (and potential cost savings) may sometimes be overlooked in business decision-making.

*Potentially Hidden Costs.* Figure 3 collects several types of environmental cost that may be potentially hidden from managers: first are **upfront environmental costs**, which are incurred prior to the operation of a process, system or facility. These can include costs related to siting, design of environmentally preferable products or processes, qualifications of suppliers, evaluation of alternative pollution control equipment, and so on. Whether classified as overhead or R&D, these costs can easily be forgotten when managers and analysts focus on operating costs of processes, systems and facilities. Second are **regulatory** and **voluntary environmental costs** incurred in *operating* a process, system or facility; because many companies traditionally have treated these costs as overhead, they may not receive appropriate attention from managers and analysts responsible for day-to-day operations and business decisions.

The magnitude of these costs may also be more difficult to determine as a result of their being pooled in overhead accounts. Third, while upfront and current operating costs may be obscured by management accounting practices, **back-end environmental costs** may not be entered into management accounting systems at all. These environmental costs of current operations are *prospective*, meaning they will occur at more or less well-defined points in the future. Examples include the *future* cost of decommissioning a laboratory that uses licensed nuclear materials, closing a landfill cell, replacing a storage tank used to hold petroleum or hazardous substances, and complying with regulations that are not yet in effect but have been promulgated. Such back-end environmental costs may be overlooked if they are not well documented or accrued in accounting systems.

Figure 3 contains a lengthy list of **'potentially hidden' environmental costs**, including examples of the costs of upfront, operational and back-end activities undertaken (1) to comply with environmental laws (i.e. regulatory costs); or (2) to go beyond compliance (i.e. voluntary costs). In bringing these costs to light, it also may be useful to distinguish between costs incurred to respond to *past pollution* not related to *ongoing operations*; to control, clean up or prevent pollution from *ongoing operations*; or to prevent or reduce pollution from *future operations*.

*Contingent Costs.* Costs that may or may not be incurred at some point in the future—here termed **'contingent costs'**—can best be described in probabilistic terms: their expected value, their range or the probability of their exceeding some dollar amount. Examples include the costs of remedying and compensating for future accidental releases of contaminants into the environment (e.g. oil spills), fines and penalties for future regulatory infractions, and future costs due to unexpected consequences of permitted or intentional releases. These costs may also be termed 'contingent liabilities' or 'contingent liability costs'. Because these costs may not currently need to be recognised for other purposes, they may not receive adequate attention in internal management accounting systems and forward-looking decisions.

*Image and Relationship Costs.* Some environmental costs are called 'less tangible' or 'intangible' because they are incurred to affect subjective (though measurable) perceptions of management, customers, employees, communities and regulators. These costs have also been termed **'corporate image'** and **'relationship'** costs. This category can include the costs of annual environmental reports and community relations activities, costs incurred voluntarily for environmental activities (e.g. tree-planting) and costs incurred for P2 award/recognition programmes. The costs themselves are not 'intangible', but the direct benefits that result from relationship/corporate image expenses often are.

### Is it an 'Environmental' Cost?

Costs incurred to comply with environmental laws are clearly environmental costs. Costs of environmental remediation, pollution control equipment, and non-compliance penalties are all unquestionably environmental costs. Other costs incurred for environmental protection are likewise clearly environmental costs, even if they are not explicitly required by regulations or go beyond regulatory compliance levels.

There are other costs, however, that may fall into a grey area in terms of being considered environmental costs. For example, should the costs of production equipment be considered 'environmental' if it is a 'clean technology?' Is an energy-efficient turbine an 'environmental' cost? Should efforts to monitor the shelf-life of raw materials and supplies in inventory be considered 'environmental' costs (if discarded, they become waste and result in environmental costs)? It may also be difficult to distinguish some environmental costs from health and safety costs or from risk management costs.

The success of environmental accounting does not depend on 'correctly' classifying all the costs a firm incurs. Rather, its goal is to ensure that relevant information is made available to those who need or can use it. To handle costs in the grey area, some firms use the following approaches:

- Allowing a cost item to be treated as 'environmental' for one purpose but not for another

- Treating part of the cost of an item or activity as 'environmental'

- Treating costs as 'environmental' for accounting purposes when a firm decides that a cost is more than 50% environmental

There are many options. Companies can define what should constitute an 'environmental cost' and how to classify it, based on their goals and intended uses for environmental accounting. For example, if a firm wants to encourage pollution prevention in capital budgeting, it might consider distinguishing (1) environmental costs that can be avoided by pollution prevention investments from (2) environmental costs related to remedying contamination that has already occurred. But for product costing purposes, such a distinction might not be necessary because both are costs of producing the good or service.

## ▌ Is There a Proper Scale and Scope for Environmental Accounting?

Environmental accounting is a flexible tool that can be applied at different scales of use and different scopes of coverage. This section describes some of the options for applying environmental accounting.

### Scale

Depending on corporate needs, interests, goals and resources, environmental accounting can be applied at different scales, including those shown in Figure 4. Specific environmental accounting issues or challenges may vary depending on the scale of its application.

### Scope

Whatever the scale, there also is an issue of scope. An initial scope question is whether environmental accounting extends beyond conventional costs to include potentially hidden, future, contingent and image/relationship costs. Another scope issue is whether companies intend to consider only those costs that directly affect their bottom-line

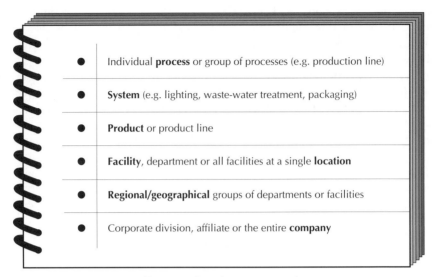

- Individual **process** or group of processes (e.g. production line)

- **System** (e.g. lighting, waste-water treatment, packaging)

- **Product** or product line

- **Facility**, department or all facilities at a single **location**

- **Regional/geographical** groups of departments or facilities

- Corporate division, affiliate or the entire **company**

**Figure 4:** Different Scales of Environmental Accounting

financial profit or loss (see, e.g., examples of costs listed in Fig. 3) or whether companies also want to recognise the environmental costs that result from their activities but for which they are not accountable, referred to as 'societal' or 'external' costs. These latter costs are described in the next section.

Thus, the scope of environmental accounting refers to the types of cost included. As the scope becomes more expansive, firms may find it more difficult to assess and measure certain environmental costs. This is illustrated by Figure 5.

## ▌ What is the Difference between Private Costs and Societal Costs?

Understanding the distinction between private and societal costs is necessary when discussing environmental accounting, because common terms are often used inconsistently

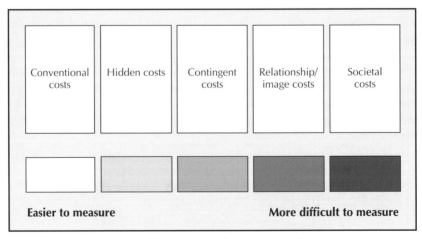

| Conventional costs | Hidden costs | Contingent costs | Relationship/ image costs | Societal costs |

**Easier to measure**                    **More difficult to measure**

**Figure 5:** The Spectrum of Environmental Costs

to refer to one or both of those cost categories. Figure 6 provides a graphical representation of the important difference between private and societal costs. It also shows that many private costs are not currently considered in decision-making. This perspective can apply to a process, product, system, facility or an entire company.

The innermost box in Figure 6, labelled 'conventional company costs' includes the many costs that businesses typically track well (e.g. capital costs, labour, material). Many of these costs may already be directly allocated to the responsible processes or products in cost accounting systems and be included in financial evaluations of capital expenditures. The larger unshaded box includes all of the potentially overlooked costs a business incurs. Examples of these costs are shown previously in Figure 3.

Together, the unshaded area represents **'private costs'**, which are the costs a business incurs or for which a business can be held accountable (i.e. legally responsible). These are the costs that can directly affect a firm's bottom line.

The outside, shaded box, labelled **'societal costs'**, represents the costs of business's impacts on the environment and society for which business is not legally accountable. (These costs are also called 'externalities' or 'external costs'.) Societal costs include both (1) environmental degradation for which firms are not legally liable; and also (2) adverse impacts on human beings, their property and their welfare (e.g. employment impacts of spills) that cannot be compensated through the legal system. For example, damage caused to a river because of polluted waste-water discharges, or to ecosystems from solid waste disposal, or to asthmatics because of air pollutant emissions, are all examples of societal costs for which a business often does not pay.

Because laws can vary from state to state, the boundary between societal and private costs may differ as well. At present, valuing societal costs is both difficult and controversial; nevertheless, some businesses are attempting to address these costs and EPA supports their efforts. A major North American power utility, Ontario Hydro, has made a corporate commitment to determine external impacts and, to the extent

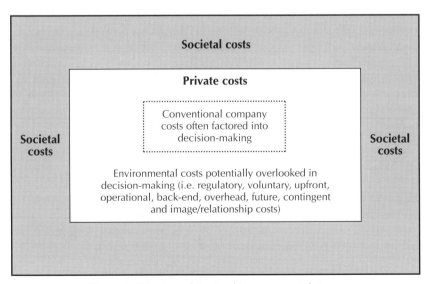

**Figure 6:** Private and Societal Environmental Costs
*Source: Adapted from White, Becker and Savage 1993*

possible, value societal costs in order to integrate them into its planning and decision-making (US EPA 1996c; see also Chapter 16, this volume). EPA urges businesses to address all private environmental costs shown in Figure 3, including hidden, future, contingent and image/relationship costs, to the extent practical. Companies are also encouraged to move beyond consideration of private costs to incorporate societal costs, at least qualitatively, into their business decisions.

### Life-cycle perspective can help to identify private and societal costs.

The life-cycle of a product, process, system or facility can refer to the suite of activities starting with acquisition (and upfront pre-acquisition activities) and concluding with back-end disposal/decommissioning that a specific firm performs or is responsible for. This life-cycle perspective can foster a thorough accounting of private costs (and potential cost savings) in addition to facilitating a more systematic and complete assessment of societal impacts and costs due to a firm's activities.

## ■ Who Can Undertake Environmental Accounting?

Environmental accounting can be employed by firms large and small, in almost every industry in both the manufacturing and services sectors. It can be applied on a large scale or a small scale, systematically or on an as-needed basis. The form it takes can reflect the goals and needs of the company using it. However, in any business, top-management support and cross-functional teams are likely to be essential for the successful implementation of environmental accounting because:

- Environmental accounting may entail a new way of looking at a company's environmental costs, performance and decisions. Top-management commitment can set a positive tone and articulate incentives for the organisation to adopt environmental accounting.

- Companies will probably want to assemble cross-functional teams to implement environmental accounting, bringing together designers, chemists, engineers, production managers, operators, financial staff, environmental managers, purchasing personnel and accountants who may not have worked together before. Because environmental accounting is not solely an accounting issue, and the information needed is split up among all of these groups, these people need to talk to each other to develop a common vision and language and make that vision a reality.

AT&T is one example of a company that has combined senior-management support and use of a cross-functional team for its environmental accounting initiative (US EPA 1995b).

Companies with formal environmental management systems may want to institutionalise environmental accounting because it is a logical decision-support tool for these systems. Similarly, many companies have begun or are exploring new business approaches in which environmental accounting can play a part:

- Activity-based costing/activity-based management
- Total quality management/total quality environmental management

- Business process re-engineering/cost reduction
- Cost of quality model/cost of environmental quality model
- Design for environment/life-cycle design
- Life-cycle assessment/life-cycle costing

All of these approaches are compatible with environmental accounting and can provide platforms for integrating environmental information into business decisions. Companies using or evaluating these approaches may want to consider explicitly adopting environmental accounting as part of these efforts.

Small businesses that may not have formal environmental management systems, or are not using any of the above approaches, have also successfully applied environmental accounting. As with larger firms, management commitment and cross-functional involvement are necessary.

## ❚ Applying Environmental Accounting to Cost Allocation

An important function of environmental accounting is to bring environmental costs to the attention of corporate stakeholders who may be able and motivated to identify ways of reducing or avoiding those costs while at the same time improving environmental quality.

This can require, for example, pulling some environmental costs out of overhead and allocating those environmental costs to the appropriate accounts. By *allocating* environmental costs to the products or processes that generate them, a company can motivate affected managers and employees to find creative pollution prevention alternatives that lower those costs and enhance profitability. For example, Caterpillar's East Peoria, Illinois, plant no longer dumps waste disposal costs into an overhead account; rather, the costs of waste disposal are allocated to responsible commodity groups, triggering efforts to improve the bottom line through pollution prevention (Owen 1995).

**Overhead** is any cost that, in a given cost accounting system, is not wholly attributed to a single process, system, product or facility. Examples can include supervisors' salaries, janitorial services, utilities and waste disposal. Many environmental costs are often treated as overhead in corporate cost accounting systems. Traditionally, an overhead cost item has been handled in either one of two ways: (1) it may be allocated on some basis to specific products; or (2) it may be left in the pool of costs that are not attributed to any specific product.

If overhead is allocated incorrectly, one product may bear an overhead allocation greater than warranted, while another may bear an allocation smaller than its actual contribution. The result is poor product costing, which can affect pricing and profitability. Alternatively, some overhead costs may not be reflected at all in product cost and price. In both instances, managers cannot perceive the true cost of producing products and thus internal accounting reports provide inadequate incentives to find creative ways of reducing those costs.

Separating environmental costs from overhead accounts where they are often hidden and allocating them to the appropriate product, process, system or facility directly responsible reveals these costs to managers, cost analysts, engineers, designers and others. This is critical not only for a business to have accurate estimates of production costs for different product lines and processes, but also to help managers target cost-reduction activities that can also improve environmental quality. The axiom 'one cannot manage what one cannot see' pertains here.

There are two general approaches to allocating environmental costs: (1) build proper cost allocation directly into cost accounting systems; or (2) handle cost allocation outside of automated accounting systems. Companies may find that the latter approach can serve as an interim measure while the former option is being implemented.

The four steps of environmental cost allocation can be described as follows:

1. Determine scale and scope
2. Identify environmental costs
3. Quantify those costs
4. Allocate environmental costs to responsible process, product, system or facility

A simple example illustrates the problem; Figure 7 depicts a traditional accounting system that assigns environmental and certain other costs to overhead. Such overhead costs are generally allocated to widgets A and B in proportion to their consumption of labour and materials.

Figure 8 highlights the misallocation of environmental costs. Suppose widget B is solely responsible for toxic waste management costs, and widget A creates no toxic waste costs. The misallocation occurs because the toxic waste management cost is lumped together in an overhead cost pool that is misallocated to both widgets A and B, even though none of the toxic waste management cost results from the production of widget A. The effect is to distort the actual costs of producing widget A and widget B.

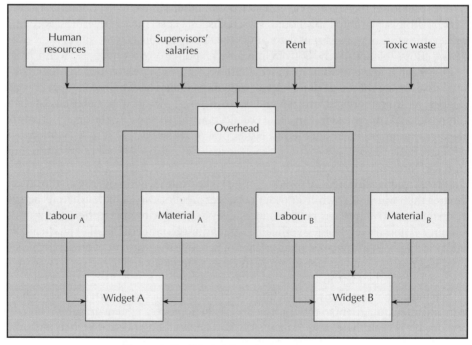

**Figure 7:** Traditional Cost Accounting System
*Source: Todd 1992*

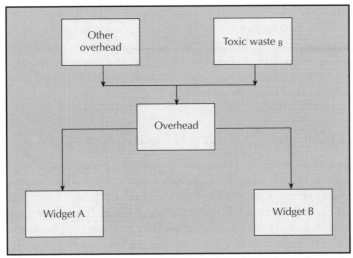

**Figure 8:** Misallocation of Environmental Costs under Traditional Cost System
*Source: Todd 1992*

Figure 9 illustrates a cost accounting system that correctly attributes the environ-mental costs of widget B only to widget B. By breaking environmental costs out of overhead and directly attributing them to products, managers will have a much clearer view of the true costs of producing widgets A and B. Alternatively, environmental costs can be allocated to responsible processes, systems or departments. Environmental costs resulting from several processes or products may need to be allocated based on

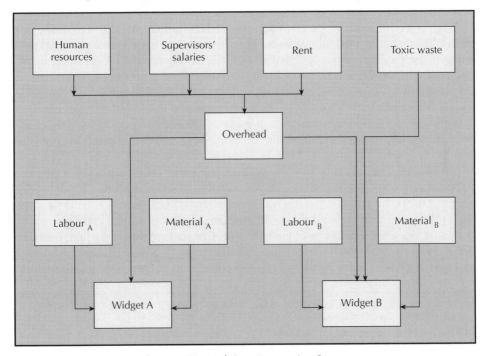

**Figure 9:** Revised Cost Accounting System
*Source: Todd 1992*

a more complex analysis. And future costs (e.g. toxic waste disposal) may need to be amortised and allocated to proper cost centres.

The preceding discussion applies equally to the appropriate crediting of revenues derived from sale or use of by-products or recyclables (e.g. raw materials and supplies). Although the focus of cost allocation is on environmental costs, environmental revenues should be treated in a parallel fashion.

## ▊ Applying Environmental Accounting to Capital Budgeting

Capital budgeting includes the process of developing a firm's planned capital investments. It typically entails comparing predicted cost and revenue streams of current operations and alternative investment projects against financial benchmarks in light of the costs of capital to a firm (White and Becker 1992). It has been quite common for financial analysis of investment alternatives to exclude many environmental costs, cost savings and revenues. As a result, corporations may not have recognised financially attractive investments in pollution prevention and 'clean technology'. This is beginning to change.

When evaluating a potential capital investment, it is important to consider fully environmental costs, cost savings and revenues to place pollution prevention investments on a level playing field with other investment choices. To do this, identify and include the types of costs (and revenues) (i.e. the 'cost inventory') that will help to demonstrate the financial viability of a cleaner technology investment. Analyse qualitatively those data and issues that cannot be easily quantified, such as the potential less tangible benefits of pollution prevention investments. Figure 3 may help in identifying potentially relevant costs (and savings).

After collecting or developing environmental data (either from the accounting system or by manual means), allocate and project costs, cost savings and potential revenues to the products, processes, systems or facilities that are the focus of the capital budgeting decision. Begin with the easiest-to-estimate costs and revenues and work toward the more difficult-to-estimate environmental costs and benefits such as contingencies and corporate image. The benefit of improved corporate image and relationships due to pollution prevention investments can impact costs and revenues in ways that may be challenging to project in dollars and cents (See Fig. 10). For example, a company selected as 'Clean Air Partner of the Year' under a Colorado partnership programme attracted several new clients from the positive publicity.[8] Information about past expenditures on corporate image may also be helpful in estimating future benefits (e.g. potential savings or reductions in those outlays resulting from the investment) for companies that want to go beyond the qualitative consideration of these benefits.

Be sure to use appropriate financial indicators that include the time value of money (i.e. a dollar today is worth more than a dollar next year). Sound financial indicators include net present value,[9] internal rate of return,[10] and other profitability indices.

---

8. Reported by representative of Majestic Metals, 22 March 1995, at EPA Regional Office training programme on pollution prevention.

9. The present value of the future cash flows of an investment less the investment's current cost. It incorporates the time value of money.

10. The discount rate at which the net present value of a project is equal to zero.

- Increased sales due to enhanced company or product image
- Better borrowing access and terms
- Equity more attractive to investors
- Health and safety cost savings
- Increased productivity and morale of employees, greater retention, reduced recruiting costs
- Faster, easier approvals of facility expansion plans or changes due to increased trust from host communities and regulators
- Enhanced image with stakeholders such as customers, employees, suppliers, lenders, stock-holders, insurers and host communities
- Improved relationships with regulators

**Figure 10:** Potential Less Tangible Benefits of Pollution Prevention Investments

Payback,[11] although commonly used, does not recognise the time value of money. Further, payback may not recognise the long-term benefits of pollution prevention investments.

Consider cash flows and the profitability of a project over a sufficiently long time-horizon (e.g. economic life of the capital investment) to capture the long-term benefits of pollution prevention investments. Finally, prepare the data and information in a format that managers and lenders can understand and find useful.[12]

The four steps for integrating environmental accounting into capital budgeting can be summarised as follows:

1. Inventory and quantify environmental costs

2. Allocate and project environmental costs and benefits

3. Use appropriate financial indicators

4. Set a reasonable time-horizon that captures the environmental benefits

Note that it may be easier to include environmental costs in capital budgeting if existing processes, systems and products are already being assigned environmental costs in cost accounting systems.

## ▮ Applying Environmental Accounting to Process/Product Design

The design of a process or product significantly affects environmental costs and performance. The design process involves balancing cost, performance, cultural, legal and environmental criteria (US EPA 1993b).

Many companies are adopting 'design for the environment' or 'life-cycle design' programmes to take environmental considerations into account at an early stage. To do so, designers need information on the environmental costs and performance of alternative product/process designs, much like the information needed in making capital

---

11. The time-period required for revenues or cost savings to equal costs; payback typically does not involve discounting.

12. For more information on integrating environmental costs into capital budgeting, see Tellus Institute 1992.

budgeting decisions. Thus, making environmental cost and performance information available to designers can facilitate the design of environmentally preferable processes and products.

For example, the Rohm & Haas Company has developed a model to estimate in R&D the environmental cost of new processes. The model includes conventional, hidden, contingent and relationship costs. In early phases of process development, the cost model prompts process researchers to select and justify process chemistries, operating conditions and equipment that embody the principles of pollution prevention. As the project progresses, the model identifies environmental cost-reduction opportunities. The model can provide financial analysts with an economic picture of the potential environmental risk of a new process prior to its commercialisation (Thomas *et al.* 1994).

The three steps for integrating environmental issues into design can be described as follows:

1. Include environmental issues in needs analysis
   - Consider environmental costs and performance in defining scope of design project
   - Establish baseline environmental cost and performance
2. Add environmental requirements to design criteria
3. Evaluate alternative design solutions taking into account environmental cost, performance, cultural and legal requirements

## ▌Key Terms and Underlying Concepts

A company that wants to use environmental accounting for management purposes may find the terminology confusing and used rather loosely. This section identifies and explains some commonly encountered terms and, most importantly, their underlying concepts. Unlike a glossary, the following discussion does not prescribe how these terms *should* be used.

The section has six parts: the first part recapitulates the three different uses of the term **environmental accounting**; the second part reviews terms such as **environmental cost accounting, full-cost accounting, total cost assessment** and related terms, highlighting critical distinctions that can clarify what people intend to mean in using these terms; the third part summarises some **life-cycle** terms and concepts that relate to environmental accounting; the fourth part comprises terms describing key applications of environmental accounting: **cost allocation, capital budgeting** and **process/product design**; the fifth part lists a series of terms used to categorise or describe **environmental costs**; and the last part presents two other terms related to environmental accounting.

### Environmental Accounting
As noted earlier, the term 'environmental accounting' has three distinct meanings:

- **Environmental accounting, in the context of national income accounting,** refers to natural resource accounting, which can entail statistics about a nation's or region's consumption, extent, quality and value of natural resources, both renewable and non-renewable.

- **Environmental accounting in the context of financial accounting** usually refers to the preparation of financial reports for external audiences using generally accepted accounting principles.
- **Environmental accounting as an aspect of management accounting** serves business managers in making capital investment decisions, costing determinations, process/product design decisions, performance evaluations, and a host of other forward-looking business decisions.

## Commonly Used Terms

Figure 11 lists nine terms that are frequently used in various ways with the same or different meanings. To understand what someone means when using these terms, it is essential to determine whether they are referring to a specific management application of environmental accounting (e.g. cost accounting, capital budgeting, process/product design) or the scope of environmental costs meant to be included (e.g. private costs only, or both private and societal costs).

Sometimes, the terms are used to refer to a specific application of environmental accounting. As noted below, **total cost assessment** is often used to refer to the act of adding environmental costs into capital budgeting, whereas **life-cycle costing** may be most frequently used to refer to incorporating environmental accounting into process and product design. Whether or not one uses these terms to refer to environmental cost allocation, capital budgeting, process/product design or other applications, there

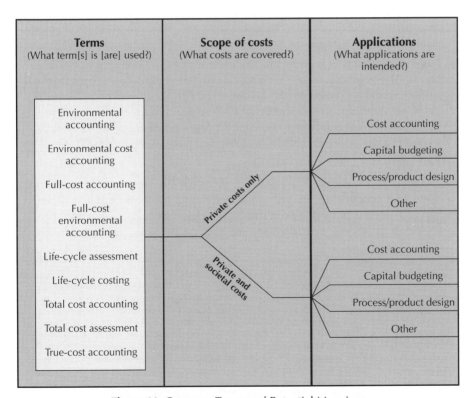

**Figure 11:** Common Terms and Potential Meanings

is another key difference in the way the terms are commonly used. Some professionals use the terms to refer to:

- A firm's private costs only (i.e. those that directly affect the firm's bottom line), or
- Both private and societal costs, some of which do not show up directly or even indirectly in the firm's bottom line.

For some people, **'full-cost accounting'**, **'full-cost environmental accounting'**, **'total cost accounting'** and the other terms refer only to *private costs*. Other people may use the terms to refer to both *private and societal costs*. Some people use one of the terms for private costs alone and another for both private and societal costs together. Understanding the basic distinction between private and societal costs makes it possible to clarify the intended meanings of the vocabulary and thereby hold a conversation with anyone interested in environmental accounting (see earlier section on the difference between private and societal costs).

This difference is at the heart of much of the confusion in environmental accounting terminology. It confuses those items that can be handled more easily—incorporation of private costs—with those that are more difficult to address: societal costs. Clarifying what someone means when using environmental accounting terms is the first step in advancing communication and co-operation.

- **Environmental cost accounting** is a term used to refer to the addition of environmental cost information into existing cost accounting procedures and/or recognising embedded environmental costs and allocating them to appropriate products or processes.

- **Full-cost accounting** is a term often used to describe desirable environmental accounting practices. In the accounting profession, 'full-cost accounting' is a concept and term used in various contexts.[13] In management accounting, 'full costing' means the allocation of all direct and indirect costs to a product or product line for the purposes of inventory valuation, profitability analysis and pricing decisions (Institute of Management Accountants 1990).

- **Full-cost environmental accounting** embodies the same concept as full-cost accounting but highlights the environmental elements.

- **Total cost accounting**, an often-used synonym for full-cost environmental accounting, is a term that seems to have origins with environmental professionals. It has no particular meaning to accountants.

---

13. For example, as required by GAAP for external financial and income tax reporting, accountants calculate the costs of goods sold and value inventory using **full absorption costing** (also called 'absorption costing') which assigns all types of manufacturing costs (direct material and labour as well as manufacturing overhead) to products. In this context, full costs per unit equals full absorption cost per unit plus selling, general and administrative, and interest expenses, per unit (see, e.g., Stickney, Weil and Davidson 1991). An alternative procedure, known as 'variable costing', is often considered superior for certain internal management purposes. A 'full-cost method of accounting' is available for oil- and gas-producing activities and this includes the capitalisation and amortisation of upfront activities (e.g. acquisition, exploration, development), estimated future expenditures of developing proven reserves, and estimated back-end costs of dismantling and abandonment. See Regulation S-X governing financial statements to be filed with the Securities and Exchange Commission (SEC), Section 50410 Rule 4-10.

- **Total cost assessment** has come to represent the process of integrating environmental costs into a capital budgeting analysis. It has been defined as the long-term, comprehensive financial analysis of the full range of private costs and savings of an investment. Adding to the confusion, the acronym for total cost assessment (TCA) is the same as the acronym for total cost accounting (TCA).

- **True-cost accounting** is a less-used synonym for full-cost accounting. The EPA Office of Solid Waste in its programme to encourage local governments to apply full-cost accounting to municipal solid waste management uses the term 'true-cost accounting' to encompass both private and societal costs, while employing the term 'full-cost accounting' to refer exclusively to costs that affect the bottom line of solid waste management activities.

## Life-Cycle Terminology

Life-cycle terms also are used in connection with environmental accounting. These terms include: 'life-cycle design', 'life-cycle assessment', 'life-cycle analysis', 'life-cycle cost assessment', 'life-cycle accounting' and 'life-cycle cost'.

- **Life-cycle design** has been defined as an approach for designing more ecologically and economically sustainable product systems, integrating environmental requirements into the earliest stages of design. In life-cycle design, environmental, performance, cost, cultural and legal requirements are balanced (US EPA 1993b).

- **Life-cycle assessment** has been described as a holistic approach to identifying the environmental consequences of a product, process or activity through its entire life-cycle and to identifying opportunities for achieving environmental improvements. EPA has specified the four major stages in the life-cycle of a product, process or activity as: raw materials acquisition; manufacturing; consumer use/re-use/maintenance; and recycle/waste management (US EPA 1993a). By itself, life-cycle assessment focuses on environmental impacts, not costs.

- **Life-cycle analysis** is sometimes used as a synonym for life-cycle assessment. EPA uses the 'life-cycle assessment' term. Neither term addresses the costs and revenues of environmental consequences and improvements, however.

- **Life-cycle cost assessment** is a term that highlights the costing aspect of life-cycle assessment. It has been termed a systematic process for evaluating the life-cycle costs of a product, product line, process, system or facility by identifying environmental consequences and assigning measures of monetary value to those consequences. Ideally, life-cycle cost assessment can be used to evaluate options for reducing total life-cycle costs and optimising the use of resources. Some people view life-cycle cost assessment as basically adding cost information to life-cycle assessments.

- **Life-cycle accounting** is a term used to describe the assignment and analysis of product-specific costs within a life-cycle framework including *usual, hidden, liability* and *less tangible* costs (US EPA 1993a: 122-29).

- **Life-cycle cost**, according to the US Office of Management and Budget, means the sum total of the direct, indirect, recurring, non-recurring and other related costs incurred, or estimated to be incurred, in the design, development, production, operation, maintenance and support of a major system over its anticipated useful life-span (US Office of Management and Budget 1976). More recently, **life-cycle cost** has been defined in an Executive Order as the amortised annual cost of a product, including capital costs, installation costs, operating costs, maintenance costs and disposal costs discounted over the lifetime of a product.[14] The term may also be used more expansively to include societal costs.

These life-cycle terms are also subject to terminological confusion. For example, some people view life-cycle costing as referring only to private costs, while others view it as including both private and societal costs. Some apply a life-cycle perspective to capital budgeting, while others apply life-cycle concepts to process and product design. As previously mentioned, the key to facilitating communication is to recognise the different uses of common terms and to be able to identify underlying concepts. A threshold question is to determine whether someone is using an environmental accounting term to include solely private or both private and societal costs. A related question is to determine what application(s) a person has in mind when using these terms.

### Scope of Costs

Because people may use environmental accounting terminology to refer to specific sets of environmental costs, or may be imprecise about what they mean, careful delineation of which types of costs are intended to be within the scope of one term or another can reduce confusion and enhance communication. There is an important distinction between costs for which a firm is accountable and costs resulting from a firm's activities that do not directly affect the firm's bottom line:

- **Private costs** are the costs a business incurs or for which a business can be held responsible. These are the costs that directly affect a firm's bottom line. Private costs are sometimes termed **'internal costs'**.

- **Societal costs** are the costs of a company's impacts on the environment and society for which the business is not financially responsible. These costs do not directly affect a firm's bottom line. Societal costs may also be referred to as **'external costs'** or **'externalities'**. These costs may be expressed, qualitatively, in physical terms (e.g. tons of releases, exposed receptors), or in dollars and cents. **Societal costs** (or externalities) are sometimes subdivided according to whether the impacts are environmental, referred to as **'environmental costs'** or **'environmental externalities'**, or social, referred to as **'social costs'** or **'social externalities'**.

- **Internal costs:** a synonym for 'private costs'.

- **External costs:** a synonym for 'societal costs'; also termed **'externalities'**.

- **Social costs** can be a synonym for 'societal costs' or can refer to a subset of external costs.

---

14. US President, Executive Order 12873 of 20 October 1993, 'Federal Acquisition, Recycling, and Waste Prevention' (Federal Register 58, no. 203, 22 October 1993): 54911-19.

- **Environmental costs** can refer to a subset of external costs or can be used as a synonym for 'environmental externalities', 'societal costs', 'private costs' or both 'private and societal costs'.

## Applications

Of the many types of forward-looking business decisions (see Fig. 2) that can benefit from environmental accounting, this chapter focuses on cost accounting, capital budgeting and process/product design:

- **Cost allocation** refers to the procedures and systems for identifying, measuring and allocating or assigning costs for internal management purposes.

- **Capital budgeting**, also known as **investment analysis** and **financial evaluation**, refers to the process of determining a company's planned capital investments.

- **Process/product design** refers to the process of developing specifications for products and processes, taking environmental costs and performance, among other factors, into account.

## Environmental Costs

Terms used to classify or categorise environmental costs are listed below:

- **Regulatory costs** are costs incurred to comply with federal, state or local environmental laws (also termed **'compliance costs'**).

- **Voluntary costs** represent costs incurred by a company that are not required or necessary for compliance with environmental laws but go beyond compliance.

- **Grey-area costs** refer to costs that are not solely or clearly 'environmental' in nature but may also be viewed, in whole or part, as health and safety costs, risk management costs, production costs, operational costs, etc.

- **Upfront costs** include pre-acquisition or pre-production costs incurred for processes, products, systems or facilities (e.g. R&D costs).

- **Operational costs** refer to costs incurred during the operating lives of processes, products, systems and facilities, as opposed to *upfront* costs and *back-end* costs.

- **Back-end costs** include environmental costs that arise following the useful life of processes, products, systems or facilities. See also *exit costs*.

- **Conventional costs** include costs typically recognised in capital budgeting exercises such as capital equipment, raw materials, supplies and equipment. They are referred to as 'usual costs' in the *EPA Pollution Prevention Benefits Manual* (US EPA 1989).

- **Direct costs** is an accounting term for costs that are clearly and exclusively associated with a product or service and treated as such in cost accounting systems.

- **Usual costs:** see *conventional costs*.

- **Hidden costs** refer to the results of assigning environmental costs to overhead pools or overlooking future and contingent costs.

- **Overhead** is often used synonymously with *indirect* or *hidden* costs as comprising all costs that are not accounted for as the *direct* costs of a particular process, system, product or facility. The underlying distinction is between (1) costs that are either pooled and allocated on the basis of some formula, or not allocated at all; and (2) costs that an accounting system treats as belonging (directly) to a process, system, product or facility (i.e. a cost centre, in accounting terminology).

- **Manufacturing or factory overhead** refers to costs that are allocated using more or less sophisticated formulae as contrasted with *general and administrative (G&A)* overhead costs that remain in pools and are not allocated.

- **General and administrative (G&A)** costs are overhead or indirect costs that are not allocated to the costs of goods and services sold.

- **Research and development (R&D)** costs can include the costs of process and product design. See also *upfront costs*.

- **Exit costs** are the costs of proper closure, decommissioning and clean-up at the end of the useful life of a process, system or facility. See also *back-end costs*.

- **Contingent costs** refer to environmental costs that are not certain to occur in the future but depend on uncertain future events (e.g. costs of remediating future spills). These are sometimes referred to as **'environmental liabilities'**, **'liability costs'** or **'contingent liabilities'**.

- **Future (or prospective) costs** refer to environmental costs that are certain to be incurred at a later date, which may or may not be known. These are sometimes referred to as **'environmental liabilities'**.

- **Environmental liabilities** is an umbrella term used to refer to different types of environmental cost, including costs for remediating existing contamination, costs of complying with new regulations, future environmental costs of current operations (also known as *back-end* or *exit costs*), and/or contingent costs.

- **Less tangible costs** refer to expenses incurred for corporate image purposes or for maintaining or enhancing relationships with regulators, customers, suppliers, host communities, investors/lenders and the general public. These are also termed **'relationship costs'** or **'image costs'**.

## Other Related Terms

Two other terms that are relevant to environmental accounting include the following:

- **Activity-based costing (ABC)** is a means of creating a system that ultimately directs an organisation's costs to the products and services that required these costs to be incurred. Using ABC, overhead costs are traced to products and services by identifying the resources, activities and their costs and quantities to produce output (Institute of Management Accounting 1993).

- **Materials accounting or materials balance** refers to an organised system of accounting for the flow, generation, consumption and accumulation of materials in a facility or process in order to identify and characterise waste streams (see, e.g., Freeman 1995). Some view a materials balance as a more rigorous form of materials accounting (INFORM 1992: 8).

## ▮ Conclusion: Moving Ahead

A successful environmental management system should have a method for accounting for full environmental costs and should integrate private environmental costs into capital budgeting, cost allocation, process/product design and other forward-looking decisions.

Companies can make progress in environmental accounting incrementally, beginning with limited scale, scope and applications. Companies can start with those costs that they know the most about and work toward the more difficult-to-estimate costs and revenues. Where private costs or revenues are difficult to estimate, and there is little management support for integrating them, then it may be best to handle them qualitatively. In many instances, it may be unnecessary to quantify the more difficult-to-estimate costs and benefits of capital investment choices because the more easily measured costs (and benefits) are sufficient to justify an investment in cleaner technologies. The same is true for process/product design, if one design direction is clearly superior to the alternatives. Ultimately, businesses will benefit from including probabilistic and difficult-to-estimate costs in cost allocation, capital investment, process/product design and other decisions. The best approach is to go as far as you can in integrating environmental costs, including hidden, future and contingent costs, into management decisions.

Efforts to integrate societal costs into business decisions will continue and expand. Most corporate information and decision systems do not currently support such proactive and prospective decision-making (US EPA 1996a). The capital markets do not yet have adequate ways of evaluating the financial performance of progressive companies who do so. Although some companies are at the leading edge of efforts to address societal costs, it will probably be some time before societal impacts and costs can be integrated into cost allocation, capital budgeting, process/product design and general business decisions. However, there is a growing body of information documenting a variety of businesses engaged in advancing the state of the art to bring societal costs into their decision-making.[15]

---

15. EPA is committed to helping businesses understand their environmental costs and integrate those costs into decision-making. The EPA's Environmental Accounting Project performs education, research, guidance and outreach on this issue. Join EPA's Environmental Accounting Network to keep informed of the latest developments by contacting EPA's Pollution Prevention Information Clearinghouse (PPIC) and asking for a Network Membership Form. PPIC can also provide a variety of materials including case studies and lists of EPA and non-EPA publications on these topics. Please telephone PPIC on +1 202 260 1023 for details and/or more information on the EPA Accounting Project.

# 3 Calculating the True Profitability of Pollution Prevention[1]

Stefan Schaltegger and Kaspar Müller

MANY FIRMS are still not aware of the potential savings that can be gained from improved pollution prevention. Three reasons account for this deficiency (see Schaltegger *et al.* 1996):

- **Management is frequently unaware of the actual amount of environment-driven costs**, as most managerial accounting systems still do not differentiate. From a managerial perspective, it makes sense to track and trace these costs to determine to what extent the company is financially affected by environmental issues.

- **Environment-related costs are considered as general overhead costs.** However, to calculate the actual profit margins of products, environment-related costs should be allocated correctly.

- **The indirect costs of pollution are neglected.** In many cases, investment appraisals of pollution prevention measure both the direct costs (capital investment) *and* indirect costs (labour, maintenance, etc.) of pollution prevention against *only* the direct costs of pollution (waste, sewage, etc.), and the indirect costs of pollution are often not considered.

This chapter discusses activity-based costing of investment appraisal of pollution prevention. Applying this approach can increase economic performance as a consequence of improved environmental protection (Schaltegger with Müller and Hindrichsen 1996). Moreover, ignoring this approach could mislead investment decisions, as it will be seen that many formerly externalised costs are now internalised into companies' accounts.

---

1. We would like to thank Frank Figge for his excellent comments.

## ▮ Tracking and Tracing of Environmental Costs

The identification of environment-related internal costs poses few problems, as long as it is solely for purposes of environmental protection. Such is the case with 'end-of-pipe' technologies, which are clean-up devices installed after the core production equipment. Scrubbers and waste water treatment plants are typical examples, which can help concentrate toxic substances and/or reduce toxic impacts. However, end-of-pipe technologies do not usually solve environmental problems at source, but rather catch emissions before they are released. Another characteristic of this kind of technology is that it often merely shifts emissions from one environmental medium to another (e.g. a scrubber shifts emissions from the air to the water and/or soil).

The example in Figure 1 shows that costs of 'joint environmental cost centres', such as incinerators, waste water treatment plants, etc., should be differentiated from other overhead costs. A manufacturer has three production steps that all produce waste. The entire waste is treated in a shared incinerator on the production site. The costs of incinerating the waste from current production are $800; the remaining overhead costs for general administration, salaries of top management, etc. are $9,000.

The identification and measurement of environment-related costs is much more difficult with 'integrated technologies' (also called 'clean technologies'), which are more efficient production technologies that reduce pollution at the source, or before it occurs (e.g. new equipment that uses 50% less energy and creates 20% less toxic effluents than the old). Environmental issues were already integrated when the equipment was

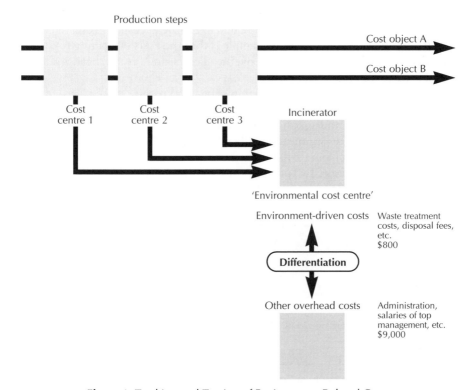

**Figure 1:** Tracking and Tracing of Environment-Related Costs

developed. Because of this, the following question arises: what part of the equipment (assets) and of the maintenance expenditure is environment-driven?

The main criterion for answering this question is the cost difference in relation to the environmentally-next-(less)-favourable solution. For example, 20% of the capital costs may be traced as environmental if the integrated technology has caused 20% extra costs relative to comparable equipment. Or the costs of depreciation of the old technology over two years may be considered as environment-driven if the integrated technology was installed two years earlier than would have been economically justified, only to comply with environmental regulations.

However, the costs should not be considered environment-related if the integrated technology is simply current state of the art, and has been installed for no other reason than the routine replacement of old equipment.

## Allocation of Environment-Driven Costs

It has been argued extensively that conventional costing methods are oriented too much towards past activities instead of present and future ones (see, for example, Johnson and Kaplan 1987). Also, direct costing, another popular approach, is less decision-oriented than activity-based costing, which concentrates on calculating the costs of specific business activities. Hence, the focus of this chapter is on activity-based costing ('ABC').

### Traditional Allocation of Environment-Related Costs

Internal environmental costs are often treated as overhead costs and divided equally between all cost drivers. A common example is that the costs of treating toxic waste of a product are included in the general overhead costs, and the overhead is allocated in equal parts to all products.

However, 'dirty' products cause more emissions and require more clean-up facilities than 'clean' products. Equal allocation of those costs therefore subsidises environmentally more harmful products. The clean products, on the other hand, are 'penalised' by this allocation rule as they bear costs that they did not cause.

A simple example in Figure 2 illustrates how equal allocation can lead to suboptimal management decisions. Two processes are compared: process A is 'clean' and does not cause any environment-driven costs for the company, while process B causes $50

|  | 'Clean' process A | 'Dirty' process B |
|---|---|---|
| Revenues | $200 | $200 |
| Production costs | $100 | $100 |
| Environmental costs | $0 | $50 |
| True profit | $100 | $50 |
|  |  |  |
| If environmental costs are overhead | $25 | $25 |
| Then the book profit is | $75 | $75 |
| Which is incorrect by | −25% | +33% |

**Figure 2:** Example of Correct and Incorrect Cost Allocation
*Source: Hamner and Stinson 1993: 3*

of extra costs because it is environmentally harmful. If these costs are assigned to general overheads and allocated equally, both processes appear to create a profit of $75. (If $50 is allocated to overhead, $25 will implicitly be allocated to each process. This leads to a profit of $75 [$200-$100-$25]). In reality, however, process A has created a profit of $100, while process B has contributed only $50 to the company's profit.

Suboptimal management decisions materially influence the pricing of products. The cross-subsidised dirty products are sold too cheaply whereas the environmentally less harmful products are sold too expensively. In consequence, market share is lost in more sustainable fields of activity and at the same time the company's position is enhanced in fields with higher risk and no business future.

### Activity-Based Costing

Whenever possible, environment-driven costs should be allocated directly to the activity that causes the costs and to the respective cost centres and cost drivers. Consequently, the costs of treating, for example, the toxic waste of a product should directly and exclusively be allocated to that product.

Many terms are used to describe this correct allocation procedure, such as 'environmentally-enlightened cost accounting', 'full-cost accounting' or 'activity-based costing' ('ABC'; see Fig. 3). In this chapter, the term 'ABC' ('activity-based costing') is adopted, although 'activity-based accounting', 'full-cost accounting' or 'process costing' is often used in practice. However, the term 'full-cost accounting' can be misleading, for (as Chapter 2 discusses), it sometimes includes external environmental costs. ABC represents a method of managerial cost accounting that allocates all internal costs to the cost centres and cost drivers on the basis of the activities that caused the costs. The activity-based costs of each product are calculated by adding the joint fixed and the joint variable costs to the direct costs of production. The strength of ABC is that it enhances the understanding of the business processes associated with each product. It reveals where value is added and where value is destroyed.

The example in Figure 4 illustrates the method of ABC. It shows two steps of allocation: first, from joint environmental cost centres to the 'responsible' cost centres (i.e. production processes); and, second, from the production cost centres to the respective cost drivers (i.e. products A and B).

Today, it is definitely wrong to include all environment-related costs in the general overhead costs; nevertheless, some remain as overheads, such as those costs clearly related to general overhead activities (e.g. new insulation of the head-office building). Also, costs of past production that are clearly related to strategic management decisions for the whole company might qualify as general overhead costs (e.g. liability costs of products that have been phased out).

At present, even in some advanced managerial accounting systems, only the visible (direct) costs of environmental cost centres are directly allocated to production

---

**Figure 3:** Activity-Based Costing

AFTER TRACKING and tracing, the costs of joint environmental cost centres, such as incinerators, sewage plants, etc., have to be allocated to the 'responsible' cost centres and cost drivers.

The total input of production is 1,000 kg, of which 200 kg is treated as waste in the incinerator. The total costs of incineration are $800. The cost key to determine the cost contribution of different kinds of waste should consider the costs of incineration that those kinds of waste cause. The treatment of one kilogramme would cost $4 if every unit of waste caused the same costs.

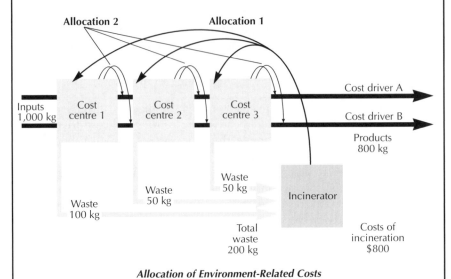

***Allocation of Environment-Related Costs***

As a first step, the costs of the incinerator have to be allocated to the three cost centres (allocation 1): $400 to cost centre 1 ($4 × 100 kg of waste); and $200 to cost centres 2 and 3 respectively ($4 × 50 kg each).

As a second step (allocation 2), the costs have to be allocated to the cost drivers (i.e. products A and B). The cost key should reflect the costs of waste treatment that the respective product has caused at each production step.

**Figure 4:** Twofold Allocation of Environment-Related Costs

cost centres and cost drivers. However, additional costs can be environment-driven even though they do not directly relate to a joint environmental cost centre (e.g. an incinerator). Yet some indirect costs could be saved if less waste was created. Waste occupies manufacturing capacities, requires labour, increases administration, and so on. If no waste was produced, the equipment would not depreciate as quickly, and less salaries would have to be paid.

For instance, in the example in Figure 5, 200 kg of the 1,000 kg of inputs were purchased only to be emitted without creating any value. Thus, the related waste has caused a 20% higher purchasing cost, higher costs of depreciation and administration, etc. Therefore, a third allocation step is necessary. As shown in Figure 5, this third allocation step can motivate management to realise huge efficiency gains by improving the environmental record at the same time!

### Allocation Keys

The choice of an adequate allocation key is crucial for obtaining correct information in cost accounting. It is important that the chosen allocation key is closely linked with

THIS EXAMPLE illustrates the third step of allocation on the basis of the example used in Figure 4. 1,000 kg of inputs has been purchased to create 800 kg of products. Of the 200 kg of waste, 100 kg are created in step 1, and 50 kg each in steps 2 and 3.

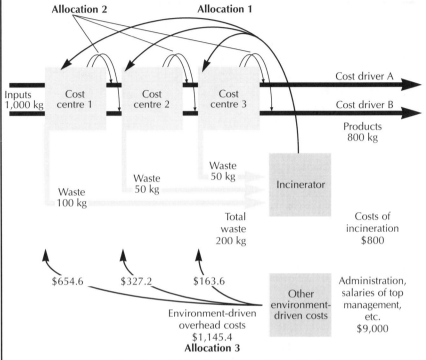

*Allocation of Other Environment-Driven Costs*

With the first and second allocation steps, the costs of the environmental cost centre ($800 for incineration) have been traced, tracked and allocated to the cost centres and drivers. However, environment-driven costs are actually much higher. Some inputs were purchased 'just to be thrown away', without having created any value. Therefore, the management should also account for other environment-driven costs, such as increased depreciation, higher costs for staff, etc. which are not directly related to joint environmental cost centres but which nevertheless vary with the amount of throughput. To take these environment-driven costs into consideration, a third allocation step is necessary.

In the case presented, it is assumed that the overhead costs of $9,000 are variable, the mass of the waste is an adequate allocation key and that the overhead costs per kilogramme are the same in all three cost centres.

The processed kilogrammes of material are 1,000 kg in cost centre 1, and 900 kg and 850 kg in cost centres 2 and 3 respectively (see table below). Possible allocation keys for the total overhead costs are 36.36% (cost centre 1), 32.73% (cost centre 2) and 30.91% (cost centre 3) if the total amount of processed material (e.g. 1,000 kg of 2,750 kg for cost centre 1) is taken as the allocation key. Thus, the total overhead costs per cost centre are $3,273 (cost centre 1), $2,945 (cost centre 2) and $2,782 (cost centre 3).

| Environment-Driven Indirect Costs | | | | |
| --- | --- | --- | --- | --- |
| | Cost centre 1 | Cost centre 2 | Cost centre 3 | Total |
| **Kilogrammes processed** | 1,000 kg | 900 kg | 850 kg | 2,750 kg |
| **As a %age of total** | 36.36% | 32.73% | 30.91% | 100% |
| **Total overhead costs per cost centre** | $3,273 | $2,945 | $2,782 | $9,000 |
| **Waste processed** | 200 kg | 100 kg | 50 kg | |
| **Waste as %age of material processed** | 20% | 11.11% | 5.88% | |
| **Waste-driven overhead costs** | $654.6 | $327.2 | $163.6 | $1,145.4 |
| **Waste-driven costs as %age of total overhead costs** | | | | 12.73% |

In this case, the environment-driven indirect (overhead) costs can be calculated as follows:

*Cost centre 1*: Physically, 100 kg of waste show up in cost centre 1. Economically, however, also the waste that later shows up in cost centres 2 and 3 already causes costs in cost centre 1. In all 200 kg (100 kg + 50 kg + 50 kg) of the 1,000 kg of inputs (= 20%) purchased causes indirect costs (e.g. production equipment is worn out faster) in cost centre 1. In this case, the additional, environment-driven indirect costs at cost centre 1 are:

   20% (200 kg of 1,000 kg) of $3,273 = $654.6

*Cost centre 2*: 900 kg of material enters cost centre 2, but only 800 kg will finally leave the company as products. Thus, 100 kg of the 900 kg (11.11%) that enters cost centre 2 only causes waste. The allocated total overhead costs to cost centre 2 are $2,945. The indirect waste costs are:

   11.1% (100 kg of 900 kg) of $2,945 = $327.2

The respective costs in *cost centre 3* are:

   5.9% (50 kg of 850 kg) of $2,782 = $163.6

In summary, calculated with ABC, the total of all environment-driven indirect costs is $1,145.4 ($654.6 + $327.2 + $163.6).

**Figure 5:** A Third Allocation Step

the actual, environment-related costs. In practice, mainly the following four groups of allocation keys are discussed for environmental issues:

- Volume of emissions, waste, etc. treated
- Toxicity of emissions and waste treated
- Environmental impact added (volume × impact per unit of volume) of the emissions treated
- Relative costs of treating different kinds of emissions

One possibility is to allocate the environment-driven costs based on the volume of hazardous waste caused by each cost driver (e.g. volume treated/hour, waste/kg of output, and emissions/working hour of equipment). This is a rather arbitrary key in cases where the capital cost (interest and depreciation of construction costs) as well as the variable costs are not related to the total volume treated. Due to higher safety and technological requirements, the construction costs and the variable costs often increase substantially with a higher degree of toxicity of the waste treated. In many cases, these additional costs are due to only a small percentage of the waste. Thus, the costs of a treatment or prevention facility are often not clearly related to the overall volume treated, but rather to the relative cleaning performance required.

Another possibility is to allocate according to the potential environmental impact added of the treated emissions. The environmental impact added is calculated by multiplying the volume by the toxicity of the emissions. However, this allocation key too is often inappropriate, as the costs of treatment do not always relate to the environmental impact added.

Thus the choice of allocation key must be adapted to the specific situation, and the costs caused by the different kinds of emissions treated should be assessed directly. Sometimes a volume-related formula best reflects the costs caused, while in other cases a key based on environmental impact is appropriate. The appropriate allocation key depends on the variety and the kind of emissions treated or prevented. Also the time of occurrence may be relevant (past, current or future costs).

## ■ Investment Appraisal

Investment appraisal (capital budgeting) is one of the most important managerial activities. The basic idea is to compare different investment alternatives. As with the evaluation of investment opportunities, there is no single correct way of incorporating environmental considerations into investment appraisal. However, the task of investment appraisal has been complicated with the increasing importance and uncertainty of environment-driven future costs (see Fig. 6). Although not discussed here, the same holds true for financial investments.

Possible steps for incorporating environmental considerations into investment appraisal are (White, Becker and Goldstein 1991):

- Expansion of cost inventory
- Correct allocation of costs
- Extension of time-horizon and use of long-term financial indicators (net present value and option value)

---

ESTIMATES OF future costs are uncertain because nobody can be fully informed about the future. Because of substantial uncertainties over future costs, management is inclined to underestimate potential future costs that are not proven.

Lack of adequate information about the future has forced many companies to phase out products that were warmly welcomed at the time of their introduction (e.g. CFCs).

However, investment decisions taken under uncertainty strongly influence future activities. Once investments are made, a big financial incentive exists to delay phasing-out as much as possible.

*Chlorofluorocarbon (CFC) Production of Du Pont\**

| 1986 | 1987 | 1988 | 1989 | 1990 | 1991 | 1992 | Goal 1995 |
|------|------|------|------|------|------|------|-----------|
| 100% | 117% | 119% | 112% | 66% | 51% | 45% | 0% |

\* 'Based on a request from the US Government, Du Pont could produce in 1995 in the US as much as 25% of its 1986 CFC production levels. Du Pont will base production on customer commitments' (Du Pont 1993: 5).

First, delayed phasing-out extends the depreciation period, which increases the short-term profit potential. Second, the last companies in the business have no other competitors and will benefit from exceptionally high profit margins (e.g. Du Pont with CFCs; Du Pont 1993). These advantages have to be weighed against a potential bad image, legal requirements and pressure from stakeholders.

---

**Figure 6:** Future Consequences of Investment Decisions

An **expanded cost inventory** considers four categories of costs:

- Direct costs (capital expenditure, operation, maintenance, expenses, revenue, waste disposal, energy, etc.)
- Indirect costs (administrative costs, regulatory compliance costs, training, monitoring, insurance, deterioration, depreciation, etc.)
- Potential liabilities (contingent liabilities, potential fees, fines, taxes, etc.)
- Less tangible costs (image, absent workers, morale, etc.)

The calculation of direct costs is part of any method of investment appraisal. However, environment-related costs are sometimes hidden in general overhead costs and therefore not considered. In particular, indirect costs, potential liabilities and less tangible costs are often difficult or impossible to identify, measure and allocate, although these can very much influence the profitability of an investment.

The rules of allocation applied to environment-related costs can substantially influence investment decisions. As discussed, three steps of **correct allocation** can be distinguished in the allocation of:

- Costs of joint environmental cost centres (e.g. incinerators) to production cost centres
- Costs of production cost centres to cost drivers
- Other environment-driven costs to production cost centres and cost drivers

The application of this three-step allocation procedure in investment appraisal has a large influence on deciding which investments, including pollution prevention investments, are economically favourable (see Fig. 7).

First empirical studies have demonstrated that between six and ten times higher savings could be realised if investment decisions for pollution prevention were based on correct allocation rules. An example from the German metal industry is given in Figure 8.

It may seem surprising that cost savings due to improved pollution prevention have not been realised. However, the collection and analysis of the related (accounting) information also incurs costs: in the past, the costs of environmental accounting were expected to be larger than the benefits of being better informed. With the growing knowledge about these savings potentials, companies that fail to realise them will face competitive disadvantages.

A third step for the inclusion of environmental considerations in investment appraisal is to **extend the time-horizon** and to **use long-term financial indicators**. Environmental investments often have longer payback periods than other investments, because the relevant benefits and losses often lie many years in the future. However, this is not always true, as the example in Figure 9 shows. Nevertheless, the application of long-term financial indicators is useful, especially with high contingent liabilities and high potential future benefits.

New regulations requiring the internalisation of previously external costs can suddenly appear. The US Superfund legislation under CERCLA (Comprehensive Environmental Response, Compensation and Liability Act of 1980) is an excellent, well-known example, as it shows the enormous, unexpected financial impacts that environmental issues can cause.

THIS EXAMPLE shows how correct allocation, compared with the mode of allocation often employed, can substantially influence the result of an investment appraisal. The calculations are based on the example that has already been used.

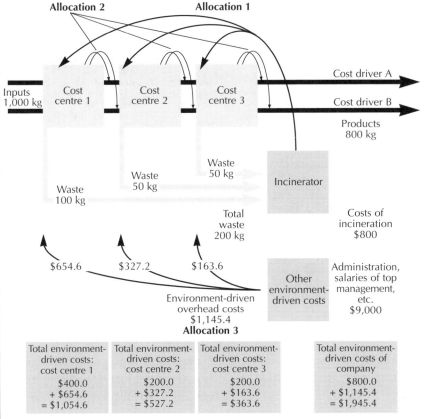

*The Correct Calculation of All Environment-Driven Costs*

The total costs of the environmental cost centre (i.e. the incinerator) are $800, whereas the total of all indirectly-caused costs is $1,145.4. The total of all environment-driven costs are shown in the above diagram for each cost centre as well as for the whole company.

Many economically profitable investments, especially for pollution prevention, would not be implemented if management relied on traditional allocation rules which consider only the direct costs of environmental centres. For example, a pollution prevention investment with annual costs of $300 would not be introduced to reduce 50 kg of waste in cost centre 1, even though it would result in savings of $363.6 ($200 direct + $163.6 [50 kg of 200 kg] indirect waste costs).

Pollution prevention activities do not create the same economic benefit in every cost centre, even if they achieve the same reduction of emissions. Also, pollution prevention investments in cost centre 3 are attractive as they reduce waste-related costs at earlier cost centres. By preventing 50 kg of waste in production step 3, costs could also be reduced in cost centre 2 (50 kg instead of 100 kg of waste processed) and cost centre 1 (150 kg instead of 200 kg of waste processed). Thus, the prevention of 50 kg of waste in cost centre 3 would reduce costs by $690.8 ($363.6 in centre 3 + $163.6 [50% of indirect waste costs in centre 2] + $163.6 in centre 1).

**Figure 7:** Investment Appraisal

| Usual method of calculating environment-driven costs | | Correct method of calculating environment-driven costs | |
|---|---:|---|---:|
| **Costs of waste disposal** | | **Costs of waste disposal** | |
| Fees | 500,000 | Fees | 500,000 |
| Disposal costs | 300,000 | Disposal costs | 300,000 |
| **Total** | **800,000** | First total | 800,000 |
| | | **Environment-driven costs in production** | |
| | | Logistics and transportation | 150,000 |
| | | Additional personnel | 250,000 |
| | | Additional depreciation | 200,000 |
| | | Storage | 100,000 |
| | | Second total | 1,500,000 |
| | | **Excess material input** | |
| | | Purchase | 4,500,000 |
| | | **Correct total** | **6,000,000** |

**Figure 8:** Example of the Significance of Correct Calculation of Environment-Driven Costs
*Source: Wagner 1995*

THE SWISS multinational company Ciba-Geigy Ltd installed a computerised environmental monitoring and internal billing system within its plant in Huningue in France. The costs were approximately SFr2.4 million. The system saved energy costing more than SFr0.8 million per year. Therefore, from a static perspective, the payback period was three years.

**Figure 9:** Short Payback of Environmental Investment at Ciba

The application of long-term financial indicators can provide a stimulus for advance consideration of future financial impacts. Two long-term financial indicators in particular—the net present value and the option value—are discussed in the context of environmental accounting.

The opportunity costs of capital (the lower value of future cash flows or the loss of value over time) are considered by the application of the discount rate of financial markets. The sum of all discounted future cash flows determines the value of a project. Projects with a positive net present value should be carried out; projects with negative net present value should be cancelled.

Some may argue that the concept of discounting is fundamentally unethical, as a lower value is assigned to the needs of future generations. This is in sharp contradiction to the conservation of assets for future generations and to sustainable development in general (see also Fig. 10). Others may argue that discounting is necessary for economics to function, and, acknowledging its flaws, propose a social discount rate for environment-related investments that is lower than the market-based interest rate. Thus, investments are given a longer time-horizon to pay off, and the time-horizon of the company is extended.

The omission of discounting and the manipulation of the discount rate are problematic as the calculated results do not reflect the actual economic situation. Furthermore, investment appraisal should indicate the economic value of alternative investment opportunities. Other, non-economic aspects might be considered separately but should not distort the economic analysis.

POLLUTION PREVENTION equipment adds an extra $4 million to capital spending on a new factory costing $20 million. New equipment could prevent soil pollution that would have to be cleaned up in ten years at estimated costs of $10 million.

The pollution prevention investment is not economic if the method of net present value is strictly applied. The discounted value of the clean-up costs is $3.9 million (discount rate of 10%), which is lower than the prevention costs of $4 million. Therefore, the value of the company is $0.1 million higher without the new equipment. However, in ten years, soil contamination might completely prevent further operations. Thus, the net present value method might lead to incorrect decisions in strategic management, if the option value is not considered as well.

**Figure 10:** Option Value versus Net Present Value
*Source: Koechlin and Müller 1992*

However, the net present value method also has flaws, as the discussion in Figure 10 shows. The **option value** takes the net present value as well as the strategic value of investments into account (see, for example, Brealey and Myers 1991; Dixit and Pindyck 1993). An option entails the right but not the obligation to do something: 'Opportunities are options—rights but not obligations to take some action in the future' (Dixit and Pindyck 1993). Thus, the option can *also* be defined as the right *not* to undertake a follow-on investment. The price of the option is determined by the price of the underlying stock (free cash flows of the project), the exercise price (the follow-on investment), the time to maturity (the date when the decision has to be made) and also by the risk of the stock and the risk-free interest rate.

The strict application of the net present value method very often ignores the value of creating or exercising options or the costs of impeding future options. Not undertaking a project with a negative net present value today might result in a follow-on project becoming too expensive or completely impossible.

Some investments create a special value in the context of other investments of the company. Sometimes an investment that appears uneconomic might be crucial if, in fact, it can create an option that enables the company to undertake other investments in the future (see, for example, Brealey and Myers 1991; Dixit and Pindyck 1993). A negative net present value shows only that the project as such (in isolation) does not pay off. However, the project could be very important in the context of other projects a company envisages. This effect is called the 'strategic value' of a project and can be expressed as a call option.

Because of new scientific evidence about environmental problems, new issues with huge effects on an industry can emerge very quickly. Many environmentally-crucial projects (e.g. the launch of a green product line or the introduction of an integrated environmental management programme in the company) are of a strategic nature because of their long time-horizon as well as their effect on public perception (signals for the general public and for customers). The capacity to adapt quickly to new circumstances also clearly represents an option value.

Just as in financial markets, the value of the strategic option increases with the variability of the project's cash flows and the risk of the project, respectively. With stricter legislation and increasing risk, investments to prevent environmental liabilities or to initiate green product lines to create new markets have an option value. They entail the option to stay in the market.

The strategic value of going further than mere compliance with regulations increases with the probability of future environmental legislation. The option value can therefore exceed the net present value of pollution prevention equipment (Koechlin and Müller 1992).

Future free cash flows of environmental management activities are often worth less than the present actual costs (see Fig. 11). If a company receives free cash flows of $100 million in fifty years but has to invest $1 million today, it will have a negative net present value. However, the respective investments might create options for future economic activities which are not taken into consideration in the calculation of the net present value.

It is not only foregone options due to high future costs that are ignored. Free cash flows in the distant future also tend to be underestimated if the option value is not taken into consideration. Short-term profits are possible, for example, if the value of the shares of a green company is less than the present value of the current assets. Thus, the management of a green company going public must inform the investors about the expected future cash flows.

It has been shown that environmental issues involve special problems in investment appraisal. Management is often unaware of these environment-driven issues. Some external stakeholders (for example, regulators and investors' associations) have therefore started to influence management to give greater consideration to environmental aspects in investment appraisal and in managerial accounting in general.

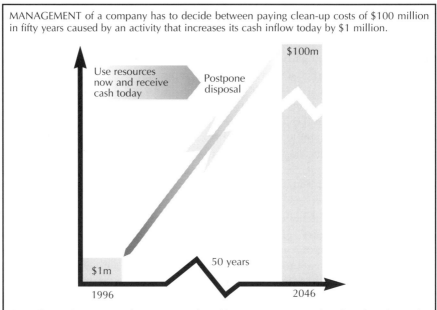

MANAGEMENT of a company has to decide between paying clean-up costs of $100 million in fifty years caused by an activity that increases its cash inflow today by $1 million.

Use resources now and receive cash today

Postpone disposal

$100m

50 years

$1m

1996

2046

According to the concept of net present value, this management must be advised to choose the option that creates the cash inflow today, because the discounted value of the costs in fifty years (discount rate 10%) is lower than the cash inflow of $1 million today.

This means that the company should use resources now, cash in today and postpone disposal. However, the damage of $100 million can be disastrous for future generations as well as for the company. The damage caused may reduce options and prevent the company from making future investments. The effects on future business options should therefore be carefully considered, too.

**Figure 11:** High Clean-Up Costs in the Distant Future

## ▌Conclusions

For most businesses, economically rational management requires the direct allocation of environment-driven costs to the 'responsible' cost centres and cost drivers.

Management accountants have several responsibilities with respect to investment appraisal of pollution prevention. First, the environment-driven costs must be tracked and traced. Second, they should be allocated as directly as possible to the related cost centres and cost drivers. With ABC, this can be done according to the activities caused. Third, no general rule for the ideal allocation key exists. The suitability of an allocation key depends on the variety and kinds of emissions prevented or treated. However, it should reflect the costs actually caused by an activity. Fourth, not only the direct but also the indirect costs of pollution should be considered in an investment appraisal of pollution prevention measures.

Implementation of these steps does not require a revolution in a company but is necessary from a purely business perspective; nevertheless, implementation may meet with opposition even though it makes logical sense. The change of allocation rules leads to a redistribution of power in a company. Line managers with currently profitable products tend to refuse allocation rules whenever they expect losses, whereas the 'company-internal lobby' for clean products and technologies is often either small or still not established.

# 4 Integrating Environmental Impacts into Capital Investment Decisions

## Marc J. Epstein and Marie-Josée Roy

APPROACHES to capital investment decision-making have changed very little in recent years. Discounted cash flow analysis was introduced decades ago and only minor refinements to that analysis have been developed. However, a broader understanding of life-cycle impacts, long-term corporate profitability and shareholder value has forced many companies to rethink the capital investment decision process. They recognise that without a broader understanding of long-term constituent and company impacts of products, services, processes and other activities, evaluations of capital investments are incomplete and improper decisions are made.

Companies in the US have, over the past decade, been burdened under Superfund with unexpected corporate environmental costs in the tens or hundreds of millions of dollars. These costs relate to production that was completed decades ago and are only now being incurred. These companies have become rightly concerned with other likely future costs related to other pending responsibilities such as product take-back (Epstein 1996a). So, the identification and measurement of environmental costs and benefits and the integration of these impacts into all capital investment projects has become critical in the proper forecasting and evaluation of long-term project success. In an increasing number of companies, decisions regarding products, processes, facilities and equipment now require the approval of, or at least consideration by, the environmental, health and safety (EH&S) department.

Making informed capital investment decisions on environmental projects is also becoming important for other reasons. External pressures often act as positive forces toward a greener process. These may include regulation, consumer pressures, competitors' environmental initiatives, increasing public scrutiny, globalisation of markets, technological availability and ethical investments (see, for example, Arora and Cason 1995). Further, numerous industry-led initiatives such as BS 7750 and ISO 14000, industry-specific standards and anticipated stringent government regulations suggest that companies will be devoting increasing attention to environmental issues in coming years. As more and more regulations, standards and technologies emerge, managers must evaluate how to respond to these environmental pressures, the long-term implications of the available options, and how these options can be effectively integrated into corporate strategy.

This chapter presents the corporate and environmental strategy implications of capital investment decisions and a framework for improving the corporate capital investment decision-making process. We also examine methods that can be used to improve measurement of the environmental costs and benefits to both the company and society and how to integrate those impacts into capital investment decisions for regulatory environmental projects, voluntary environmental projects and general capital projects, including equipment purchase and facility location decisions.

## ▌ Strategic Implications of Capital Investment Decisions

There are many ways in which firms may choose to respond to environmental pressures. Among these are end-of-pipe solutions, product/process modifications and managerial system modifications (Klassen and Whybark 1995). These can be minor changes of existing organisational routines or radical new ways of doing business. In the US, the Clean Air Act Amendments of 1990, which attempt to set overall emissions levels, clearly illustrate the type of decisions that managers are now facing. Utilities are required to choose among six compliance options:

1. Fuel switching and/or blending
2. Obtaining additional allowances
3. Installing flue gas desulphurisation equipment
4. Using previously implemented controls
5. Retiring facilities
6. Boiler repowering

What elements should guide managers as they integrate these environmental considerations into capital investment decisions and how should they choose between the various options?

Examining environmental programmes, Hunt and Auster (1990) suggest that organisations differ on two main aspects: (1) commitment of the organisation (i.e. general mindset of corporate managers, resource commitment, support and involvement of top management); and (2) the programme design (i.e. performance objectives, level of integration/involvement within the organisation and reporting structure [both internal and external]). They suggest that a firm's commitment to environmental performance can be structured according to a stage model,[1] where the intensity of the efforts, the type of

---

1. There are several typologies describing environmental behaviour found in the literature (Post and Altman 1994; Roome 1994). They all reflect a progression from a reactive strategy to a proactive one.

efforts, and the method of implementation vary according to those stages. The capital investment decision-making process, and, more specifically, its ability to handle the environmental aspects of the decisions, will include substantially different methods and procedures of analysis as a firm moves from a reactive stage to a proactive one.

As they impact the firm's long-term profitability, capital investment decisions are strategic in nature. Few business decisions are more important. The decisions translate a firm's chosen direction in terms of opportunities it wishes to pursue or abandon (Hayes *et al.* 1988). Past capital investment decisions pave the way for future decisions, and through their capital investment decisions managers commit an organisation to its future environmental performance.

The strategic nature of the environmental challenge is indeed firmly grounded in the management literature. Several authors have suggested that environmental leadership can lead to competitive advantage (Porter and van der Linde 1995a; Shrivastava 1995). The path to competitive advantage may be through cost reduction benefits, increased market share or technological leadership. An important strategic issue concerning environmental leadership as a competitive weapon is the **early-mover** advantage. Skea (1995) proposes that this may:

- Provide longer timetables for developing more innovative solutions
- Influence regulatory authorities to set tighter standards for the industry
- Lead to possible licensing of technological solutions

Skea also acknowledges the risks involved in this strategy: second-movers could benefit from a wait-and-see strategy as cleaner and potentially cheaper technologies could become available. The optimal timing of the adoption of any new technology is a complex issue. In the specific case of environmental technologies, the level of uncertainty is very high. Indeed, scientific evidence relating to the potential environmental impacts of various substances is a continual subject for debate. Whom should we believe? Newer and better technologies are emerging all the time. Which one is the best?

Though research on the relationship between environmental performance and financial performance is not clear, there is substantial evidence that poor environmental performance may affect stock price, access to capital and cost of capital (see, for example, Cormier, Mangan and Morard 1993, 1994; Epstein 1996c; Freedman and Jaggi 1992). In part, this may be affected by a reputational effect, financial analysts' concern over lingering environmental liabilities, and the interests of environmental investors and investment funds.

Thus, managers need to evaluate how environmental considerations will affect their long-term positioning. Corbett and Van Wassenhove (1995) review how environmental issues affect the five traditional competitive priorities: cost, quality, dependability, flexibility and innovativeness. Some of these elements are presented in Figure 1. The authors describe how these environmental pressures can endanger and reshape a firm's competitive basis and how, given these same pressures, a firm needs to develop new skills and knowledge.

Strategic decisions will also impact the range of environmental responses considered by the firm. Competitive priorities pursued by a firm could influence the scope of environmental initiatives available. Indeed, compatibility issues are of paramount importance given past technological investments in product and process design. These investments in equipment, knowledge, staff, etc. are made in light of a particular corporate priority such as cost leadership or quality.

| Competitive priorities | Possible impacts |
|---|---|
| Cost | Cleaner technology can be cost-efficient. |
| | High cost of maintaining current activities at a profitable level, given legislation. |
| Quality | Environmental products can be perceived as having higher quality. |
| Dependability | Environmental problems can affect the quality of the inputs, endangering dependability. |
| Flexibility | Firms must respond quickly to uncertain and ever-changing regulations and market pressures. |
| Innovativeness | Market pressures are forcing firms to introduce new products more frequently. |

**Figure 1:** Impact of Environmental Pressures on Competitive Priorities
*Source: Adapted from Corbett and Van Wassenhove 1995*

Are firms responding to environmental pressures with environmental initiatives that maintain their competitive priorities or are they changing or reshaping their competitive basis? Firms often need to develop new skills and knowledge to respond to this new environment. For example, as product take-back legislation and standards become more commonplace, which firms would be best positioned to meet this new challenge? Are firms with capabilities for innovation and flexibility more likely to respond creatively and positively to environmental pressures? In the case of cost leadership, cost-driven firms could choose to produce a cleaner more expensive product as some consumers may be willing to pay a little more for clean products (Roberts 1996). But, often, cost-driven companies are less likely to invest in new manufacturing technologies. Some already have major investments in highly specialised and expensive production equipment and are unwilling to re-invest in newer technologies. Thus, technical capabilities may be an important barrier to environmental changes.

> Managers of such projects face a paradox: core capabilities simultaneously enhance and inhibit development (Leonard-Barton 1992: 112).

As they choose to sustain their technical capabilities, managers must realise that the choices they are making today could inhibit their ability to compete with other firms on what will be an important competitive dimension: environmental performance.

## ▮ Capital Investment Decision Process

While environmental pressures are intensifying, most corporations still do not have the proper capital investment decision infrastructure to improve decision-making. Indeed, most companies have acknowledged that they do not have a system that adequately identifies and tracks environmental costs (Epstein 1996b, 1996d). Thus many environmental costs are hidden in overhead and general administrative accounts, and no assignment is made to the activities or products that caused those costs. Relevant cost and benefit information, appropriate time-frames and risk and uncertainty associated with projects are evidently key components of capital investment analysis that are too often ignored.

The proper integration of environmental issues into the decision-making process is a significant challenge for managers. There are, in fact, many barriers to overcome. Roome (1994) proposes that environmental problems are complex because they cut across many organisational, functional and disciplinary frontiers. Further, Post and Altman (1994) propose two types of barriers to environmental change: industry and organisational barriers. Industry barriers include technical availability and knowledge, information, and regulatory constraints. Attitudes, both from staff and top management and from past management practices, are among organisational barriers. Past procedures regarding capital investment decisions could also be an important barrier to environmental change.

In the case of energy-saving technologies and the US Environmental Protection Agency's 'Green Lights' programme, DeCanio (1993) suggests that the barriers to corporate adoption of energy-saving technologies are primarily behavioural and institutional, not technical. He describes four factors that cause under-investment and are therefore barriers to profitable opportunities. These are:

- Myopia (short payback periods)
- High internal hurdle rates
- Managerial compensation tied to short-term performance
- Low priority (for top management) for relatively small cost-cutting projects

Managers must address all of these factors to improve the capital investment decision process.

So far, the evidence suggests that in many cases, corporate environmental response consists of relatively small projects focusing on adjustments to the existing production system rather than important technological investments. For example, the average savings per project in 3M's 'Pollution Prevention Pays' programme are about $180,000 (Skea 1995). (These analyses are dependent, in part, on the definition of environmental investments and the difficulty in separating environmental investments from capital investments to improve product quality or corporate productivity.) But significant improvement in environmental performance will demand larger investments and, therefore, a more thorough examination. As the firm moves closer to 'zero emissions', reductions will become more capital-intensive and may require broader modifications in product design and technology (Walley and Whitehead 1994). A framework that integrates environmental impacts into capital investment decisions is needed in forward-looking organisations. Presently, in most firms, environmental projects and non-environmental projects do not, in fact, go through the same screening process.

The current screening process for three types of capital investments is described in Figure 2: general, environmental/compliance-driven, and environmental/voluntary. Figure 2 suggests that these projects will not receive the same type of screening. Regulatory pressures remain the driving force behind most environmental initiatives currently undertaken by firms (Green, McMeekin and Irwin 1994). Many environmental investments are compliance-driven. Given their 'must-do' or 'profit-sustaining' categories,

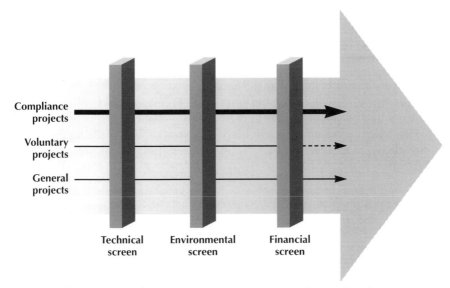

**Figure 2:** Capital Investment Screening Process: Current Practices

firms approach these investments differently. The projects are not perceived as discretionary and managers do not believe there is a need for thorough analysis (White, Becker and Savage 1993). Usually, firms evaluate a limited number of options and do not evaluate their full environmental impact. Instead, they check that environmental standards and norms, such as a prescribed emission level, are observed. The financial screen's scope is very limited. Financial analysis techniques such as discounted cash flow (DCF) are not typically used. The objective is to meet the environmental requirements in the least expensive manner. Voluntary environmental projects must typically pass through an important barrier: the financial screen. Well-developed financial analysis techniques normally used to analyse general capital projects are still not deployed in the case of voluntary environmental projects. Budget constraints are common, and required hurdle rates are high. As with compliance projects, there is not typically a complete evaluation of environmental costs and benefits. Many companies do not believe that there are potential benefits to environmental investments nor do they have the proper infrastructure to evaluate environmental costs. Yet there is abundant anecdotal evidence to support the benefits of environmental performance and the value of integrating environmental impacts into management decisions (Epstein 1996c, 1996e). Michael Porter's *Scientific American* article (1991) was one of the first significant contributions to this position.

In later work, Porter and van der Linde (1995a: 101) suggest that 'innovation' can be a by-product of environmental performance, and offer examples of both product and process offsets. Examples of product offsets include:

> higher-quality products, lower product costs (perhaps from material substitution or less packaging), products with higher resale or scrap value (because of ease in recycling or disassembly) (Porter and van der Linde 1995a: 101).

Process offsets include:

better utilisation of by-products, lower energy consumption during the production process, reduced material storage and handling costs, conversion of
waste into valuable forms, reduced waste disposal (Porter and van der Linde
1995a: 101).

If managers do not believe that the benefits exist, or perceive them as immeasur-
able or intangible, like improving the firm's image (Lefebvre, Lefebvre and Roy 1995),
they are not likely to include them in their decision-making process. White and Savage (1995: 54) conclude that current accounting and capital investment methods do
not provide the proper framework to analyse the environmental impacts, both costs
and benefits:

> While most companies quantify the most obvious and measurable environ-
> mental costs, substantially fewer have grappled with costs that are uncer-
> tain, less tangible, and difficult to quantify.

## ▌ Proposed Framework

To improve capital investment decisions, managers must have the proper framework to
help them evaluate all of the impacts of these decisions on the performance of the firm,
from a financial, operational and environmental perspective. The same framework should
be used for general capital projects and environmental capital projects, both regulatory
and voluntary. This requires projections of future revenues and costs and the likely changes
in environmental regulations, technology, and the cost of technology.

Figure 3 suggests a three-screen decision-making process[2] for capital investment decisions: a technical, environmental and financial screen. The first element of the screening process is the type and number of projects that will be entering the process. A firm's
skills and knowledge will have an important bearing on the investments considered:

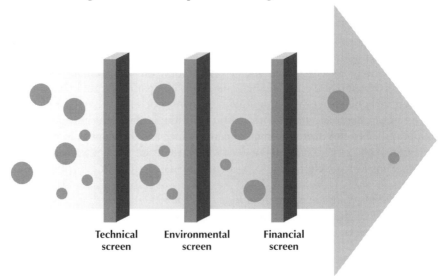

Technical    Environmental    Financial
screen          screen          screen

**Figure 3:** Capital Investment Screening Process

2. A related framework is the development funnel described by Clark and Wheelwright (1993). It
   provides an innovative framework for examining the generation and screening process for
   research and development projects.

> The ability of a firm to recognise the value of new, external information,
> assimilate it, and apply it to commercial ends is critical to its innovative
> capabilities (Leonard-Barton 1995: 136).

This is the concept of absorptive capacity and is an important determinant of a firm's capacity to exploit new or outside knowledge (Cohen and Levinthal 1994). It will have a significant impact on how a firm can actually respond to environmental pressures. Does the company have the skills and expertise to exploit a new and cleaner technology? Does it have a clear understanding of all of the options, technical and managerial, available to it? Firms should recognise the importance of building their absorptive capacity through technical training and extensive monitoring of technical literature in the field of the environment.

Our proposed framework emphasises that, through the broader definition of life-cycle costs and benefits, these impacts will be analysed and included in the decision process. These learning opportunities could create future competitive advantage and cannot be ignored.

The government plays an important role in this process and has a significant impact on the amount and type of options considered by the firm. It can facilitate this process through numerous programmes, incentives and databank access to information. Furthermore, its choice of environmental policy instruments will also induce certain type of behaviour. Market-based environmental policy instruments will encourage companies to adopt new and better technologies and processes (Jaffe *et al.* 1995). When choosing policy instruments, regulators must be responsive to costs and efficiency considerations and should allow for flexibility and innovation (Rice 1994).

## Technical Screen

The first screen of the decision-making process is the technical screen. All investments should be carefully analysed to ensure that they meet technical requirements. From a technical perspective, capital investments impact many aspects of the business value chain, including inbound logistics, operations, outbound logistics and after-sale service (Porter 1985). As capacity decisions are often critical, marketing and sales data will also be included in this screening process. For each of these activities, a proper evaluation of how the investment may affect product performance and sales must be included.

The existing technical capabilities and their features will have a significant influence on the screening process. Indeed, compatibility issues are of significant importance in any adoption decision (Rogers 1983). Therefore, the age of the equipment, type of production process and type of product could become either barriers or driving forces in the screening process.

## Environmental Screen

Choices among process improvements, product improvements and capital improvements can be evaluated properly only by completely identifying and measuring all environmental costs and benefits. The next section presents some cost assessment methods to help in measuring these costs.

*Life-cycle Assessment and Life-cycle Costing.* 'Life-cycle' refers to the scope of physical operations involved in producing, selling, using and disposing of a product—from raw material extraction through to final disposition. Life-cycle assessment (LCA) is a technique involving comprehensive evaluation of the environmental impacts of a particular product, throughout its life-cycle.

LCA provides managers with a more complete picture of the environmental impacts linked to a particular product or process, beyond those evident in the manufacturing stage. Managers who comprehend the full scope of a product's environmental consequences are then better equipped to make intelligent investments in pollution prevention, rather than looking only at the 'end of the pipe' for solutions. As a result, many companies have realised significant bottom-line benefits through the inclusion of life-cycle principles in their capital decision-making process.

LCA involves three stages:

- **Inventory analysis:** a quantitative inventory of all energy and raw materials inputs and environmental releases throughout the life-cycle

- **Impact assessment:** an evaluation of the inventory's findings in terms of environmental and human health impacts, as well as economic costs and benefits

- **Improvement analysis:** an analysis of opportunities for improvement in each stage of the life-cycle, including changes in product design, raw material usage, industrial processes or consumer waste

Conducting an LCA first requires boundaries to be defined. Given the incomplete nature of current methods of evaluating environmental impacts, as well as practical constraints of time and money, a complete life-cycle profile for decision-making could be a very complex undertaking.

Identifying and evaluating environmental impacts, by itself, may not yield results that are readily meaningful to business decision-makers. For LCA to be an effective decision-support tool for managers, impacts must be translated into metrics that business managers understand—namely, dollars and cents. Accurate comprehensive identification of environmental costs puts environmental investments on an even playing field with other uses of corporate funds, allowing the company to make the best financial decisions rather than separating 'environmental' benefits from 'financial' benefits.

Life-cycle costs can be classified into three categories. Conventional costs include direct product or process costs, such as labour and materials, as well as one-off capital expenditures. These costs are typically accounted for in process control, product costing, investment analysis and capital budgeting. Less tangible, hidden, and indirect costs also directly affect the bottom line. However, they are often probabilistic in nature or hidden in overhead accounts and thus are overlooked or miscalculated in assessing product costs. These include costs of regulatory compliance, waste disposal, as well as less easily quantifiable liabilities such as the effects of non-compliance on market share.

External costs are defined as

> those for which a company, at a specified point in time, is not responsible in the sense that neither the marketplace nor regulations assign such costs to the firm (Curvan 1995).

The boundary between internal and external, however, is constantly shifting. For example, the re-authorisation of the Clean Air Act imposed regulatory burdens on facilities with substantial emissions of nitrogen oxide ($NO_x$), a primary component of smog. Before 1970, the cost for $NO_x$ pollution and its resulting environmental and human health impacts was not borne privately: it was external. However, manufacturers must now pay for contributing to air quality degradation, through investments in pollution prevention or through fines for non-compliance. It has become an internal cost, with real impact on profitability.

Should manufacturers have invested in pollution prevention or control before the Clean Air Act? More generally, should companies consider external costs when calculating total life-cycle costs? The shifting nature of the boundary between internal and external suggests that ignoring external costs entirely is a poor long-term strategy. The general trend in the past decade towards more stringent environmental regulation suggests that, at a minimum, impacts likely to be subject to future regulation should be considered in terms of their probable future cost (Curvan 1995). The firm must also consider the growing public sensitivity to corporate environmental activities. What may appear to be external costs at first glance may have real effects on market share and profitability, as consumers increasingly base their decisions on a new set of criteria.

Quantifying less tangible and external costs is not a straightforward process. Putting a value on environmental impacts, especially those that do not carry regulatory penalties, involves a fair amount of judgement and estimation. Accounting for future regulatory trends, and other forces that cause the internal–external cost boundary to shift, often relies on extrapolation and probabilistic forecasting. However, decision-making under uncertainty is hardly foreign to the business world. Ignoring costs that are difficult to quantify or are probabilistic in nature implicitly assigns them a value of zero, which is misleading if their values are likely to be positive (Epstein 1996c). A number of techniques exists that can help both with monetising environmental externalities and with accounting for uncertainty and risk.

*Monetising Environmental Externalities.* There are two main approaches to monetising environmental externalities: cost of control and damage costing. The cost of control approach uses the cost of reducing or avoiding environmental damage before it occurs to approximate the cost of the damage itself. For example, the cost of polluting a waterway with dioxin would be estimated by the cost of the equipment necessary to control its presence in effluent, or the cost of process improvements that would eliminate the need for such controls. The cost of control approach avoids difficult-to-determine actual costs of environmental damage by replacing them with more easily estimated costs of installing, operating and maintaining environmental control technologies.

In contrast, damage costing attempts to assess the actual cost incurred from environmental damage. The loss of value attributable to the damage is estimated by the public's willingness to pay to avoid the damage. Ontario Hydro (see Chapter 16) suggests four different approaches to damage costing:

1. The market price method uses information on market prices of, for example, crops that have been damaged or lost due to toxic emissions.
2. The hedonic-pricing method uses geographical differences in real-estate values or wage rates, assuming that such differences are attributable to relative environmental quality.

3. The travel cost method uses information on cost of travel from polluted areas to recreational sites not affected by environmental damage.

4. The contingent valuation (CV) uses survey responses on willingness-to-pay or willingness-to-accept from 'perpetrators' and 'victims' of environmental impact to express physical manifestations of environmental impact in dollar terms.

These methods can be used to evaluate both environmental benefits and environmental costs. For example, CV is a survey-based method that attempts to determine how individuals value non-marketed goods. Similar to market research, it attempts to predict how people will make decisions. The concept has been in use for more than thirty years and is now used widely in both government and industry. Most federal, state and regulatory agencies responsible for natural resources or environmental assessments use CV. One of the recent uses of this concept was in the valuation of the environmental impact of the Exxon *Valdez* disaster. But the method also has been used to value environmental benefits created by, for example, the development of recreational facilities by a corporation as a by-product of its commercial activities. Though there remains some controversy over the reliability of its estimates, contingent valuation does provide useful information about the value of non-marketed goods that would not otherwise be available.

None of these methods is a perfect proxy for the cost of environmental damage. However, each approach provides information about non-market values that decision-makers otherwise would not have. Companies can use a combination of damage costing or cost of control methods to develop a reasonable range of values for the environmental impacts of their activities. This can lead to better decisions than would the current assumption that externalities have no cost at all. Though measurement difficulties remain, the reliability of the measures is now as high as many other estimates used in business decisions.

### Financial Screen

For decades, business schools have been teaching, and modern corporations have been using, discounted cash flow techniques for capital investment decision-making. But companies still commonly make use of inadequate tools for these decisions (see Fig. 4).

Continued use of the payback method rather than discounted cash flow is inadequate in most situations. Analyses that include only current costs and ignore current

| Methods | US | Japan | UK | Canada |
|---|---|---|---|---|
| Payback | 59% | 52% | 76% | 50% |
| Internal rate of return (IRR) | 52% | 4% | 39% | 62% |
| Net present value (NPV) | 28% | 6% | 38% | 41% |
| Accrual accounting rate of return | 13% | 36% | 28% | 17% |
| Other | 44% | 5% | 7% | 8% |

**Figure 4:** International Comparison of Capital Budgeting Methods
*Source: Horngren et al. 1994: 703*

benefits, future costs and future benefits are also inadequate. The complete examination of all impacts is required. Some managers continue to make decisions that minimise current capital costs related to project construction and ignore other long-term impacts. They also often ignore how changing regulations may require more rapid capital replacement. Other managers invest in capital projects that maximise short-term operating profits and have a quick payback but ignore longer-term impacts that may dramatically increase the company's long-term costs and ultimately produce a negative net present value. Discounted cash flow techniques such as net present value (NPV) and internal rate of return (IRR) do provide for an analysis of relevant cash flows (costs and benefits) over the economic life of the project, adjusting for the time-value of money and riskiness of the project through an appropriate discount rate. However, estimates of future environmental impacts must be estimated and included.

***Social Discount Rate.*** The terms 'uncertainty' and 'risk' are often used interchangeably, but are distinctly different. Uncertainty relates to a situation in which the probability distribution is not known. Risk relates to a situation in which the probability distribution of an event *is* known. To assess risk in environmental situations, it often is suggested that adjustments be made to the cost and benefit profiles rather than to the discount rate. A better approach to this problem is to test the sensitivity of the outcome of project evaluations to variation in the key parameters (Kula 1992).

Traditional capital budgeting techniques make use of an internal discount rate to compare alternatives in investment decisions. The choice of the discount rate obviously affects the outcome of the decision.

> Most companies discount all future cash flows at the same rate, usually the firm's cost of capital. In most cases this is appropriate, however there may be compelling arguments why the costs and benefits of some pollution prevention projects should be discounted at a lower rate (Environmental Law Institute 1993: 29).

Many of the more traditional analyses use short-term payback methods, while environmental decisions may require a longer time-horizon.

Gray (1990) and Gray, Babbington and Walters (1993) propose that 'all new investments should meet environmental criteria' as well as other corporate criteria in the investment decision process. They suggest the use of an 'environmental hurdle rate' through which all investment decisions should be filtered. They acknowledge that this interest rate is more difficult to ascertain because of the long-term nature of environmental decisions.

Environmental capital budgeting must use the same techniques for decision-making that other areas use throughout modern corporations. Payback is inadequate, so discounted cash flow methods should be used. Risk analysis should be incorporated into the decision and appropriate discount rates used. Whereas certain hurdle rates are generally established for all other capital investments, they are usually ignored for environmental capital investments because companies believe incorrectly that a positive return on investment is not achievable. The biggest improvement in environmental capital investment decisions, however, probably will be made through a broader identification and measurement of costs and benefits. By including all environmental impacts, companies can apply the same discount rates as they do for other projects. They can complete proper analyses that consider all stakeholders and all impacts in an attempt to minimise environmental impacts and maximise corporate profits. Through this approach, benefits are often seen to be substantially greater than previously believed and return on investments substantially higher.

Many companies have already recognised the significant value that can be added by fully integrating the identification and measurement of environmental impacts into capital investment decisions or, at a minimum, requiring an EH&S review of capital expenditures. Pharmaceutical and consumer products manufacturer Bristol-Myers Squibb includes EH&S in all of its capital investment decisions (see Datar, Epstein and White 1996). Capital appropriations requests (CARs) are required for all new processes, products and facilities, and each facility has its own environmental co-ordinator, who must review all facility CARs. Capital expenditures of more than $4 million, as well as purchases and sales of land, must be approved by the corporate general finance committee, of which the corporate vice-president of EH&S is a member.

***Option Assessment, Option Screening, and Scenario Forecasting.*** To measure environmental costs and benefits adequately, estimates of likely future effects are critical and approaches to the analysis are helpful. Option assessment, option screening and scenario forecasting are among techniques that can be used to assist in this analysis. Option assessments and option screenings are designed to provide decision-makers with all of the available options regarding alternatives.

> There is a three-phase methodology developed that helps decision makers assess, and act on, the relative attractiveness of options to reduce the environmental impact of substance chains (Winsemius and Hahn 1992: 252).

The first phase is to generate options. This phase includes four steps:

- Drawing a flow diagram
- Identifying the major environmental issues
- Defining the options
- Selecting the most likely options for further evaluation

This selection step is based on cost-effectiveness, relevance for decision-makers, and environmental impact.

The next phase prioritises the options by determining an economic and environmental profile of the effects. These effects are quantified in monetary terms and typically include the net changes in operating and capital costs. The options then are positioned on an 'option map' based on the relative weight and importance of the costs and benefits of each option. The last phase requires the establishment of targets, resources and responsibilities (Winsemius and Hahn 1992).

The Niagara Mohawk Power Company (NMPC) of Syracuse, New York, uses option screening to compare potential environmental scenarios and associated costs of environmental considerations. It implemented a system to identify and measure the options related to both the demand and supply side of electric power usage. The company uses option screening to determine the optimum mix of demand and supply strategies that provide electrical energy services at the lowest cost, within a set of various constraints. It used focus groups to determine the appropriate options and assign probabilities to the most likely scenarios.

Some companies are also using scenario forecasting techniques to help them examine the likely impacts on the total environmental costs of changing regulations, changing technologies, and the changing cost of technologies. In companies with high levels of uncertainty, where change is imminent and diversity of opinion exists, scenario forecasting can be useful to identify clearly the various choices for decision-makers. Some have suggested that scenario forecasting aids in assessing and managing risk, broadens corporate thinking, and makes managers focus on the long-term impacts of their decisions.[3]

***Decision Trees and Monte Carlo Simulation.*** Other ways to deal with uncertainty include decision trees and Monte Carlo simulation. A decision tree is a visual portrayal of the structure of a decision problem, displaying the alternative courses of action, all possible outcomes, and the probability values of each decision.

> Monte Carlo analysis is a simulation technique that permits the probability distributions of outcomes to be calculated for complex decision trees. A computer is used to simulate the outcome of a series of probable events, over and over again, very rapidly (Deloitte & Touche 1991: 4).

Many companies have applied Monte Carlo analysis to the problem of comparing the possible costs of alternative environmental remediation options. Using Monte Carlo random sampling from an option's cost probability distribution, the probability that one option will cost more than another can be estimated and the most likely costs of each operation can be compared. Probabilities (i.e. confidence levels) can be assigned to a range of possible costs, leading to more credible and dependable comparisons.

Monte Carlo simulation assigns a probability distribution to environmental risk that can increase or decrease depending on changes to environmental legislation. Once probability distributions are established for all inputs required for an NPV analysis, the Monte Carlo simulation is initiated. The computer goes through the decision tree, drawing a sample from the relevant probability distributions at each point where an event occurs, and then applies simple logic to determine how to proceed through the tree. Where alternative technologies are available, the computer model will determine the probability distributions of the possible costs of the technologies and then choose the least costly option. If the decision tree has different possible events, the computer will model each event and the possible outcomes. This process is repeated until meaningful probability distributions can be established.

All of these techniques can be used to improve the capital investment decision process. Companies need to use the same techniques for environmental investments as they do for other capital investment decisions. Thus discounted cash flow techniques should be used and estimates of future impacts should be included as they are in other capital investment decisions. Further, cross-functional integration is necessary to facilitate the consideration of these impacts. Companies must bring together staff from operations, engineering, accounting, design, legal, and marketing with EH&S for the

---

3. Building on the same logic and potential benefits of this framework, Mercer (1995) proposes a simpler approach to this methodology.

identification and measurement of life-cycle impacts. Otherwise, capital investment decisions will be incomplete. Companies must complete full analyses of both voluntary and regulatory environmental projects. Only then, will proper project selections be made. Until that time, companies cannot move from reactive to proactive environmental management.

## ▌ Conclusion

Implementing a decision-making process for capital investments that integrates environmental impacts is an important challenge for managers. There are many barriers to its implementation. As the regulatory climate changes and government agencies become increasingly flexible regarding acceptable alternative technologies, the complete analysis of projects becomes even more critical. Companies need to incorporate environmental considerations into their capital investment decisions to provide additional perspectives with which to make those decisions. All environmental costs and benefits need to be identified and measured before investment decisions are made and strategy implemented. A trend is emerging whereby companies have gone beyond regulatory requirements in meeting their environmental obligations. They are beginning to understand that being proactive and planning for pollution prevention rather than reactive and settling for mere compliance is the most cost-effective and responsible strategy.

Some organisations are already familiar with more sophisticated techniques that include risk analysis. Other organisations will require more changes to their existing decision-making process to provide a framework that includes the long-term impacts of decisions. Though not without their limitations, these approaches represent significant improvements for environmental management over the simplified model used presently in most corporations. They will definitely improve management decisions. Our proposed framework can also be adopted and implemented by companies in high-impact and low-impact industries, in one location or globally, small or large, with large EH&S staff or no full-time EH&S personnel. However, it will require the involvement and integration of many expertises: namely, engineers, EH&S and financial analysts. Though the proposed framework seems to suggest sequential screens, the underlying analysis relies on continuous consideration of these impacts throughout the decision process. Only in this way can companies improve investment decisions to reduce environmental impacts and improve profitability.

# 5 Valuing Potential Environmental Liabilities for Managerial Decision-Making

## A Review of Available Techniques[1]

### US Environmental Protection Agency

## ▋ Introduction

THE ESTIMATION of environmental liability costs is an area fraught with uncertainty. Managers may feel that they cannot estimate these costs with a sufficient degree of accuracy to merit inclusion in decision-making calculations.[2] They may be unaware of existing environmental liability valuation techniques. They may also be unsure of the relevance of the various techniques to different types of liability. As of yet, accounting for environmental liability costs receives relatively little attention, with the exception of accounting and disclosure requirements under Securities and Exchange Commission (SEC) rules, Financial Accounting Standards Board (FASB) statements, and a recently issued Statement of Position from the American Institute

---

1. This chapter is an edited version of a longer report of the same title published by the Environmental Protection Agency (US EPA 1996b). It was prepared by ICF, Inc. for the Agency's Environmental Accounting Project. The full report can be read on the Project's website at *http://www.epa.gov/opptintr/acctg* or obtained free of charge from the EPA's Pollution Prevention Information Clearinghouse by telephoning +1 202 260 1023.

2. In a survey of US manufacturing firms conducted by the Tellus Institute and commissioned by EPA's Environmental Accounting Project, an average of approximately one-third of the respondents indicated that, for the purposes of investment financial evaluations, they regularly calculate cost values for such items as environmental fines and penalties, personal injury, future regulatory compliance costs and natural resource damages. With respect to Superfund remediation liabilities, it was found that only about 10% of all respondents regularly determine the liability costs for inclusion in project financial analyses. The most commonly cited barriers to calculating Superfund liability costs were the difficulties associated with estimating the likelihood, magnitude and timing of the liability costs (Tellus Institute 1995).

of Certified Public Accountants (AICPA). Regardless of financial accounting requirements, however, managers are naturally concerned about minimising the costs associated with environmental liabilities, just as they are concerned about minimising other business expenses. Thus, as the costs of environmental liabilities mount, due both to expansion in the nature and types of environmental liability and to increased costs for each type of liability, business managers have greater incentives to consider carefully the extent to which existing and proposed activities give rise to environmental liability costs.

The Environmental Protection Agency (EPA) prepared the report, of which this chapter is an edited version, as a step toward helping companies assess and manage their environmental liability costs. It describes valuation approaches and tools that have been specifically developed or adapted for estimating environmental liability costs for consideration in business management decisions such as capital investments, process/input substitutions, product retention and mix, facility-siting and waste management. The emphasis is on techniques for placing a monetary value on *potential, preventable* environmental liabilities. Sources of information on estimating environmental liability costs are somewhat obscure; EPA thus offers this summary of documented valuation techniques to managers who are interested in estimating potential environmental liability costs, but are unaware of techniques for doing so. While estimation techniques are still under development, and possibly controversial, this chapter is intended to assist organisations in estimating future and/or potential environmental liability costs within reasonable limits of accuracy such that a manager is comfortable using the estimations when making decisions.

The research uncovered over three dozen references with potential applicability to forward-looking management decisions. The unedited version of the report (see footnote 1) identifies the valuation approach(es) employed by each of the techniques, assesses their data requirements and describes their scale of application. The full report also categorises the techniques by the types of liability that each was developed to address. Additionally, the full report provides an annotated bibliography that contains availability information for each of 37 references and detailed profiles for 24 of the references. For reasons of space, these details are not included here.

## ▮ Environmental Liabilities: Definitions and Categories

The term 'environmental liabilities' crops up in many discussions of environmental issues. Yet there is much confusion about the term. Often, 'environmental liabilities' is used to refer to the potential for fines, penalties and jail terms for violations of environmental laws. 'Environmental liabilities' also frequently serves as short-hand to refer to the clean-up obligations under the federal Superfund and state counterpart laws for contaminated sites. Another common usage is to label the costs involved in complying with regulations as 'environmental liabilities'. In contrast, when companies perform or commission 'environmental liability' assessments, they want to know their exposure to potential environmental liabilities even when they are in complete compliance with regulatory standards.

Clearly, 'environmental liability' is an umbrella term. The following pages present a brief definition of the term and describe the major types of environmental liability in order to establish a framework for reviewing approaches and tools identified for expressing these liabilities in monetary terms. More information about the timing, likelihood and uncertainty characteristics of the various forms of environmental liabilities is provided in the appendix to this chapter.

## Defining 'Environmental Liability'

The term 'liability' has important accounting and legal dimensions. Accounting institutions define liability as a 'probable future sacrifice of economic benefits arising from present obligations to transfer assets or provide services in the future as a result of past transactions or events' (Financial Accounting Standards Board, Concept Statement No. 6, Paragraph 35 [1985]; Institute of Management Accountants 1990). More simply, a liability is a *present* obligation to make an expenditure or to provide a product or service in the *future*.

Liability has an important legal dimension as well. A liability is a legally enforceable obligation, whether it is voluntarily entered into as a contractual obligation, or is imposed unilaterally, such as the liability to pay taxes. The law both establishes liabilities and determines who is responsible for discharging them.

For the purposes of this chapter, an **environmental liability** is a legal obligation to make a future expenditure due to the *past* or *ongoing* manufacture, use, release or threatened release of a particular substance, or other activities that adversely affect the environment. A *potential* **environmental liability** is a *potential* legal obligation to make a future expenditure due to the *ongoing* or *future* manufacture, use, release or threatened release of a particular substance, or other activities that adversely affect the environment. An obligation is *potential* when it depends on future events or when a law or regulation creating the liability is not yet in effect. A 'potential environmental liability' differs from an 'environmental liability' because an organisation has an opportunity to prevent the liability from occurring by altering its own practices or adopting new practices in order to avoid or reduce adverse environmental impact.

## Types of Environmental Liability

Environmental liabilities arise from a variety of sources. Federal, state and local environmental statutes, regulations and ordinances, whether enforced by public agencies or through private citizens' suits, give rise to many types of environmental liability. Another legal source of these liabilities is 'common law' (i.e. judge-made law) that can vary from state to state. A detailed list of environmental liabilities would be very lengthy. Thus, this chapter distinguishes the following broad categories of environmental liability:

- **Compliance** obligations related to laws and regulations that apply to the manufacture, use, disposal and release of chemical substances and to other activities that adversely affect the environment

- **Remediation** obligations (existing and future) related to contaminated real property

- Obligations to pay civil and criminal **fines and penalties** for statutory or regulatory non-compliance

- Obligations to **compensate** private parties for personal injury, property damage and economic loss

- Obligations to pay **'punitive damages'** for grossly negligent conduct
- Obligations to pay for **natural resource damages**

The following paragraphs elaborate on each of these types of environmental liability.

*Compliance Obligations.* As laws and regulations are enacted that apply to the manufacture, use or release of regulated substances, companies find themselves facing future compliance costs. In evaluating business plans, some companies may also consider the possibility that new laws and regulations will be enacted. Additionally, a company may discover that it is not in compliance with existing laws and regulations. The costs of coming into compliance can range from modest outlays required to conform to administrative requirements (e.g. record-keeping, reporting, labelling, training) to more substantial outlays, including capital costs (e.g. to pre-treat wastes prior to land disposal or release to surface waters, to contain spills, to 'scrub' air emissions). Laws and regulations also impose 'exit costs' (e.g. to close waste disposal sites properly and provide for post-closure care, and to decommission nuclear power reactors at the end of their useful lives).

*Remediation Obligations.* These are sometimes subsumed under 'compliance' because some property clean-up requirements have been enacted as part of regulatory programmes applicable to operating facilities under, for example, the Resource Conservation and Recovery Act (RCRA) and the Safe Drinking Water Act's Underground Injection Control programme. Also, it is easy to blur the distinction between the compliance obligation of routine closure of facilities at the end of their useful lives and the remediation obligation for cleaning up pollution posing a risk to human health and the environment. And meeting current compliance obligations may help minimise future remediation obligations. Nevertheless, remediation obligations are considered a separate category in this chapter because of some distinguishing characteristics of the liability and the attention that has been paid to this category of environmental liability. Remediation tends to be expensive, ranging up to many millions of dollars, and can include excavation, drilling, construction, pumping, soil and water treatment, and monitoring, and can include the response costs incurred by regulatory agencies. Remediation costs also can include the provision of alternative drinking water supplies for affected community residents and, in some circumstances, purchase of properties and relocation expenses. Technical studies and the expenditure of management, professional and legal resources add to the cost of remediation.

The remediation obligation is distinctive because a company may face remediation obligations due to contamination at inactive sites that are otherwise unregulated; at property formerly but not currently owned or used; at property it never owned or used, but to which its wastes were sent; and at property it acquired but did not contaminate (e.g. in 'Superfund liability' scenarios). Because many dollars will be needed in the near term to remediate existing environmental contamination, particularly at inactive and abandoned sites, these liabilities often dominate (and can distort) a firm's assessment of its environmental liabilities. Therefore, it is helpful to distinguish between remediation obligations for existing contamination and potential remediation obligations for future contamination because managers can have more impact on ongoing and future activities and releases—whether accidental or not—that may trigger future remediation obligations.

**Fines and Penalties.** Companies that are not in compliance with applicable requirements may be subject to civil or criminal fines or penalties for non-compliance and/or expenses for projects agreed to as part of a settlement for non-compliance. Such payments fulfil punitive and deterrent functions and are in addition to the costs of coming into compliance. Fines and penalties (and related outlays for supplemental environmental projects) can range from modest amounts to a few million dollars per violation. Generally, a civil penalty is assessed that is at least equal to the costs a company saved through non-compliance, thus removing any financial incentive to ignore a law. Other factors may add to or reduce the penalty amount assessed for a violation.

**Compensation Obligations.** Under common law and some state and federal statutes, companies may be obligated to pay for compensation of 'damages' suffered by individuals, their property and businesses due to use or release of toxic substances or other pollutants.[3] These liabilities may occur even if a company is in compliance with all applicable environmental standards.

Distinct sub-categories of compensation liability include personal injury (e.g. 'wrongful death', bodily injury, medical monitoring, pain and suffering), property damage (e.g. diminished value of real estate, buildings or automobiles; loss of crops) and economic loss (e.g. lost profits, cost of renting substitute premises or equipment). Compensation costs can be fairly minor or quite substantial, depending on the number of claimants and the nature of their claims. Often, legal defence costs (potentially including technical, scientific, economic and medical studies) can be substantial in handling such claims, even when the claims are ultimately determined to be without merit. Moreover, responding to compensation claims can consume management time and require expenditures in order to control damage to corporate image. Compensation liabilities may involve costs for remediation of contaminated property as well as provision of alternate water supplies, thus somewhat overlapping the remediation category.

Because of workers' compensation and employer liability laws, payments to compensate employees for occupational exposure and injury from hazardous or toxic substances are not generally determined through litigation against the employer or considered environmental liabilities. However, occupational claims sometimes may be brought against another party who is not the employer; for example, workers responding to a train wreck have sued the shipper of hazardous wastes released at the scene of the wreck; for the shipper, these claims can be viewed as environmental liabilities. Managers will want to understand the potential costs of occupational exposure and injuries, because actions taken to prevent or reduce environmental liabilities may also eliminate or reduce occupational liabilities.

---

3. This is also known as 'toxic tort' liability where the word 'tort' is a legal term meaning 'a wrong' under common law. Most 'toxic tort' cases do not relate to environmental liabilities but fall under the realm of 'product liability' where the issues are the danger posed by a 'product', such as a pharmaceutical, pesticide, household chemical or industrial product (e.g. asbestos insulation), and whether there was adequate warning or disclosure of the risk. In contrast, 'environmental torts' are most often associated with emissions from a facility, waste disposal sites and accidental releases.

*Punitive Damages.* To supplement compensatory payments to those harmed by the actions of others, the law allows the imposition of what are called 'punitive damages' to punish and deter conduct viewed as showing a callous disregard for others. Unlike compensatory liability, the measure of punitive damages is not directly tied to the actual injuries sustained. Punitive damages are often many times larger than the costs of compensation; although rarely assessed, punitive damages in environmental litigation usually exceed $1 million. Punitive damages tend to be more common in product liability than environmental liability cases; the most notable recent imposition of punitive damages in the environmental context arose from the Exxon *Valdez* spill.

*Natural Resource Damages.* A relatively new category of environmental liability is best termed 'natural resource damages'. Established by state and federal statutes, notably Section 311 of the Clean Water Act, Section 107 of the Comprehensive Environmental Response, Compensation and Liability Act (CERCLA or 'Superfund'), and Section 1006 of the Oil Pollution Act (OPA), this liability generally relates to injury, destruction, loss or loss of use of natural resources that do *not* constitute private property. Rather, the resources must belong to or be controlled by federal, state, local, foreign or tribal governments. Such resources include flora, fauna, land, air and water resources. The liability can arise from accidental releases (e.g. during transport) as well as lawful releases to air, water and soil. To date, most natural resource damage payments have been relatively small.[4]

## ▌ Research Findings

### General Overview

Discussions of environmental accounting and accounting for environmental liabilities can be found in a variety of accounting, managerial, environmental and legal publications. Much of the available literature focuses on SEC requirements and generally accepted accounting principles (GAAP) in connection with the reporting of environmental remediation liabilities in external financial statements.[5] It is clear from this literature that rules and practices for measuring (as well as recognising and reporting) environmental liabilities for a company's *financial statements* are still evolving. Some of the tools and approaches developed to measure environmental liabilities for these purposes may also be useful for addressing potential environmental liabilities outside of the external reporting context. However, in general, the requirements and limitations (e.g. focus on expenditures and materiality at the level of the firm as a whole) of this branch of environmental accounting result in data that do not meet the needs of most managers for disaggregated, forward-looking information for planning, decision-making and operations.

---

4. In June 1995, the General Accounting Office (US GAO 1996) testified before Congress that natural resource damage claims had been settled for relatively small amounts: of 98 cases settled by federal trustees, 48 cases were settled for zero dollars; 36 cases for less than $500,000 each; nine cases for between $500,000 and $5 million; and five cases for greater than or equal to $12 million. GAO also identified four states where seventeen claims had been settled by non-federal trustees for an aggregate of $23.4 million.

5. A parallel body of literature addresses tax treatment of environmental remediation expenditures.

Environmental accounting literature oriented towards *managerial decision-making* focuses on: (1) methods for evaluating pollution prevention investments; (2) examples of financially and environmentally attractive actions taken by companies; and (3) internal recognition and proper allocation of environmentally driven expenditures. This literature often refers to the value or importance of considering the dollar magnitude of environmental liabilities, but tends to stop there. Or, the literature describes elements to be considered in valuing environmental liabilities without providing numerical illustrations, detailed methodologies, databases or default values. Some of the literature claims that environmental liabilities cannot be estimated, and some provides qualitative scoring approaches only. But more of the authors make the point that even an uncertain monetary estimate may be better than ignoring a potential environmental liability, which implicitly equates to a monetary value of zero. In failing to place a value on environmental liabilities, managers may reject pollution prevention actions that would be seen as financially attractive if the potential reduction in liability costs were valued. EPA's *Pollution Prevention Benefits Manual* (EPA 1989) recognises that accounting for conventional costs (e.g. labour, materials and utilities) and potentially hidden environmental costs in project financial analyses alone will often suffice to identify financially attractive pollution prevention actions. It suggests a middle course: value environmental liabilities only when they might make a difference in the cost–benefit calculus or assess what the value of avoided liability costs would have to be for the pollution prevention action to make financial sense.

## Valuation Approaches

There are a variety of general *approaches* to valuing environmental liabilities, such as actuarial techniques, professional judgement, engineering cost estimation, decision analysis/statistical techniques, modelling, scenario techniques and valuation methods. These approaches have been developed and applied, usually in combination, in specific tools for valuing particular types of environmental liability that can arise from certain situations. Although full explanations of these general approaches go beyond the scope of this chapter, they can be described as follows.

**Actuarial techniques** involve the statistical analysis of historical data on the costs and/or occurrence of environmental liabilities or events (such as accidents) or consequences (such as adverse health outcomes) that can lead to environmental liabilities.

**Professional judgement** includes the expert judgement of engineers, scientists, lawyers, environmental specialists and other professionals (see, e.g., Cooke 1991). Specifically, engineering judgement can be used to identify appropriate compliance and remediation activities and estimate the likelihood of accidental releases; scientific judgement can be used to assess hazards, the transport and fate of substances released to the environment, and the potential responses of exposed plants, animals, human beings, their property and ecosystems; and legal judgement can be used to assess legal bases for liability and potentially recoverable damages.

**Engineering cost estimation** develops costs (e.g. for remediation, restoration, compliance, provision of replacement water supplies) by systematically identifying required implementation activities (termed 'activity-based estimating') and corresponding units, unit costs (e.g. labour, materials, utilities) and contingency factors or through parametric cost estimation which uses cost equations, either individually or grouped into more complex models, developed by analysing the correlation between cost drivers and costs (see, e.g., Humphreys 1984). A *life-cycle cost* estimate encompasses all costs, including design, development, operation, maintenance and final disposition over the anticipated life-span of a process, product, facility or system.

**Decision analysis techniques** (e.g. event trees, probability distributions, level of confidence calculations) are used in structuring expert judgement, reflecting (and quantifying) uncertainties in liability valuation, and in characterising and presenting the results of environmental liability valuation (see, e.g., Ross 1985; Schaeffer and McClave 1990). Uncertainties regarding the magnitude, likelihood and timing of potential environmental liabilities can be explicitly addressed, producing a set of liability values and their associated likelihoods.

**Modelling** is used as an alternative or supplement to professional judgement when historic data are limited or not available and cost or occurrence values must be simulated due to many uncertain variables or complex interactions. (A 'model' is a set of equations and associated rules for their applicability and interaction.) Modelling typically draws on available data; professional judgement; quantitative expressions of pollutant release, transport, fate, exposure and consequences; and statistical analysis techniques.

**Scenario techniques** are used to describe and address future situations that can affect environmental liabilities, such as changes in regulatory requirements, remediation policy, legal standards for compensation and natural resource damages, and enforcement policy. A few scenarios can bracket a wide range of possibilities, represent diverse views and challenge managerial thinking. The scenario development process involves the formal elicitation of expert judgements about future environmental scenarios (Schoemaker 1991, 1995).

**Valuation methods** include a variety of legal rules and economic techniques for putting monetary values on environmental consequences for compensation and natural resource damage liabilities, respectively. Legal approaches to valuation of injuries to people, their property and their businesses comprise the accepted practices that have been developed and used to put monetary values on compensation claims in litigation. Different types of claim (e.g. fear of cancer, increased risk, clinical impairment, morbidity, pain and suffering, mortality, diminished market value, lost profits) will each have their own valuation approaches, which can vary across legal jurisdictions. Economic approaches to valuing the services provided by natural resources that are not privately owned constitute a set of techniques intended to value, directly or indirectly, the various use and non-use services injured or lost due to releases of pollutants.

### Overview of Valuation Tools

This research has documented a diverse set of specific tools that employ one or more approaches to valuing environmental liabilities. Some of the tools can help the user calculate potential liability costs and/or provide monetary values, while other tools describe either a more general process for developing, or a framework for using, such monetary estimates; this latter group may or may not provide clear illustrations or examples. The level and clarity of documentation vary considerably, both within and across tools. For some of the tools, the source(s) of key information (e.g. liability costs per unit, probability of accident) are not provided; other tools may require the user to estimate such variables, but provide little in the way of guidance or benchmarks to help that process. The diversity of features, approaches and orientations in the tools makes them very difficult to classify.

For the most part, the tools have been developed to address specific situations or answer specific questions. Often, the results or methods may be applied to other situations or questions. Linking the results or approaches of different tools may be feasible for some situations but not for others. The tools uncovered in this research do not address all potentially relevant situations; and the research uncovered no comprehensive, up-to-date compilations of data that could be used in assessing magnitude, likelihood and timing of all potential liabilities.

Currently, the most common applications of environmental liability valuation tools relate to waste management and disposal, and releases from underground storage tanks of petroleum, with fewer tools applied to releases from fixed facilities.

Environmental liability valuation approaches and, hence, tools that use those approaches, also differ in the scale of their application; a single process (e.g. waste disposal) or unit (e.g. underground storage tank), an entire facility, a population of units, or an entire company. Liabilities can be valued through bottom-up or top-down approaches. The former focuses on the liabilities of individual units of a firm or classes of liabilities and can serve as stepping stones to estimate firm-wide aggregate environmental liability. The latter approach starts with an aggregate liability estimate and apportions that liability to industries and firms (and, by extension, subdivisions and operations within firms).

Most of the identified tools appear to take a bottom-up approach, some characterise a population's environmental liabilities, and examples of the top-down approach are few. Some of the tools focus on waste management approaches, others focus on the industrial process level, while others deal with incidents, and a small number of tools relates to specific chemical substances.

## ▍Appendix: Timing, Likelihood and Uncertainty Issues

When incorporating the monetary values of environmental liabilities into planning, decision-making and operations, organisations may want to express or adjust the estimated dollar value magnitudes to reflect the timing, likelihood and/or uncertainty associated with such future expenditures. This appendix introduces these topics.

### 1. Timing
By definition, all environmental liabilities involve *future* costs. A common approach to evaluating future payments is to calculate their net present value. This requires an estimate of both the magnitude and the *timing* of future outlays (as well as selection of an appropriate discount rate). Sometimes, particularly for compliance and remediation obligations, the outlays may occur within three to five years. Such obligations, however, also may not occur until a point quite remote in the future. For example, the cost of the compliance obligation to decommission a nuclear power plant or close a landfill following the end of its useful life may be 25 or more years away. Similarly, a remediation obligation to clean up a contaminated site may not arise for decades. Compliance and remediation obligations also may be both years away and continuous in nature: in most cases, regulations require companies to provide for at least thirty years of 'post-closure care' following the closure of certain waste disposal facilities. To remediate contaminated groundwater in some locations could require fifty or more

years of removal and treatment. Thus, these obligations may well entail a stream of future payments extending over time, as opposed to a lump sum outlay.

Compensation and natural resource damage liabilities can arise in the near term or can have long time-frames as well. Compensation and natural resource damage liability can be triggered by sudden accidents that can occur at any time. Conversely, existing or future releases of contaminants and exposure of persons and property can continue over many years before any adverse environmental aspects or health effects, such as many forms of cancer, manifest themselves (termed the 'latency period'). Thus, compensation and natural resource damage claims may not materialise until years in the future after the use or release of a chemical or waste. And the process of investigating and litigating claims (e.g. linking health effects to prior exposures) can take several more years. On the other hand, claims for medical monitoring costs may be brought following mere exposure and in the absence of clear health impacts, meaning that some portion of the potential liability may arise earlier than other components.

In addition to understanding the potential timing of environmental liabilities, it is important to distinguish between (1) future costs that cannot be avoided or prevented due to events that have already taken place; and (2) future costs that can be avoided or prevented. Most reported accounting and insurance discussions of environmental liability relate to the first category, while most planning and risk management articles apply to the second category. The terms 'retrospective' and 'prospective' liability are sometimes used as labels for these two different situations. A related term 'retroactive liability' has a special connotation and usage.[6]

Understanding the timing of environmental liabilities is important in weighing the costs of taking actions to reduce or prevent those liabilities. Although there may be some opportunities to manage future costs resulting from activities and releases that have already occurred, different ways of doing business now and in the future can greatly affect a company's other environmental liabilities. In thinking about (and valuing) environmental liabilities, it may be useful to distinguish between:

1. Liabilities or potential liabilities due to past activities that have ceased

2. Liabilities or potential liabilities due to current activities

3. Potential liabilities due to future activities that have not yet begun

The latter two categories represent opportunities for companies to prevent the incurrence of environmental liability costs. By considering potential liability costs in forward-looking business decisions, planners, managers and managerial accountants may realise the financial benefits of altering current practices or of avoiding new activities that might cause the incurrence of environmental liabilities.

### 2. Likelihood
Managers may want to consider *likelihood* when weighing the monetary value of their potential environmental liabilities against the costs of potential management options

---

6. 'Retroactive liability' is a term most often used in connection with Superfund to refer to liability resulting from past activities that ceased before establishment of the law creating the liability. 'Retrospective liability' is a broader term that encompasses liability for past deeds that cannot be avoided or prevented. While all retroactive liabilities are retrospective in nature, retrospective liabilities are not necessarily retroactive. It is important to note that Superfund liability (and other laws) also applies *prospectively* and, in so doing, creates incentives for proper management of hazardous substances.

for preventing (or reducing the likelihood of) those liabilities.[7] A common approach to evaluating a potential environmental liability is to calculate its expected value, which is the product of its forecasted magnitude and its likelihood of occurrence.[8] In this way, the monetary estimate is adjusted (i.e. reduced) to reflect the probability that a future expenditure will be made. For example, if a potential environmental liability has an *estimated value* of $1 million and the likelihood of its occurrence is 10%, then the *expected value* of that liability is $100,000. The likelihood that a future expenditure will be made depends on whether the potential liability is certain to be incurred or, alternatively, depends on the uncertain occurrence or non-occurrence of one or more future events.

What complicates management's evaluation of the likelihood of environmental liability costs is that the uncertainties can include both *factual* questions (e.g. Will the tank car leak? Will children be exposed to substance x? Will they be harmed? Will groundwater be contaminated?) and *legal* questions (e.g. Does this law apply to this facility? Will there be a requirement to remediate this property? Are these damage claims compensable? Who is or will be legally responsible?) Thus, estimating the likelihood of a potential liability may require engineering, scientific, statistical and legal analysis and judgement. Both the factual and legal questions may be difficult to answer with certainty, particularly when managers are making projections for forward-looking decisions; although, with time, better data tend to make the factual questions easier to address and legal precedents tend to make the legal questions easier to answer.

**Likelihood of Compliance Liabilities.** Of all the environmental liabilities discussed here, compliance liabilities have the lowest level of associated uncertainty. If a law applies to a facility, activity or substance, expenditures to come into compliance must be made by a specified date. Therefore, there may be no need to adjust the estimated cost of compliance to reflect the likelihood of the incurrence of the liability.[9] For example, regulations require that those who open a hazardous waste surface impoundment must, at some point in the future, properly close it. There is no way to avoid that closure obligation; it is not contingent on any future event. However, if the closure does not meet 'clean closure' standards, then post-closure care must be provided; this post-closure care requirement is contingent on the results of the closure procedure. In this example, because the likelihood of closure is certain, its estimated cost need not be adjusted, but the potential cost of post-closure care can be adjusted to reflect the probability that post-closure care costs will be incurred. Another example: to avoid reporting and emissions control requirements that will apply to users of a particular ozone-depleting substance, a business may switch to an alternative substance (or process) that is not regulated. This is an example of a compliance obligation that is not contingent on a future event, but, rather, becomes inapplicable (i.e. avoidable) once use of the substance stops. In this example, because the future compliance expenditures

---

7. Where managers decide not to adjust the estimated liabilities to reflect their likelihoods of occurrence, as described below, adjustments to reflect the timing of future expenditures may nevertheless be made.

8. More precisely, the expected value is the probability-weighted product of an array of forecasted magnitudes and the corresponding likelihoods of those magnitudes.

9. As discussed earlier in this appendix, managers may choose to adjust the estimated cost of compliance to reflect the timing of the future expenditures.

will certainly be incurred, it would be inappropriate to adjust the estimated compliance cost for likelihood; the full cost of compliance should be considered when evaluating alternative practices.

**Likelihood of Potential Remediation Liabilities.** Potential remediation liabilities for future contamination have different types of associated uncertainty. Potential factual uncertainties could include, at a minimum, whether there will be a release or threat of release to the environment, transport through the environment, and/or discovery of the release. Depending on the situation, potential legal uncertainties could include whether there will be an obligation to remediate and whether that liability will be shared with others (often termed 'joint and several' liability). The latter contingency may be relevant for a company that sends its waste off-site for disposal at a commercial facility that also accepts wastes from others. (On the other hand, liability for remediation of contamination resulting from on-site waste disposal would probably not be shared with others unless the property is subsequently sold.) In these situations, the estimated monetary magnitude of the potential remediation liability could be adjusted to reflect both the likelihood that the liability will arise and the proportionate share of any liability for which other companies might be legally responsible.

However, it may be appropriate to view potential remediation liabilities for future contamination as certain to arise for such practices as the underground storage of petroleum products or other substances; the underground injection of wastes; or placing solid and liquid wastes into landfills and surface impoundments. Over time, leaks from containment may become increasingly likely. Given design limitations, releases may become quite probable. A conservative management approach might treat such future remediation expenditures as a certainty by not adjusting the magnitude of the potential liability by the likelihood of its occurring; this would provide managers with greater incentives to find alternatives to the use of such practices because a higher monetary value would be associated with the potential liability.[10]

**Likelihood of Compensation Liabilities.** Compensation liabilities are not usually considered certainties due to the many factual and legal hurdles that must be crossed prior to the incurrence of a liability. As with remediation liabilities, compensation liabilities may depend on such factors as the existence of releases (e.g. accidental, intermittent or continuous), exposure of human beings or their property to releases, and prevailing legal rules. For a compensation claim to become a compensation obligation, either the parties must agree to a settlement or one or more courts must find that the claim satisfied a variety of legal and factual tests. In evaluating such potential liabilities, companies can adjust their liability cost estimations to reflect their probabilities of occurrence. For example, the expected value of a potential compensation liability of $10 million with a 1% likelihood is $100,000. Alternatively, just as a company in its planning and decision-making might want to treat certain remediation liabilities for future contamination (e.g. from underground storage or disposal of substances) as eventual certainties, a company with a conservative management approach might choose to treat selected compensation liabilities (e.g. medical monitoring for people exposed

---

10. However, as discussed earlier in this appendix, managers may choose to adjust the estimated cost of remediation to reflect the expected timing of the future expenditures.

to *regular releases* of hazardous air pollutants) as certain to arise in order to provide a greater financial incentive for developing alternative practices that would eliminate or reduce such regular releases.[11]

**Likelihood of Penalties and Fines for Non-Compliance.** Penalties and fines are doubly uncertain. First, for most companies, non-compliance itself is an uncertain and unlikely event, unless a firm has a corporate policy deliberately to flout environmental laws. Non-compliance may be likened to accidental releases; neither is explicitly intended but management can influence to some degree (e.g. through training, technology) the probabilities that these events will occur. Second, penalties and fines are triggered only in the event of discovery (such as through an official inspection or review of required submissions of information) of non-compliance. Thus, the magnitude of the potential liability due to fines and penalties can be adjusted (1) to reflect the likelihood of being out of compliance; and (2) to reflect the probability of discovery of the non-compliance. If a company, however, prefers to base its planning and decision-making on the presumption that any non-compliance will be discovered, it would choose not to make any adjustments for the likelihood of discovery.

**Likelihood of Natural Resource Damages.** Natural resource damages that result from accidental releases, such as spills during transport, are by their nature unlikely events. Similarly, natural resource damages caused by releases of hazardous substances due to fires, explosions or earthquakes are uncommon events. Thus, it may be appropriate to adjust the potential magnitude of such natural resource damage liabilities to reflect the probability of their occurrence. Finally, because natural resource damage liability is so new, claims are likely to be litigated, perhaps further colouring natural resource damages as unlikely to occur. On the other hand, natural resource damages may be caused by the regular release of pollutants into the environment; although any resulting natural resource damage may be unintentional, the releases themselves are not accidental in these cases. For such intentional releases, a company might decide to view liability for natural resource damages as the certain outcome of its practices and recognise those costs as fully avoidable (i.e. no adjustment for likelihood) when evaluating alternative pollution prevention and control practices.[12]

### 3. Uncertainty

Estimates or projections of future costs have *unavoidable uncertainties*. There may be uncertainties regarding (1) magnitude; (2) probability of occurrence; and/or (3) timing of a liability. These uncertainties may be due to inherent variability or to a lack of information. Several of the liability valuation approaches can develop results in ways that make uncertainties more transparent. For example, levels of estimated liability costs can be paired with their respective probabilities of occurrence; this is termed a 'probability distribution'. Developing a probability distribution provides information on the potential range of liability costs and facilitates the calculation of the 'expected value' of a liability costs, which may differ from the 'most likely' value. Such a probability distribution can easily be converted into a 'cumulative frequency distribution',

---

11. The comment in footnote 10 applies as well to compensation liabilities.
12. See footnotes 10 and 11.

which indicates the likelihood that liability costs will be at or above a given value. For example, to reflect uncertainties in the potential for accidental releases and the resulting cost of remediation, decision analysis techniques have been used as a means for assessing the likelihoods of remediation costs exceeding various dollar levels. Compensation liabilities have been treated in a similar fashion. The approach is also applicable to liability for fines and penalties. This statistical approach provides more information than a single-point 'best estimate' of the cost or a multiple-point estimate of 'worst-case', 'most likely' and 'best-case' estimates. Managers can select their own levels of risk-averseness from such distributions.

# 6 The Italian Method of Environmental Accounting[1]

Matteo Bartolomeo

ITALIAN INDUSTRY is characterised by a large number of SMEs and a few large corporations, which are partly owned by the Italian Government. Large conglomerates such as Eni in the oil, gas, engineering and chemical sector; Enel in energy; Stet in telecommunications; IRI (controlling Alitalia, Stet and RAI); and FS (the national rail corporation) are state-owned and operate in a near-monopoly position.

This scenario means that environmental management and environmental accounting is a different matter compared to many other European countries. Large corporations in a non-competitive market pay little attention to the reduction of environmental impacts or to cost savings from environmental protection. Small and medium-sized companies are mainly in a defensive position as regards environmental issues and are not able, for reasons of culture and scale, to develop sophisticated management accounting systems.

In a country in which market forces are weak and in which the financial community—apart from a few isolated situations—does not play a very important role in driving Italian industry towards eco-efficiency, environmental pressures come mainly from legislation (command and control) and local community action. So it is these kinds of pressures that have had the greatest influence on environmental accounting practices, the objectives of which can be considered more focused on external stakeholders rather than on internal management. Environmental accounting has therefore been shaped to inform local communities and public administration and to assess

1. I would like to thank Giorgio Vicini and Stefania Borghini (both researchers at the Fondazione Eni Enrico Mattei) for their useful help in preparing this chapter.

compliance with existing and future legislation by identifying current environmental performance.

On the other hand, objectives such as cost savings and market benefits have not, in the past, played an important role in stimulating companies towards better environmental management and improved environmental accounting.

The following sections will focus on three areas of management accounting that are or will be particularly relevant for Italian companies: environmental reporting (internally and externally); environmental costing; and capital budgeting. These areas have been monitored in the Italian context by the Fondazione Eni Enrico Mattei as part of several research projects.

## ▍ Environmental Reports and *bilancio ambientale*

Before explaining the peculiarities of the Italian method of environmental accounting, it is necessary to clarify the meaning for Italian businesses of terms such as 'environmental accounting', 'environmental report' and 'environmental balance sheet'. 'Environmental report' (*rapporto ambientale*) is often confused with 'environmental balance sheet' (*bilancio ambientale*), not just by external stakeholders but also by companies using the term *bilancio ambientale* as the title of reports published as dialogue instruments with the general public. This misuse of the term *bilancio ambientale* can be explained by the focus of current environmental accounting practices, the disclosure of information, and is not just pure carelessness.

But what is the *bilancio ambientale*? *Bilancio ambientale* can be considered as a very basic and simple strategy (see Fig. 1) tailored to track (physical) figures as inputs and outputs of a certain production process, site or corporation. Outputs include all media emissions that are related to environmental expenditure, while inputs and products are related to their economic value (purchase and selling price). The success of this simple strategy, developed and disseminated by the Fondazione Eni Enrico Mattei, can be explained only by considering the starting point (very poor information on the basics) and the need to assess compliance with regulation and to disclose information to public authorities and local communities.

Not surprisingly, in response to the question 'What do you think environmental accounting is?', Italian environmental managers often mention only the *bilancio ambientale* and (when the difference is clear) the environmental report. Other issues, such as the improvement of the decision-making process, cost identification, cost allocation and capital budgeting, are not considered very relevant from an environmental management point of view. The reason for this lies again in the lack of a highly-competitive market for many large companies in the industrial sector together with the small scale of many businesses.

The term *bilancio* (balance) is not used here in a strict sense and should not lead one to expect a genuine balance between inputs and outputs. The *bilancio ambientale*, even though it is conceptually similar, is also neither a mass balance nor an *Ökobilanz*, but, likewise, is an information system to monitor flows in and flows out and to build environmental performance indicators. Misunderstandings also arise with accountants who approach the word *bilancio* as they are accustomed to do; and this does not help in improving communication with environmental managers.

| | Year 199x | | Year 199y | |
|---|---|---|---|---|
| | Quantity (tons and m³) | Values (1,000 Lira) | Quantity (tons and m³) | Values (1,000 Lira) |
| **Resources** | | | | |
| Crude oil (tons) | 3,166,983 | 464,531,133 | 3,577,024 | 564,167,409 |
| Lead (tons) | 69 | 623,343 | 73 | 658,866 |
| Sea water (m³) | 56,078,911 | | 63,339,654 | |
| Water (m³) | 1,672,529 | | 1,889,077 | |
| **Products** | | | | |
| LPG (Liquefied petroleum gas) | 77,133 | 19,291,272 | 87,120 | 22,165,803 |
| Naphtha | 33,549 | 6,913,605 | 37,893 | 7,943,779 |
| Premium petrol | 346,302 | 109,006,145 | 366,037 | 117,210,764 |
| Unleaded petrol | 232,787 | 78,257,680 | 288,029 | 98,503,238 |
| Kerosene | 24,173 | 6,955,174 | 28,558 | 8,358,902 |
| Automotive diesel | 1,269,183 | 371,001,468 | 1,432,254 | 425,909,981 |
| Fluid fuel oil | 111,597 | 18,355,054 | 124,790 | 20,880,077 |
| Fuel oil (high sulphur) | 91,370 | 9,270,276 | 104,455 | 10,781,153 |
| Fuel oil (low sulphur) | 700,940 | 100,836,911 | 797,969 | 116,780,698 |
| Bitumen | 116,435 | 16,706,203 | 125,235 | 18,279,557 |
| Liquid sulphur (S) | 10,232 | 463,281 | 11,557 | 532,313 |
| **Total products** (tons and values) | **3,013,700** | **737,057,069** | **3,403,895** | **847,346,265** |

| **Pollutants** | | | | | **Environmental expenditures 199x** | | |
|---|---|---|---|---|---|---|---|
| | | | | | Current | Investment | Total expenditure |
| **Air** | | | | | | Million Lira | |
| Carbon dioxide6 $(CO_2)$ | 75,000 | | 702,000 | | 0 | 0 | 0 |
| Sulphur dioxide $(SO_2)$ | 4,050 | | 3,690 | | 720 | 1,140 | 1,860 |
| Oxides of nitrogen $(NO_x)$ | 1,260 | | 1,278 | | 0 | 0 | 0 |
| Volatile organic compounds (VOC) | 1,692 | | 1,350 | | 66 | 480 | 546 |
| Total suspended particulates (TSP) | 162 | | 165 | | 24 | 12 | 36 |
| **Total** | **682,164** | | **708,483** | | **810** | **1,632** | **2,442** |
| **Water** | | | | | | | |
| Biological oxygen demand (BOD) | 6,390 | | 6,750 | | | | |
| Chemical oxygen demand (COD) | 50,670 | | 54,000 | | | | |
| Total Suspended Solids (TSS) | 24,300 | | 27,000 | | | | |
| Oils | 1,170 | | 1,350 | | | | |
| Phenols | 126 | | 135 | | | | |
| Nitrogen (N) (Ammonia [NH₄]) | 2,340 | | 2,700 | | | | |
| **Total** | **84,996** | | **91,935** | | **1,300** | **300** | **1,600** |
| **Wastes** | | | | | | | |
| Municipal | 180 | | 171 | | 30 | 0 | 25 |
| Special | 990 | | 900 | | 516 | 0 | 430 |
| Toxic and hazardous | 1 | | 1 | | 0 | 0 | 0 |
| **Total** | **1,171.26** | | **1,072.35** | | **546** | **0** | **455** |
| Total | | | | | **1,846** | **300** | **2,055** |

**Figure 1:** An Example of *bilancio ambientale* for a Refinery

*(Figures are representative and not actual.)*

Fiat, Enichem, Agip Petroli, IBM Semea, Eni, SNAM, Enel, Favini, Ciba Additives, Montecatini, Edison and Solvay Italia are some of the companies operating in Italy that have published environmental reports. The large majority of these companies named their environmental report *bilancio ambientale*, thus showing the weak borderline between the external dialogue instrument (report) and internal management controlling tool (*bilancio*). Furthermore, environmental reports of companies such as IBM Semea and SNAM have been published—also called *bilancio ambientale*—which are in line with international best practices and from which they do not differ either in terms of structure or of information disclosed.

Agip Petroli preferred first to build up their environmental information system (on the basis of the *bilancio ambientale* methodology), and second to publish their environmental report. Eni, SNAM, and Enichem, on the other hand, have used the environmental report as a starting point to establish a comprehensive *bilancio ambientale*. The management of Eni in particular—during the first roadshow to float part of the corporation's stocks on the New York, London and Milan markets—was requested to disclose more information on environmental issues: namely, environmental expenditures, liabilities and performance. The environmental report has been the driver here to establish and link (via the *bilancio ambientale*) different environmental information systems currently in operation mainly at site level.

Figure 2, derived from the Agip experience, clearly explains the integration of the environmental report with the *bilancio ambientale*. In this example, the *bilancio ambientale* (consisting of physical and financial components and of environmental expenditure components) is part of an external reporting system. These accounting strategies are very similar to those in Figure 1, and together they create a complete picture of significant environmental flows of a site, division or corporation.

For Agip—and also for other companies that have implemented a *bilancio ambientale* in a strict sense—management benefits include improved knowledge of the situation and a better identification of poor performances and their causes. The *bilancio ambientale* for a multi-site company such as Agip, Agip Petroli or Italiana Petroli is normally assisted by software which, while facilitating data entry and analysis, also supports the proper use of corporate terminology by line management.

## ▐ The Cost of Non-Environment

In Italy, identification of environmental expenditures is normally carried out with the purpose of informing top management and dialoguing with public administrations; this has always been the motive behind including this information in environmental reports. As regards the term 'environmental expenditures', companies refer to the definition accepted by the European Statistical Office (Eurostat) and therefore only to those expenditures 'deliberately and mainly undertaken for environmental reasons' (for example, this definition is used in the *bilancio ambientale* of Agip Petroli refineries in Fig. 1).

Such a narrow definition, determined in order to ensure homogeneity of data in different sectors and countries, does not in fact help managers improve the internal decision-making process. The true cost of environmental protection is deeply underestimated and is claimed to account for just 2% or 3% of operating costs (depending

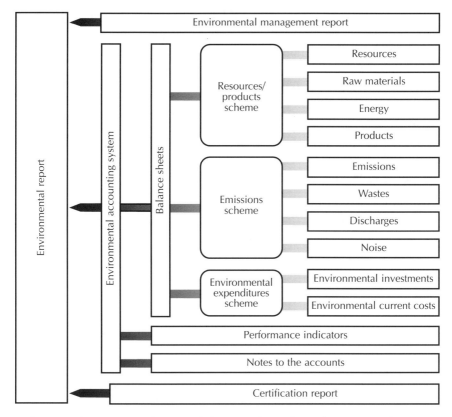

**Figure 2:** The Integration between *bilancio ambientale,* Performance Evaluation and Environmental Report for Agip

on the sector). By adopting the statistical office definition, managers track only a small proportion of environmental figures: namely, the conventional environmental cost defined by the US EPA (Chapter 2 and US EPA 1995a). Meanwhile, they fail to consider the so-called potentially hidden cost, potential and contingent cost and the image and relationship cost; nor do they consider the cost of inefficiency which, according to some US-based companies, represents the larger part of environment-related cost.

While the majority of Italian companies do not track environmental cost at all, and a small number of them refer only to end-of-pipe environmental expenditures, a few innovative organisations have started tracking the hidden part of environmental cost and the cost of inefficiency. This approach is becoming quite common in companies controlled by multinationals, such as IBM Semea, Ciba and the French–Italian joint venture SGS-Thomson.

Ciba Additives, at the Pontecchio Marconi site, using a wider definition of environmental cost, has, for example, accounted for a decrease in environmental cost from 11% to 7% of operating cost, while SGS-Thomson is now rapidly moving towards the integration of the cost of non-quality and cost of non-environment (or cost of ineffi-ciency). Such integration will induce the company to shift from failure cost (internal and external) related to the environment to prevention and appraisal cost. Telecom

Italia, as part of a new commitment towards sustainable development, has recently organised a successful workshop focused on cost of environmental inefficiency that will form the basis for a set of initiatives in line with best British Telecom (BT) practices.

Identification of the cost of inefficiency, still in its infancy, will probably become quite common in companies that have adopted the philosophy of Total Quality Management, that are familiar with the concept of the 'cost of non-quality' and therefore likely to identify the 'cost of non-environment'. (Fig. 3 shows the position of some Italian companies in relation to the scope of cost identification.)

## ▮ The Decision-Making Process and the Environment

On the capital budgeting side, Italian businesses are more followers than leaders. Even if, in many of the interviewed companies, the decision-making process routinely involves environmental managers, environmental concerns are not really integrated into the decision-making process, nor are investment-screening procedures normally applied to environmental investments. Environmental investments therefore run along different lines and are accepted mainly in order to comply with existing legislation rather than to cope with economic efficiency requirements.

Only a few innovative cases can truly be considered as integration between environmental management and the traditional decision-making process via an improved

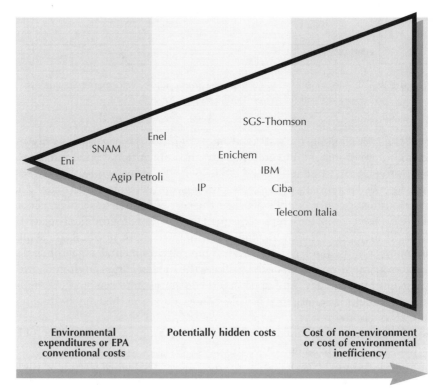

N.B. The position of Telecom Italia and SGS-Thomson represents company objectives and not the current situation.

**Figure 3:** The Scope of Environment Cost Identification in Some Italian Companies

environmental accounting system. The two cases described briefly below refer to companies where the environmental challenge has genuinely become part of the business, by virtue of environmental legislation and public concern.

## ▌Accounting for Decommissioning at Agip

Agip is the leading upstream Italian oil company,[2] operating in four continents and 24 countries. In order to acquire a correct quantitative description of the relationship between the company and the environment, Agip has implemented an environmental balance sheet designed to control flows (physical and financial) in a given period of time and with a retrospective approach.

As illustrated before, this balance sheet, while providing relevant management benefits, does not assist in the capital budgeting process. Like all upstream oil companies, Agip faces the problem of decommissioning drilling facilities, which has an increasing impact on the future economic position of the company. The Brent Spar case in particular has highlighted important economic threats for upstream oil companies and has induced management to accelerate the process of reshaping strategic planning, environmental management options and accounting practices.

There are two major accounting systems (not mutually exclusive) employed as a response to the decommissioning problem: from a financial accounting point of view, oil companies actually make provisions for this kind of future cost; accounting bodies are on the way to defining accepted criteria for such provisions. From a management accounting point of view, Agip, like other oil companies, is developing new accounting instruments to assist the strategic decision-making process, the design of facilities and the selection of environmental management options. From a financial accounting point of view, annual provisions are calculated as shown in Figure 4. The management accounting side is very relevant for Agip, where the capital budgeting process now includes back-end environmental cost in cash flow analysis for all the new production projects.

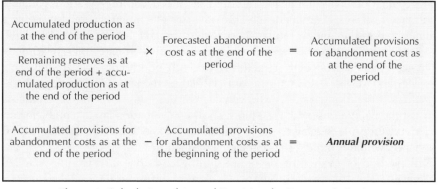

**Figure 4:** Calculation of Annual Provision for Decommissioning

---

2. The term 'upstream' is part of the language of the oil business, and refers to the exploration and production of oil and gas. Likewise, the term 'downstream' relates to the refining process and the marketing of oil derivatives.

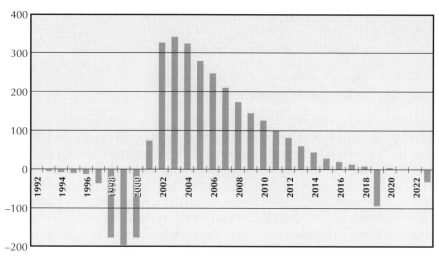

**Figure 5:** An Example of Cash Flow Distribution along the Whole Life-Cycle of an Oil Production Field

Decommissioning is the process of shutting down operations at the end of a field's life. It has four distinct stages: first, the operator has to stop the extraction of oil or gas, plug the wells deep below the surface and make them safe; secondly, the installations have to be cleaned; thirdly, all or part (the superstructure can be removed without the substructure) of the installations usually have to be removed from the site; finally, the parts removed have to be disposed of or recycled.

Decommissioning costs are considered in economic analysis as operating costs with a negative cash flow at the end of economic lifespan. Figure 5 shows cash flows (the amount on the vertical axis) during the economic life of an oil field (horizontal axis). As seen, the first part of the oil platforms will be removed in 2019 and a second part in 2024; related costs have been identified and represent a significant negative cash flow. In the recent past, they would not have appeared in a cash flow analysis to this magnitude.

The evaluation of decommissioning costs can differ according to various parameters and its precision is highly influenced by the time-horizon, by anticipated legislative and community pressures and by future technologies. Agip has therefore developed different decommissioning cost categories and well-plugging cost categories, relating costs to physical characteristics of the structures (sea depth, number of conductor piles, number of legs, and weight). After removal, disposal of the structures is considered in licensed, permanent waste disposal sites. Although decommissioning costs have been included in investment appraisals for the last ten years, their evaluation is still a problem due to obvious uncertainties related to the long time-period and to current decommissioning experience, which is limited to the removal of small structures.

This method of evaluating decommissioning costs has its advantages and disadvantages. Using this method, Agip avoids many of the uncertainties linked to decommissioning costs evaluation. On the other hand, because the method is so rigid, it does not allow the proper inclusion of cost evaluation in the design of new facilities. Furthermore, the use of financial indicators such as net present value, which take into account discounted cash flows, reduces the relevance of those costs that appear at the end of the lifespan.

The first important step, i.e. considering environmental back-end costs in investments evaluation, has been made; nevertheless, the identified categories of decommissioning costs should be considered not only as given costs but also as cost that can be reduced with a sophisticated facility design.

## ▌ Accounting for Risk Management at Italiana Petroli

Italian Petroli (IP) operates a distribution network for oil products with more than 2,800 petrol stations and a number of storage terminals disseminated throughout Italy. In response to recent environmental regulations and privatisation plans for the company, IP has embarked on a risk management programme, an environmental balance sheet and an environmental report.

The environmental risk management programme in particular is a very interesting example of integration between environmental management and the capital budgeting process. 'Environmental risk' refers to deposits and petrol stations where product containers are considered a major source of soil contamination and therefore environmental liabilities. The project is implemented in two different phases:

- **Risk assessment.** IP wished to identify the causative factors determining spills from underground tanks, and therefore implemented database software that includes the structural characteristics of 2,830 petrol stations (age and kind of tanks, capacity and products, stratum, lithology, wells, and other characteristics). Ad hoc audits on a representative sample of network stations have been carried out to identify causative factors that are most likely to influence the frequency of oil spills; these causative factors have been used to generate normalised risk scores for the entire network. This phase has generated information on the petrol stations that are most likely to face the problem of spills from underground tanks.

- **Economic assessment.** As risk is determined, costs for assessment, corrective action and monitoring of sites can be evaluated. This evaluation takes into account different cost categories related to: prevention (audit, tank level measurement, environmental audit, tank test, technical procedures, soil gas survey); substitution and repair (tank substitution, cathodal protection, internal verification, double skin tank); and monitoring and reclaiming (spill monitoring and control, hydrogeological control, reclaim actions, land control). These costs—together with maintenance and repair action, reclaiming costs, fines and penalties—represent the core of IP's capital budgeting process for environmental risk management.

Different environmental management options have therefore been investigated (described in Fig. 6) that show the optimal level of prevention policy: prevention policy will be implemented only if savings from avoided reclaims surpass or equal prevention cost. IP has calculated the total number and quality of petrol stations that will be involved in the prevention policy.

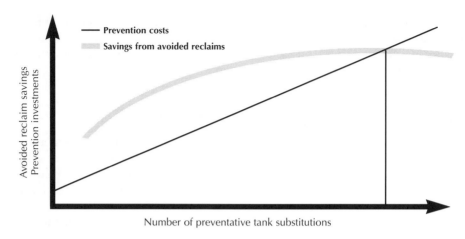

**Figure 6:** The Optimal Level of Prevention Policy for IP

## ▌ Conclusions

The Italian method of environmental accounting is only partly original in terms of instruments, although it can be considered different from those of other northern European countries in relation to business pressures and business strategies.

These pressures have not yet created sufficient incentive for companies to improve environment-related inefficiency. Nevertheless, despite the small number of companies that have included environmental accounting activities as part of their environmental programmes, a few organisations are making important steps in the right direction and can claim to be contributing to the diffusion of a green accounting philosophy among industry and service sectors.

Business networks and industry associations can play an important role in the dissemination of accounting practices that complement an environmental policy that creates an incentive to be eco-efficient instead of simply to comply with legislation.

# $7$ Environmental Management Accounting in the Netherlands[1]

Jan Jaap Bouma

## ∎ Introduction

THE NETHERLANDS experiences considerable environmental pressure as a result of having one of the world's highest population densities, a large concentration of heavy industry, such as oil-refining and chemicals, a highly intensive agricultural system and a geographical situation that creates transboundary pollution, for example, carried down in the Rhine. As a result—and also because of high levels of awareness and concern among the population and political parties—the country has ambitious environmental targets. Increasingly, financial measures such as eco-taxes are being introduced to achieve these targets.

For these reasons, there is considerable academic and business interest and action in the area of environmental management accounting in the Netherlands.[2] However, much of this is available only in the Dutch language and has not yet influenced English-language discussions. The main purpose of this chapter is therefore to bring some of this work to a wider audience. To achieve this, the next section discusses the Dutch context for environmental management accounting in further detail. The subsequent sections then outline three specific areas of work: on defining environmental costs, valuing the externalities created by Dutch railways, and incorporating projections of future environmental cost increases into building investment

---

1. The author is grateful to Peter James and Martin Bennett for their valuable comments on earlier versions of this paper.

2. Research has also been fertilised by a large body of work on the costs and benefits of major infrastructure projects such as a new runway at Schiphol Airport and flood-prevention schemes.

appraisal. The final section provides conclusions and discusses the extent to which approaches to environmental management accounting are influenced by national differences.

## ▌ The Dutch Context for Environmental Management Accounting

Dutch environmental targets have been set out in the National Environmental Policy Plan ('Nationaal Milieubeleids Plan' [NMP], 1989) and the National Environmental Policy Plan Plus (NMP+, 1990). The NMP and NMP+ provide environmental objectives for several sectors of the Dutch economy: agriculture, industry, transportation and consumers.

The environmental policy objectives are to be achieved in three main ways (Winsemius 1986: 78):

- Direct regulation, i.e. through 'command-and-control' measures
- Financial regulation through eco-taxes, subsidies and other means
- Self-regulation, based on agreements between the Government and individual sectors or companies to achieve specified environmental targets

### Direct Regulation

With regard to direct regulation, the approach of the governmental organisations (especially the agencies responsible for providing environmental permits and controlling the firms with regard to their compliance to permits) is important to environmental management at the level of the firm. At the beginning of the 1970s, the governmental agencies followed the approach of 'strict command and control'. In this approach, firms received permits with detailed prescriptions on the technical means to be implemented in order to comply with environmental standards. For a firm with complex and unique technical production processes, the permits were often the outcome of intense negotiations between firms and government.

Since 1985, an increasing number of firms have implemented environmental management systems, which have influenced the process of permitting. The so-called 'flexible permits' have appeared, by which companies gained more freedom in how they could achieve environmental objectives and standards, though these objectives and standards themselves were more exacting. The environmental objectives and time schedules to achieve these objectives were prescribed in the permit, but only to a lesser extent the technical means of how to comply. The focus of the Government shifted from detailed control over production processes towards monitoring the output of these processes.

The cost implications of direct regulation are illustrated by the clean-up costs of soil contamination. Figure 1 shows the development of investments by companies with more than twenty employees to prevent a future need to clean-up soil contamination in the Netherlands. The Dutch Air Force alone has estimated clean-up costs of 500 million guilders—equivalent to approximately US$250 million or £150 million (Bouma 1995).

In the Dutch context, detailed cost information is crucial in negotiations over the severity of regulation, time-scales for introduction and other matters. This was true of the negotiations between industry and government over the phasing-out of ozone-depleting chemicals, when detailed data on cost implications helped to produce a more

| | 1975 | 1980 | 1985 | 1988 | 1989 | 1990 | 1991 | 1992 | 1993 |
|---|---|---|---|---|---|---|---|---|---|
| **Total investments with regard to soil** | 11 | 13 | 25 | 32 | 34 | 40 | 41 | 101 | 96 |

**Figure 1:** The Development of Investments (in million Dutch guilders) by Companies with more than 20 Employees (excluding the building sector) to Prevent the Clean-up of Soil Contamination in the Netherlands (1975–93)

*Source: CBS 1996*

manageable timetable. Similarly, when stringent legislation was proposed to deal with volatile organic compound (VOC) emissions from the transportation of liquid fuels, the industry-based trade organisation that deals with the interests of this sector performed a study that demonstrated its high costs and the possibility that it would bankrupt some companies. The result was that the proposed legislation was shelved.

Another company used cost information to settle a dispute with neighbours. This company demonstrated the extent of future expenses if it were forced to reduce noise. Such cost information is of particular importance when environmental measures are prescribed in environmental licences. The licensing staff may demand that those measures be implemented in accordance with the principle of ALARA (that emissions should be reduced 'As Low As Reasonably Achievable').[3] Under ALARA, the economic feasibility of a technique is tested to determine if a technical option to achieve environmental targets is feasible. In the dialogue between a corporation and the licensing staff, the discussion often focuses on the interpretation of the term 'reasonable'. The amount of environmental costs and the ability of the corporation to bear these costs will largely determine the outcome of the dialogue.

However, a research project at the Erasmus Centre for Environmental Studies (ECES) has found that, while bold estimates of environmental costs by companies are often accepted as a starting-point for negotiations, policy-makers are becoming more experienced at checking these estimates and revealing any weaknesses. Hence, companies will need to develop more sophisticated accounting systems and procedures in future.

### Financial Regulation

Dutch financial regulation aims not only to raise costs—through eco-taxes and other means of internalising externalities—but also to create positive incentives for action. For example, the management of Dutch firms increasingly wanted to be informed about the cost linked to effluent charges. In the past, these costs were externalised, but, as the result of an increase in the levies payable for emissions to surface water, these costs became relevant to decision-makers. Figure 2 shows the tremendous increase in the costs that companies faced due to their emissions of effluents to sewage, and also the increase of costs to the transportation system due to emissions of noise from aeroplanes. These cost increases can be explained not only by an increase of internalisation (raising

---

3. Within Dutch legislation, ALARA is explicitly mentioned in article 8.11 of the General Environmental Control Act (GECA). In this context, ALARA is not an impediment for the adoption of best applicable techniques at the level of the firm, since financial costs are considered at the level of the industry sector rather than of the individual firm. Only in those cases where an individual firm reflects the whole of a sector is ALARA practised at a firm level (e.g. in the case of the railway company).

| | 1985 | 1989 | 1990 | 1991 | 1992 | 1993 | 1994 |
|---|---|---|---|---|---|---|---|
| Charges on emissions to sewage | 313 | 467 | 542 | 624 | 707 | 795 | 864 |
| Charges on noise emissions of aeroplanes | 7 | 17 | 23 | 17 | 20 | 19 | 52 |

**Figure 2:** Costs (in million Dutch guilders) of the Taxation of Emissions to Sewage and of Noise Emissions (derived from the total revenues that the Dutch Government receives from the taxation of these emissions)
*Source: CBS 1996*

taxes on emissions) but also due to an increase in the activities that result in emissions of effluent and noise.

### Self-Regulation

Besides direct regulation, the Government makes agreements with industry to reduce environmental impacts. Such agreements, or 'covenants', are the outcome of intensive negotiations between the Government and industry-based trade associations, though it is presently still unclear to what extent individual firms are obliged to comply with the agreements reached between the Government and trade associations. Cost information can considerably influence the outcome of covenant negotiations. The reduction of environmental impacts usually has to be realised under a set of constraints, and the economic position of the industry is often one of these constraints. Cost information is frequently used to identify the necessity of a reasonable economic position under which industry can bear the investments in environmental improvements.

Covenants are one of a series of communicative instruments used by the Government in the process of formulating a policy of self-regulation. Other policy instruments in this process include information and education, demonstration projects and environmental management systems. The main characteristic of these communicative instruments is that they depend on voluntary actions by firms to provide a follow-up to these instruments. These self-regulation instruments (including covenants) are used in combination with other policy instruments and are not regarded as substitutes for direct regulation. The covenants are regarded as a clarification of how both the Government and the target-group of the covenant (industry, transport sector, etc.) plan to achieve environmental objectives. Sometimes this approach results in confusion, since it may be unclear if a covenant offsets the necessity for firms to be confronted with prescriptions in permits. Presently, it seems that the attitude of governmental agencies towards a firm heavily depends on the state of the art of its environmental management. The implementation, certification and verification of a firm's environmental management systems is an important factor in assessing the level of quality of its overall environmental management.

The Dutch emphasis on self-regulation also encompasses strong pressures on industry associations and large companies to report on their environmental performance and their environmental costs. In this respect, an increasing number of companies are publishing environmental reports. To generate the information needed to formulate its environmental report, the chemical company Akzo Nobel is exploring the possibilities of using an accounting approach in reporting on its financial performance, and also reports on its environmental investments in its environmental report (Akzo Nobel 1996). The financial information (the amount of environmental investments) is accompanied by an extensive explanation of the nature of the investments, which is useful, since financial figures alone would have a low information content for the average reader of an environmental report.

The verification of environmental reports is a topic increasingly discussed and studied in the Netherlands as well in other European countries (FEE 1996). However, in only some cases are cost figures used to communicate the effort that a company puts into its environmental management. For example, a Dutch electricity producer demonstrated in its external environmental report that in 1991 more than 3.6 million guilders were spent on environmental care (EPON 1991), divided into three categories: (1) large environmental investments; (2) minor environmental expenses; and (3) staff expenses.

Increasingly, companies are also developing environmental cost estimates for more proactive purposes, such as to support pollution prevention rather than end-of-pipe approaches. Such approaches can be extended across the supply chain. For example, one Dutch company who wanted to cut its waste management expenses used cost information in its dialogue with one of its suppliers, to demonstrate the total amount of waste management costs that arose due to the packaging that the supplier used. This dialogue persuaded the supplier to change its packaging system.

Environmental issues are also becoming increasingly important when take-overs take place. Dutch legislation means that companies can often have large potential costs or liabilities arising from the clean-up of sites, the storage of hazardous wastes and the need to renew or update environmental licences. For example, when a British company wanted to buy a Dutch construction company, environmental aspects highly influenced the buying price. A 'due diligence' audit showed that the Dutch company had emitted a large quantity of contaminated mud into surface water. Additionally, contaminated mud was stored on the facility that still had be treated at high costs. Those future environmental costs represented 7% of the take-over price. The management of the Dutch company was not aware of these costs before the British company informed them about the results of their due diligence audit. In the end, the take-over did take place but only after a downward renegotiation of the price.

## ▌ Defining and Calculating 'Environmental Costs' in the Netherlands

In national statistics, the central purpose of calculating the environmental costs is to provide an overview of the costs to business of environmental management. Although this purpose refers to accounting procedures at a macro level, it has some important links to accounting procedures at the micro level in the Netherlands.

The Dutch Bureau of Statistics (CBS) has registered environmental costs from a corporate perspective since 1979, i.e. the statistics of industrial costs involved in protecting the environment (CBS 1994). In these statistics, environmental costs are defined as the costs of environmental care. Environmental care is defined as environmental activities that are undertaken with the intention of preventing negative effects on the environment of the facility. Additionally, the primary purpose of the activities must not be labour safety or other safety reasons (for example, the safety of neighbouring housing). Moreover, under this definition, any environmental activities that result in a profit are excluded.

CBS makes a distinction between two categories of environmental activities: end-of-pipe activities and process-integrated activities. End-of-pipe activities are those that are undertaken only because of environmental considerations; process-integrated activities, on the other hand, relate to installations or components of a production process that are inherent to the process, without which the production process could not be operational. For process-integrated environmental activities, the environmental costs are equal to the incremental costs, i.e. the difference between the costs of the production process designed *with*, less the costs of the production process designed *without*, the consideration of environmental care.[4] The environmental costs of end-of-pipe activities are equal to the total costs of these activities.

Despite the limited definition of environmental activities, the CBS method provides a useful overview of different components of the costs of environmental activities. The total environmental costs are the sum of:

- Capital costs (interest and depreciation)
- Ongoing costs (costs of personnel, electricity costs and costs of delivered services)
- Material costs
- Clean-up costs of soil-contamination
- Extra costs for fuels that contain less sulphur
- Costs of external environmental studies
- Costs of the co-ordination of environmental activities (costs of an environmental management system)

CBS distinguishes between 'environmental costs' and 'net environmental costs'. 'Net environmental costs' consist of the costs of the company's own environmental activities and costs, levies paid and payments for environmental activities contracted out, minus any environmental subsidies that have been received. The difference between environmental costs and the net environmental costs is important, since the effect of environmental subsidies can have a significant effect on the final financial burden on a company of its environmental activities.

The most important consequence of the macro-level accounting procedure for micro-level accounting is that it establishes a definition and a calculation rule that enjoy a certain degree of consensus on what environmental costs might include. The advantage of the CBS method is that it gives a clear definition of (net) environmental costs. This makes it possible to compare the environmental costs of different corporations and sectors. A disadvantage is that the information is not primarily designed for the purpose of company managements, who may prefer to include in their own definitions costs arising from activities that are excluded from the CBS definition, especially with activities that are implemented in order to reduce environmental impacts but which are also profitable.

---

4. Although conceptually the determination of incremental costs should be possible, in practice calculation problems may occur when there is no clear technical alternative.

The CBS definition of environmental costs is narrow. Facilities that benefit the environment and have a positive net financial consequence are excluded. Also, costs of measures that are primarily undertaken in order to improve worker safety are excluded. This narrow definition of environmental costs can be explained from the history of the statistics, which were originally developed in order to show the charges that environmental legislation imposes on industry.

The CBS definition of environmental costs is most likely to be relevant for management in a company that acts defensively and reactively towards environmental legislation. In such a case, the environmental management will be similar to what is asked for by environmental legislation. However, when environmental management is more proactive, then environmental costs should be defined differently, to account for activities that have a pay-off or which are integrated with worker safety. Proactive environmental management asks for:

- The calculation of financial costs and benefits of environmental investments
- The integration of the protection of the environment, worker safety and product quality
- Information on environmental impacts during the total economic life-cycle of a product
- The prevention of environmental impacts in preference to end-of-pipe solutions

From a proactive environmental management perspective, environmental costs will generally be defined differently. The Dutch Ministry for the Environment (VROM) is aware of the shortcomings of the present basis. Consequently, ongoing research is being undertaken at the ministry to improve the rules on calculating environmental costs, and to reduce the mismatch between the macro-level accounting rules and such rules on a micro level. This research is not primarily motivated by a corporate management perspective, so it could be argued that research for improving national statistics on environmental costs is not a part of environmental management accounting. However, the agencies responsible for the national statistics request that companies report on environmental costs according to the agencies' rules, so that companies' accounting systems should be in a position to generate the requested cost information. This is often not the case, as an internal research project at the multinational Philips indicated.

Philips wants to improve its cost-accounting system, so that environmental costs are made transparent. The calculation of its environmental costs is not only in order to submit this information to external stakeholders such as CBS, but also to provide cost information for internal control. Several of Philips's plants are interested in a better understanding of their environmental costs in order to achieve cost reductions and reduce environmental impacts at the same time, and several plants belonging to different parts of the company have started internal projects to define and identify their environmental costs. The definition and analysis of this is considered to be very plant-specific, since different plants have different environmental problems and thus different environmental costs.

It is therefore concluded that, at a macro level, there is one definition of environmental costs and accordingly a technique to calculate environmental costs. At the micro level, a range of definitions of environmental costs is found. Only in some specific cases can the micro level benefit from using the definition and calculation rule of the macro level for its own purposes. In other cases, companies need to use their own definitions and calculation rules to generate tailor-made cost information.

## ▌ Environmental Cost Accounting at the Dutch Railway Company

In the Netherlands, the railway company NV Nederlandse Spoorwegen (NS) faces environmental standards for noise. These standards are part of the environmental permit requirements and impose environmental costs on the corporation. Part of NS's noise emissions are considered as 'industrial noise', for which the ALARA principle applies. Noise is one of the main environmental problems associated with trains, and occurs especially at railway yards. Noise is an environmental problem the scale of which is heavily determined by the extent of the nuisance or damage that it causes to people. If noise emissions occur in an area where there is no residential housing and hence no distress for people living in the direct neighbourhood, the environmental effects of noise are given a lower priority in environmental policies at a corporate and governmental level. However, NS has railway yards in residential areas with much neighbouring housing, so that its noise emissions result in severe environmental problems and costs have to be incurred to reduce the problem. NS refers to these costs as 'environmental costs', defining these broadly to include the costs of both end-of-pipe and preventative activities. However, costs of not being in compliance (e.g. legal penalties incurred) are excluded, as are the costs of effluent charges and subsidies.

In order to operationalise this principle, a study was conducted by ECES, with two objectives. The first was to calculate environmental costs and then allocate these to the different units of the company, which are financially independent. The relevant units are:

- The unit responsible for passengers (NS Reizigers BV)
- The unit dealing with the transportation of containers (NS Cargo NV)
- The unit dealing with the management of the railway infrastructure (NS Railinfrabeheer BV)
- The unit that manages the railway carriages and locomotives (NS Materieel BV)

NS Reizigers sells tickets to passengers and organises the timetables, hiring the railway carriages and locomotives and railway infrastructure. NS Cargo does the same for the transportation of goods. Although the NS is to become privatised, there are not yet any such plans for NS Cargo. The environmental costs result from activities at NS Materieel (e.g. the use of other locomotives) and NS Infrastructure (e.g. changes in the location of railway yards). NS Reizigers BV and NS Cargo NV ought to allocate environmental costs in proportion to their demand (e.g. use of railway and locomotives) from the other units of NS.

The second objective of the preliminary study is the execution of a general economic assessment of NS, to evaluate the influence of potential environmental costs on its financial position (liquidity, solvency and profitability). The study shows that potential environmental costs increase according to the level of noise reduction.

However, there is a maximum level of acceptable environmental costs, though this is difficult to determine. Information on environmental costs should be generated by the management accounting system utilised by the company but, at the moment, it is not able to do this. An ad hoc study to obtain the information was necessary and different sources for data were used (for example, technical reports for data on the capital expenditure on environmental measures). Additional information on depreciation methods and changes in operational costs could be largely derived from present accounting systems. The present management accounting system provided information on:

1. **The future allocation of the environmental costs.** The expected environmental costs could be allocated according to the traditional cost-drivers (the use of railway, railway carriages and locomotives). This allocation of environmental costs is necessary for the general economic assessment of NS, since the financial situation and competitiveness of NS's separate units are different from each other. Consequently, the units differ from each other in the extent to which environmental costs are economically acceptable.

2. **The minimum average return-on-investment criteria.** Because the environmental costs reduce the average return on investment of the different units of NS, the minimum average acceptable return on investments of the units had to be provided.

3. **Transfer prices.** Because the financially independent units within NS deliver services to each other, information is needed to support transfer-pricing. By using transfer prices, some units will be confronted with an increase in environmental costs. These transfer prices are of great importance, since they determine the profitability of the units within NS, particularly since some units have been privatised whereas other are still state-owned.

The management accounting system can be useful for the calculation and allocation of environmental costs.[5] Some of the reasons why this calculation and allocation can be useful to the management of the NS are:

- Those units that are responsible for generating noise are made aware of the costs that they impose on NS.

- Those units that are responsible for reducing noise can reallocate these costs to the parties who are responsible for incurring these costs (e.g. the unit that operates the railway infrastructure reallocates costs of noise to those parties who use that infrastructure). These users are not only the units that belong to the NS, but also foreign railway companies who use the infrastructure and should pay accordingly for the environmental measures that NS takes.

However, with regard to the average return on investment, the information was problematic because these values of average returns on investment have to be accepted by

---

5. At the present time it is unclear what activities are to be implemented to reduce noise, so it is not possible to assess the behavioural effects on management of having enhanced transparency in environmental costs.

the Ministry of Housing, Physical Planning and the Environment (VROM).[6] Only when these values are accepted can the noise emissions be reduced to a level on which both company and licensing staff agree, so information from this external source is necessary, which management accounting is unable to generate. Nevertheless, the results of the study at NS indicate that management accounting systems play an important role in the generation of relevant information on environmental costs and the economic feasibility of noise reduction measures. The management accounting information refers to the value of specific cost categories, the allocation of environmental costs to different units within the corporation, the necessary return on investment and transfer prices. At the moment, NS is considering the development of environmental performance indicators that provide insight into the efficiency of environmental measures. An important link between the accounting department and environmental management at the firm would be established by using such performance indicators.

## ▌ Calculating the Environmental Costs of Government Buildings

Since 1995, ECES has assisted the Dutch governmental agency responsible for the design, construction and maintenance of government buildings (Rijksgebouwendienst [Rgd]) in attaching a monetary value to the environmental impact of their buildings. The monetarised environmental impacts are referred to as 'environmental costs'. The purpose of calculating the environmental costs of buildings for the Rgd is twofold. First, by monetarising environmental impacts, different kinds of impacts (for example, emissions of $CO_2$ and water use) can be summarised to indicate the total environmental impact of a building. Second, expressing environmental impacts in a monetary unit makes the environmental effects easier to communicate to stakeholders.

The objective of the ECES research project is to develop a calculation technique that attaches a monetary value to the environmental impact quantified by a life-cycle analysis (LCA). At the start of the project (1995), the Rgd used a method that summed environmental impacts by using 'environmental impact points' (EIPs). An EIP is a rating without a dimension and is subjective. For example, the environmental impact of the building material concrete is 4.9 EIP and aluminium 1,248.5 EIP (Ministerie van Volkshuisvesting, Ruimtelijke Ordening en Milieubeheer 1995; for a further explanation of the method of calculation of the environmental impact of materials, see Haas 1996). Subsequently, a simple rule for monetarising was used. Each point had the value of 110 Dutch guilders and would be used for all building materials. With a price for an environmental impact point, the environmental impacts of building materials can be given a monetary value. Building materials with large environmental impacts that result in negative externalities have lower market prices than they would have if all environmental impacts were internalised. Therefore, the market price is adjusted by valuing the externalities. The total of environmental impacts points are multiplied

---

6. The value of the minimum acceptable return on investment according to NS was 10%, though VROM argue that this should be 7.38%. This latter value is the outcome of a method described in guidelines to calculate environmental costs (Ministerie van Volkshuisvesting, Ruimtelijke Ordening en Milieubeheer 1994). The percentage is calculated by the sum of the interest percentage for governmental bonds and a risk premium of 1%. The consequence of having different returns on investment resulted in negotiations over a minimum return on investment that will be used in the licensing process. These negotiations are ongoing at the present time.

by 110 Dutch guilders. It is expected that the total monetary value of a building material will be higher if calculated in this way than by using the market price. The difference in monetary value is regarded by the Rgd as those externalities that are to be internalised in the future. This internalisation will be the consequence of the 'polluter pays' principle and makes it attractive to consider the environmental impact of the building material in the decision-making process. The calculation technique is briefly described in Figure 3. The Rgd expects that those building options with lowest environmental impacts will be the most economical in the long term; in fact, monetarisation is only regarded as an attractive way of expressing the effect of environmental impacts to the Rgd in the longer term.

ECES evaluated the rule for deriving the monetary value of the environmental impacts of building materials, and concluded that it had severe shortcomings:

1. The method does not account for the differences in monetary value of environmental impacts related to different themes (ozone depletion, greenhouse effect, etc.). For example, an environmental impact point related to emissions of ozone-depleting materials has the same value as an environmental impact point related to $CO_2$ emissions.

2. To obtain the value of an environmental impact point related to one theme would require valuation according to an acceptable method. Using the market prices of a number of nine materials is insufficient.

Consequently, the method is to be adapted. First, the environmental impact of a building material should be determined using LCA. The outcome of the LCA is not condensed into a single indicator, but the different environmental impacts are expressed

---

The monetarisation of the environmental impact of building materials follows this rule, formulated by the Rgd:

$$EP(i) = TEIP(i) \times P(E)$$

where

EP(i) = Environmental price of building material (i)
TEIP(i) = Total environmental impact points of building material (i)
P(E)   = Value of an environmental impact point

If **P(E) = fl.110.00**, then
$$EP(i) = TEIP(i) \times fl.110.00$$

P(E) is derived from the assumption that the market price of nine building materials accounts for all environmental impacts (i.e. no externalities).* Using the outcomes of LCAs, the total environmental impact is calculated for these nine materials (expressed in environmental impact points). The value of an environmental impact point is calculated for each material by dividing the market price of the materials by the total of the environmental impact points of a material. P(E) (fl.110.00, here) is the average of these values of the environmental impact points.

......................................................................................................................

* The assumption that 'environmentally friendly' building materials have no externalities seems contrary to the basis of most of the work on externalities. When evaluating the rule described in this figure, ECES emphasised this point.

**Figure 3:** A Rule for Calculating the Environmental Costs of Building Materials
*Source: Ministerie van Volkshuisvesting, Ruimtelijke Ordening en Milieubeheer 1995.*

in different terms. Also, using an LCA broadens the scope of environmental impacts (environmental impacts such as effects on health and energy consumption of a building are to be included). For example, the impact of a material on global warming is expressed in $CO_2$ emissions. Based on these insights, the impacts are valued according to the prevention or compensation costs. For example, total kilogrammes of $CO_2$ emissions are multiplied by the cost of preventing 1 kg of $CO_2$ emissions.

Currently, the research project is focusing on acquiring data on prevention and compensation costs related to the environmental impacts of building materials. The prevention costs are those private costs that are incurred in order to prevent environmental effects from occurring. Compensation costs refer to costs that occur as a consequence of the environmental effects that take place because of the environmental impacts that were released. This includes the following

- Costs of compensation to persons or companies as a consequence of environmental impacts
- Costs of environmental liabilities of environmental damage
- Costs of reducing the total effect of an environmental impact (for example, the effect of $CO_2$ emissions from the production of a building material can be offset by planting trees)

Both the prevention and compensation costs are captured in statistics that are managed by a Dutch governmental organisation (Rijksinstituut voor Volksgezondheid en Milieu [RIVM]). Clearly, the compensation and prevention costs that occur in the Netherlands may be very different from these costs that occur for the same emissions in other countries. Therefore, the outcome of the calculation rule for a design is specific to the country in which the method is used, in accordance with the different priority and technical options in each country.

Although the approach is not finalised, it can already be used by the Rgd. The positive environmental effects of sustainable building are quantifiable using financial information on the prevention and compensation costs related to less sustainable building alternatives. The anticipated behavioural effects are large. One analysis carried out—on the design of a governmental building for the tax agency (Rijksgebouwendienst 1996)—showed that the total cost, including environmental costs, of a design that is merely in compliance with environmental legislation was higher than that of a more environmentally friendly design. Hence, the Rgd regards the more environmentally friendly option as more attractive, assuming that all the negative externalities of environmental impacts will be internalised over the lifetime of the building. This assumption of internalisation is crucial in accepting the positive difference as material to the decision.

## ▌ Conclusions

In this chapter, some research projects in the field of environmental management accounting were discussed, each with its own objectives for defining and calculating environmental costs. It is concluded that there is no generally accepted definition of 'environmental costs'. The research projects show that the definition and, hence, calculation of environmental costs depends on the specific purpose. Nonetheless, all research projects and practical initiatives must have a common approach on how environmental costs can be determined, namely:

1. The need for information on environmental costs must be identified.
2. The environmental costs or specific environmental cost categories should be defined.
3. Specific accounting systems and concepts are to be selected to generate the relevant cost information.

For some specific purposes (such as national statistics), there seems to be a consensus about the definition of environmental costs. However, this definition is very narrow. If the definition of environmental costs is to be broadened, one can refer to the literature on environmental accounting (Epstein 1996c; Schaltegger, Müller and Hindrichsen 1996; Haasis 1996; Roth 1992; Wagner 1993). In most cases, companies need to use their own definitions and calculation rules to generate tailor-made cost information.

However, the broad definition of environmental costs lacks a direct link to the purpose of generating cost information. For situations where there is a clear desire to be informed about environmental costs, new environmental accounting systems and techniques exist for both profit and non-profit organisations. However, it is shown that financial information about environmental aspects of investments requires new management accounting systems and techniques (for example, the calculation technique for monetarisation of environmental impacts of buildings). Other contexts indicate the application of existing management accounting systems (for example, the calculation of environmental costs at the Dutch railway company) or modified versions of existing systems and techniques. Research projects show that the level to which negative external effects are internalised is a major determinant of the appropriateness of techniques such as total cost assessment (TCA).

It can be argued that environmental management accounting will always differ greatly between countries and that care needs to be taken when introducing systems and techniques developed in one national context into another. This is confirmed by an ongoing study being conducted for the United Nations Environment Programme by the present author and IVAM Environmental Research, University of Amsterdam, of the capital budgeting process in companies located in Guatemala, Nicaragua, Tanzania, Zimbabwe and Vietnam. This study is finding that many of the TCA categories and assumptions—for example, of an onerous environmental liabilities regime or of green consumers' willingness to discriminate between products on environmental grounds—are not relevant for many such countries. It is also finding that factors such as the role of governmental agencies and commercial banks, and of national preferences for different investment appraisal procedures, could greatly affect outcomes. The examples presented in this chapter show that, within the Netherlands, there are different contexts that require different environmental cost definitions. However, it can be expected that the impact of country-specific variables (e.g. legislation, traditions with regard to national statistics) will strongly affect the developments of environmental management accounting in a particular country.

# 8 Cost Allocation

## An Active Tool for Environmental Management Accounting?

### Roger L. Burritt

THE CANADIAN Institute of Chartered Accountants (CICA) aptly summarises the importance of environmental performance indicators in environmental reporting as follows:

> The development of environmental performance indicators is an essential aspect of environmental reporting. They may be financial or non-financial, objective or subjective, depending on the data and the information they capture. Each type of indicator can convey useful information. The number of indicators should be kept to a minimum to avoid overloading the audience with information, but there should be enough of them to provide a balanced view of an organization's environmental performance (CICA 1994: xvi).

The key points in this summary are that:

- Environmental performance can be represented by different forms of indicators, some of which are financial, some physical, some descriptive.
- Each indicator has the potential to be of use both singly and in combination with other indicators.
- There is some level above which too many indicators will confuse the reader, and below which there will be an unrepresentative view of the environmental performance of an organisation.

Drawing upon this starting point, CICA (1994) makes two further points which throw down the gauntlet to those wishing to track progress towards environmental goals established by organisations. First, the selection of understandable, informative indicators that are consistent with environmental objectives and that respond to the audience's information requirements is a challenging task which requires some innovative thinking. Second, stakeholders and organisations need to co-operate in the process of developing useful indicators for disclosure if reporting is to be credible and successful.

In the above context, this chapter examines one element of environmental performance indicators: the process of cost allocation in the financial evaluation of environmental performance. The chapter proceeds in three stages. First, by way of introduction, it considers the reasons for cost allocation—contrasting information accuracy and behaviour-influencing views; second, it examines the areas where cost allocation may have an impact in corporate environmental performance measures; and, finally, it assesses whether and how cost allocation can be used in a pro-environmentally-benign manner by some corporations.

## ▌ The Reasons for Cost Allocation

Cost allocation is a process whereby costs are linked to a cost object, such as a particular product produced by an organisation. Some costs can be directly linked through a causal chain, while many costs can only be indirectly linked with a cost objective. Costs that can be directly linked to products are said to be *traced* to the cost object, while indirect costs have to be *allocated* to the cost object because there is no direct causal reason for making the link. The US Environmental Protection Agency identifies a hierarchy of environmental costs linked to pollution prevention (US EPA 1989). These include:

- Direct environmental costs associated with preventing pollution. These include costs of changing to environmentally-benign raw materials, operating and maintenance costs, and capital expenditures.

- Hidden costs of environmental regulation. Included are monitoring and reporting costs associated with pollution prevention—in particular labour costs and costs of these accounting systems.

- Contingent liabilities arising from the need to remediate contaminated sites, and penalties for non-compliance with legislation and regulations.

- Intangible costs and benefits of pollution prevention associated with stakeholder attitudes and perceptions of the organisation (Chapter 9 and Ditz, Ranganathan and Banks 1995).

Some of these environmental costs can be traced to the activities identified and then to products and departments, while many of them can only be allocated to the particular activity, such as monitoring, and can then only be re-allocated to products and departments because there is no direct causal link.

Cost allocation occurs for a number of reasons. One important reason is because of a desire by managers to obtain the best possible information in order to assess, in financial terms, operational performance of an organisation through product profitability, and the contribution of departments and divisions to organisational profits. This information provides the basis for future decisions. Another important reason for cost allocation is the need to motivate managers and employees to improve their performance. Subsidiary reasons include the justification of a cost charged to an outside party such as a government, or to set the basis for reimbursement of costs. Cost allocation is necessary for the measurement of income and assets in order to report this information to outside parties (Horngren, Foster and Datar 1994). Finally, cost allocation may also be required by regulators, such as the US Environmental Protection Agency (EPA) and industry associations, seeking specific information about environmental costs.

Two of these reasons for allocating costs merit further attention—the provision of accurate information for decision-making, and the use of information to motivate managers and employees. The two are somewhat at odds because the former elicits the need for accurate information, while the latter plays up the importance of influencing behaviour at the expense of accuracy of information. An emphasis on accuracy of information for decision-making has led some authors to suggest that cost allocation should not take place at all because of the arbitrary or incorrigible basis for such allocations (Chambers 1966; Thomas 1969, 1974). Others have proposed schemes for cost allocations that have a neutral impact on decisions made (Hamlen, Hamlen and Tschirhart 1977), only later to be discredited (Thomas 1982).

Recently the suggestion has been made that technology, by providing the power of easier calculation, will solve the practical problems of cost allocation (Brown and Killough 1988). Unfortunately, the theoretical objections to cost allocation as a means of improving accuracy of reported cost information remain. Finally, activity-based costing ('ABC') has become the vogue to improve competitiveness in business organisations as attempts are being made to adopt new ways of allocating overhead costs when calculating the costs of products or any other cost object (Johnson and Kaplan 1987; Cooper and Kaplan 1988). However, even the proponents of this approach have realised that there has been too much of a focus on removing the cost distortions in accounting data and not enough focus on goals that matter, such as continuous improvement, rather than goals that count (Johnson 1992: 28; 1994: 260). In Johnson's words:

> I will be…disappointed if people associate [my ideas] with the currently popular belief, erroneous in my opinion, that a surefire way to restore competitiveness to American business is to reform the way accounts trace overhead to cost objects. (Johnson 1994: 260).

Cost allocation appears to be an issue that is recycled periodically without it ever being resolved or disposed of (Sterling 1979). It is argued here that current moves to identify environmental costs and liabilities as part of a desire to make the environmental performance of corporations more transparent to decision-makers are likely to rekindle the problems of cost allocation if too much focus is placed on the identification of accurate environmental costing.

An alternative possibility is related to the potential use of cost allocation to manipulate financial environmental performance figures with the intention of changing attitudes and behaviour of managers in favour of environmentally-benign activities, processes, products and organisational structures represented by ecological modernisation of the organisation (Huber 1991). Another stream of literature addresses this issue.

Japanese management practice is cited as using cost allocation to help achieve their long-term strategic corporate objectives (Hiromoto 1988). Management accounting is claimed to have a 'behaviour-influencing' role, rather than an 'information' role in the management process. As Ito (1995) indicates, it entails a political or intentional use of cost information for the sake of organisational reform or change. It appears that Japanese companies are less concerned with whether the allocation of indirect costs reflects the precise demands that each activity makes on resources than about how that allocation affects cost-reduction priorities of middle and operating line managers. Hence, as Morgan and Weerakoon observe, Japanese managers '...use cost allocation techniques that many Western academics, and indeed managers, criticise as distorting product cost information. The secret is that they use them for different purposes' (Morgan and Weerakoon 1993: 201).

For example, Japanese companies continue to allocate indirect costs to individual products based on the number of direct labour hours used in the production of the products, when this measure has been shunned by proponents of ABC.

> As a result of this approach, the production function and process become overloaded with labour-based overhead 'burden'. Therefore, as the labour content of operations is reduced, so the burden is progressively lightened. The motivation...to reduce [direct labour] cost is obvious and pressing' (Morgan and Weerakoon 1993: 202).

The strategic goal is to encourage automation (capital-intensive production) in the organisation. One way of encouraging such a move is to load up the costs of managers who encourage labour-intensive production with indirect costs. The cost allocation method facilitates this loading. A second example is cited whereby indirect costs are allocated to products on the basis of the number of parts used in production. This encourages managers to focus on simpler designs with fewer parts and to promote the use of standard parts that help reduce the cost of production.

Johnson (1994) is critical of managers in the US for abdicating from their strategic responsibilities. He advocates a need to move towards the Japanese approach. First, he suggests a move towards adoption of the principles of Total Quality Management (TQM), and an emphasis on continuous improvement. Second, he recommends that financial accounting information be removed as the sole object of attention, with greater emphasis placed on non-financial performance and associated measures. However, although he promotes the TQM management philosophy, Johnson does not suggest that financial accounting information be used to influence behaviour in the way Japanese companies do.

The idea of using cost allocation to influence behaviour is not new. Indeed, it is unlikely that anyone would deny that cost allocations are used to affect the behaviour of managers in the areas of pricing, use of shared services, and corporate infrastructure (Wells 1980).

Zimmerman (1979) suggests two ways in which cost allocation could be so used. First, he explores the idea that allocated costs can act as a lump-sum tax which leads to a reduction of the consumption of perquisites (discretionary expenditure) by managers. This is intuitively appealing because the direction of the impact of cost allocation on a manager's behaviour can be predicted. A higher lump-sum cost allocation leads to a lower consumption of perquisites. If cost allocation is linked to profits, the direction is less easy to predict. Second, he examines the proposal that it is cost effective to use cost allocations to ration the use of scarce resources internally because the information is already being produced for tax and external reporting purposes. (See Burritt, Craswell and Wells 1980 for some difficulties with this approach.)

What is suggested here is that cost allocation could be used to encourage environmentally benign behaviour by managers in organisations that have accepted the need for a strategic posture towards the environment to be developed. In particular, cost allocation may be used to encourage behaviour that facilitates ecologically sustainable development (ESD).

## ▌Areas where Cost Allocation has an Impact on Corporate Environmental Performance

In the context of alerting readers to the information distortions inherent in traditional management accounting, Ditz, Ranganathan and Banks make the following observation: '…the common practice of pooling overhead costs can conceal and distort critical information on environmental and other costs' (Chapter 9 and Ditz, Ranganathan and Banks 1995: 1). From this, they conclude that better accounting for environmental costs is crucial to long-term business sustainability. Implicit is the need to overcome the cost allocation process. Considerable effort is now being placed on the identification of corporate environmental costs as a component of total costs of certain products and processes. If it can be shown that these costs are material, and likely to increase, then a strong case can be mounted to formalise their identification, measurement and reporting.

Following a series of nine case studies undertaken in the US, Ditz, Ranganathan and Banks (1995: 8) argue that environmental costs 'may account for 20 percent of total costs' for some products and facilities. This could be considered a worst-case perspective, as the study provides detailed studies on five large organisations where environmental costs were likely to be high—Amoco, Ciba-Geigy, Dow Chemical, Du Pont and Johnson Wax. The Australian Society of Certified Practising Accountants is in the process of replicating the Ditz, Ranganathan and Banks study for Australian companies. There is every reason to think that similar results should be expected.

Ditz, Ranganathan and Banks suggest that the relative size of corporate environmental costs is sufficiently important to merit internal management of these costs, particularly those that are indirect and long term. However, identification of costs as environmental is problematic.

First, consider a cost incurred because of a long-term investment. Where such costs are capitalised and then depreciated, the allocation of costs to different time-periods is required. Corporations appear not to be keen to identify such investments as environmental.

Typical is the example provided in the environmental performance report of ICI plc for 1994. Two of the four reported areas where ICI spend money on the environment involve capital expenditure:

- **Developing environmentally-sound products and processes.** 'We put considerable effort into developing products and processes which meet the fast-growing demand for goods with a sound environmental performance. We spend money on research, technology and engineering, building plants and providing technical services to customers' (ICI 1994). ICI does not label any of these costs as environmental expenditure.

- **New plants.** 'Whenever new plants are built anywhere in the world they are designed to keep their impact on the environment to a minimum. This makes good business and environmental sense because a clean plant will be more efficient, produce less waste and therefore cost less to run. This will eventually reduce our annual running costs for environmental protection' (ICI 1994). ICI claims that up to a quarter of the building costs can be attributed to safety, health and the environment. No dollar figures are provided and no breakdown is given. Recognition is given to the environmental component of expenditure which will later be expensed through depreciation, but figures that allocate the identifiable environmental costs are not provided.

The 1994–95 *Environmental Progress Report* for WMC (Western Mining Corporation) in Australia finds similar difficulties in classification. The environmental budget is said to exclude '$168 million for the Sulphuric Acid Plant and operating facilities at the Kalgoorlie Nickel Smelter during 1994–96' (WMC 1995: 5). From the report, it is not possible to establish whether depreciation of the above expenditures would be considered environmental costs, yet there is a clear implication that some of these capital expenditures are environmental.

The main problem with the identification of environmental components of capital expenditure is how to allocate the expenditure between new expenditure for normal commercial operational purposes and environmental expenditure which has been incurred to avoid future environmental risks and liabilities. This is a classic joint-cost problem which can only have an arbitrary or incorrigible solution. Information about the depreciation of environmental components of capital is likely to be distorted and will distort resulting decisions.

Second comes the situation examined by Kreuze and Newell (1994). They observed that arbitrary cost allocation leads to the incorrect accumulation of product costs. As a result, when environmental costs are not separately recognised, but are aggregated in a common overhead pool and then allocated to products, those products that actually have lower environmental costs provide a cross-subsidy to the products with higher environmental costs. Hence, products that are relatively more environmentally benign are discriminated against through the cost allocation procedures. This is the exact opposite result from that favoured by risk-averse managers with an eye on the reduction of potential environmental liabilities through precautionary actions to avoid environmental impacts.

In both of these cases, the focus on trying to remove distortions in cost information is doomed to failure. Instead, a reformulation of the issue on hand is needed. Here it is suggested that a move towards direct use of cost allocations to change behaviour of managers towards the reduction of environmental impacts is in accordance with the sub-principles of ESD.

## ▌ Whether and how Cost Allocation can be Used in a Pro-Environmentally Benign Manner by Some Corporations

For corporate activities in which environmental impacts are significant, physical environmental performance indicators can be designed to provide a benchmark against which to assess progress. As reported by CICA (1994: 87-88), these should be few in number, cover an organisation's most significant environmental performance issues, portray the costs and benefits of pursuing the environmental objectives and provide a balanced view of performance. It is suggested here that financial indicators of environmental performance can complement such progress, particularly if they are linked to an appropriate incentive mechanism.

Consider two situations: first, a situation where a more appropriate classification of cost can influence decisions that affect corporate environmental performance; second, a situation where cost allocation can actively be used to help encourage a better outcome for the environment.

As an example of the first, consider a situation where cost allocation can influence the choice of best environmentally benign technology. The case of Du Pont's Yorktown refinery illustrates the point (see Chapter 9 and Ditz, Ranganathan and Banks 1995). Given the choice between two waste management alternatives—deep-well injection and bio-treatment—Du Pont's Yorktown refinery manager favoured the former because the management accounting system showed it had the lowest relative measure of performance—full cost per pound of waste being treated (0.09 cents per lb compared with 0.11 cents per lb; see Fig. 1).

| Costs per lb of waste disposed (cents) | | | |
|---|---|---|---|
| **Method of waste disposal** | **Variable costs** | **Fixed costs** | **Total costs** |
| Deep-well injection | 0.08 | 0.01 | 0.09 |
| Bio-treatment | 0.05 | 0.06 | 0.11 |
| **Total costs per lb of waste disposed ($)** | | | |
| **Deep-well injection** | **10,000 lbs processed** $ | **20,000 lbs processed** $ | |
| Variable costs | (10,000 × 0.08)  800 | (20,000 × 0.08)  1,600 | |
| Fixed costs | 100 | 100 | |
| Total costs | 900 | 1,700 | |
| **Bio-treatment** | | | |
| Variable costs | (10,000 × 0.05)  500 | (20,000 × 0.05)  1,000 | |
| Fixed costs | 600 | 600 | |
| Total costs | 1,100 | 1,600 | |

**Figure 1:** Use of a Finer Cost Classification to Improve Decision-Making

*Source: Based on Ditz, Ranganathan and Banks 1995*

Several criteria, financial and non-financial, formed the basis of decisions made about waste treatment in Du Pont. These factors included the cost per pound of waste treated, the effect on the Toxic Release Inventory (a national register of toxic chemical emissions/transfers in the US), the risk of future liability, and the degree to which the method of disposal of waste is acceptable to the stakeholders. The presence of these multiple criteria meant that, although Yorktown favoured deep-well injection, both deep-well injection and bio-treatment continued to be used by Du Pont, as the non-cost criteria meant that some operating managers still had to use bio-treatment. In these circumstances, it was actually cheaper for Du Pont to channel as much waste into bio-treatment rather than deep-well treatment, as the out-of-pocket costs of bio-treatment were lower (5 cents per lb compared with 8 cents per lb). Figure 1 illustrates the point.

Given that Du Pont had to maintain a bio-treatment process for waste disposal in other parts of the organisation, it would have been appropriate for the Yorktown manager to make a decision based on only the variable cost of the process, rather than on full (fixed and variable) costs. Fixed costs of the bio-treatment process were 'sunk' and irrelevant to the decision as they continued to be incurred whatever the decision made by Yorktown. Unfortunately, as waste treatment was diverted from bio-treatment to deep-well injection, the fixed costs of bio-treatment had to be allocated over a smaller volume of waste being processed. This makes bio-treatment look progressively less attractive on a total-cost basis. By way of illustration, as volume treated declines from 20,000 lbs to 10,000 lbs, the allocation of fixed costs makes bio-treatment even more expensive on a full-cost basis than deep-well injection. The problem is caused by cost allocation of overheads to derive a full cost.

By concentrating on a traditional accounting calculation—the cost per pound for waste treatment—an incorrect decision resulted. In practice, the total fixed costs of both methods of waste treatment had to be incurred as both methods continued to be used. Hence, the only costs that were relevant to the decision of which method to favour were the variable (out-of-pocket) costs. With the present financial information being based on the allocation of fixed costs of waste treatment to each pound of waste treated, and on the assumption that bio-treatment is regarded as the most environmentally benign of the two forms of waste treatment, the decision made by Yorktown could be viewed as the worst one for the environment and the worst for the financial performance of Du Pont. Hence, it would have been beneficial in both financial and environmental terms to discover, classify, measure and report the more detailed variable/fixed-cost information as a basis for decision-making. Yet, to achieve this result, the cost allocation problem would have to be solved.

The second situation to consider is whether cost allocation could have a positive impact on corporate environmental performance through a direct strategic commitment to influence the behaviour of a manager. This would be in accordance with the suggestions made by Maunders and Burritt (1991) to:

- Decondition accountants in relation to the values and beliefs they hold with respect to ecological issues

- Recondition accountants to change the *Weltanschauungen* (world-view) of their measurement systems towards an ecocentric approach that: encourages activities that are in harmony with nature; recognises that all nature has intrinsic worth; promotes elegantly simple material needs; recommends use of appropriate technology rather than high technology solutions; perceives the virtue in doing with enough; and places focus on bio-regions

The significance of such a change cannot be underestimated, given that accounting information, in a financial form, does occupy a dominant role in corporate information systems because it quantifies and simplifies, as well as serving an ideological function.

Using a lump-sum cost allocation, as suggested by Zimmerman (1979), it would be a feasible process to identify projects and activities that should suffer a cost penalty because they encourage environmental impacts that are potentially detrimental to either the financial position or the ecological profile of the organisation—such as the encouragement of global warming, destruction of critical natural capital, or the exhaustion of non-renewable resources. For example, the activity may lead to future environmental liabilities and could be penalised by the application of an ecological risk premium. This could be calculated based on scientific evidence, or when unavailable, the precautionary principle (see Fig. 2).

Hence if, at the head office of a hotel chain with buildings mainly centred on the beaches of tropical islands, it was thought strategically important to encourage a long-term change towards wind-power sourcing because of the lower impact on global warming, then a hotel management accountant could look to use cost allocation to cross-subsidise the environmentally benign method. On the assumption that rewards were based on reported profitability, an absolute cost allocation would be used in a proactive, precautionary way to change behaviour, thereby helping to reduce global warming, and avoid future hotel closures and financial penalties caused by flooding. The effect would be similar to a tax imposed by government, but discretion would lie in the hands of the corporation. As illustrated, the allocation scheme could be environmentally risk-adjusted with a risk premium being identified and added to the cost of activities that have potential future financial or ecological liabilities attached.

Likewise, if a relative cost allocation was made at a specified amount per unit of input, again assuming an incentive system based on reported profitability, a favourable outcome from an ecological perspective could be ascertained. Returning to the Du Pont example, Figure 3 illustrates how differential variable allocated costs could

| Electricity costs per hotel ($) | | | | |
|---|---|---|---|---|
| | Ecological risk premium weighting factor | Variable costs ($) | Allocation of ecological lump-sum fixed-cost risk premium ($) | Total costs ($) |
| **Method of electricity generation** | | | | |
| Coal-fired | 1.25 | 800 | 1,000 | 1,800 |
| Hydro-electric | 0.5 | 600 | 300 | 900 |
| Wind-powered | 0 | 1,600 | 0 | 1,600 |

**Figure 2:** Use of Lump-Sum Cost Allocation to Encourage Environmentally Benign Decision-Making

| Total costs per kg of waste disposed ($) | | | | |
|---|---|---|---|---|
| | **10,000 kg processed** | **15,000 kg processed** | **20,000 kg processed** | **25,000 kg processed** |
| **Deep-well injection** | | | | |
| Direct costs | 800 | 900 | 1,000 | 1,100 |
| Allocated costs | 600 | 650 | 700 | 750 |
| Total costs | 1,400 | 1,550 | 1,700 | 1,850 |
| **Bio-treatment** | | | | |
| Direct costs | 800 | 900 | 1,000 | 1,100 |
| Allocated costs | 600 | 500 | 400 | 300 |
| Total costs | 1,400 | 1,400 | 1,400 | 1,400 |

**Figure 3:** Use of Proportional Cost Allocation to Encourage Environmentally Benign Decision-Making

be used to discriminate between the two waste treatment methods. Deep-well injection suffers an increasing penalty, while bio-treatment is favoured by a smaller variable cost allocation.

## ▌ Conclusion

The examples provided above to illustrate use of cost allocation to encourage environmentally-benign behaviour are both contrived and simplistic. However, the principle can be applied by managers who wish to recognise environmental risks stemming from their corporate activities, to continue to use financial numbers as part of their incentive systems, and to strive for ecological modernisation. The allocation scheme can be environmentally risk-adjusted with a risk premium being identified and added to the cost of activities that have potential future liabilities attached.

In conclusion, cost allocation schemes can be used in combination with non-financial environmental performance measures to influence corporate environmental behaviour in a pro-environmental manner consistent with ESD principles. Further research is needed on the combination of non-financial environmental performance indicators with cost allocation strategies in the process of developing generic environmental risk classes to which cost allocation weights can be linked. In this way, cost allocation can be turned from a financial bane to a boon, for managers and proponents of ESD, through the integration of environmental performance indicators and accounting information systems.

# Part Two

## Empirical Studies

# 9 Green Ledgers

## An Overview[1]

### Daryl Ditz, Janet Ranganathan and R. Darryl Banks

## ▌ Background

ENVIRONMENTAL LIMITS don't simply constrain business. Rather, companies are finding, environmental considerations increasingly infuse everything from product design to marketing, from purchasing to product stewardship, from employee relations to executive compensation. Now the challenge for corporations is to fully integrate environmental thinking into corporate decision-making—in other words, to translate their environmental concerns into the language of business.

Environmental costs are dispersed throughout most businesses, and can appear long after decisions are made. Unfortunately, conventional accounting practices—developed to serve financial reporting requirements—rarely illuminate environmental costs or stimulate better environmental performance. Managerial accounting's traditional dependence on discrete, historical transactions, and the common practice of pooling overhead costs can conceal and distort critical information on environmental and other costs.

Not only is it achievable, but better accounting for environmental costs is crucial to long-term business sustainability. In 1993, the World Resources Institute (WRI) began exploring how firms account for environmental costs. Working with three teams of academic investigators, WRI initiated nine case studies to improve understanding of environmental accounting through the participation of companies facing real business and environmental challenges.

---

1. This chapter is an edited version of *Green Ledgers: Case Studies in Corporate Environmental Accounting* (Ditz, Ranganathan and Banks, 1995).

The firms include five major corporations: Amoco Oil, Ciba-Geigy, Dow Chemical, E.I. Du Pont de Nemours and S.C. Johnson Wax. These diverse firms span several industries. In addition, one case study was undertaken of four mid-sized firms in Washington state—Cascade Cabinet, Eldec, Heath Tecna and Spectrum Glass—which are obliged by state regulatory requirements to assess the costs and benefits of pollution prevention. Case highlights of each of the six case studies are presented at the end of this chapter. Each firm was encouraged to focus on its internal costs, not on externalities. Otherwise, they were free to define environmental costs for themselves. No single definition emerged, but most identified environmental costs as expenditures that arise from either compulsory or voluntary actions to achieve environmental objectives. For example, capital costs for pollution control equipment are environmental costs, as are salaries of corporate environmental staff. Furthermore, the incremental cost of cleaner process technologies, extra processing steps and less-polluting materials can be considered environmental. The failure to meet environmental goals also has costs, so regulatory fines, remediation costs and even losses in sales were considered environmental costs.

These case studies provide insights into how firms account for environmental costs and show that, even where environmental costs are sizeable, they are often systematically underappreciated. More important, they illustrate how environmental costs can influence a wide variety of business decisions. These findings provide a strong rationale for asking how well a firm's managerial accounting system is serving these needs.

## ▌ Magnitudes and Sources of Environmental Costs

Just how large are environmental costs? Corporate executives and other readers who want to go straight to the bottom line can turn to Figure 1 for a glimpse at the gross magnitude of environmental costs uncovered at the five large firms. These numbers might satisfy initial curiosity about how large environmental costs can be. The double-digit percentages could well spur other firms to analyse their own environmental costs. But the magnitude of costs is not the whole story. The real value of environmental accounting lies behind these aggregate figures. By digging more deeply into the composition of the totals, the 'behaviour' of these costs and other underlying factors, firms can link cost reduction to significant improvements in environmental performance. Since these costs refer only to the specific products, product-lines or facilities studied, they should not be extrapolated to the companies at large. A quick glance at the figures in Figure 1 could easily mislead. Consider the S.C. Johnson and Dow cases. The numbers suggest that the relatively modest environmental cost percentages associated with these two products means that environmental issues are not that important. But a closer look at the S.C. Johnson case reveals that environmental costs exceed the operating profit for this product. In the case of Dow, the use of a seemingly inexpensive solvent was creating environmental challenges that jeopardised an entire product-line.

On the other hand, estimated environmental costs may actually exaggerate the net economic impact on firms. Costs that are arguably environmental (e.g. equipment for closed-loop recycling or leak-detection programmes) sometimes provide multiple benefits such as greater occupational safety or enhanced product quality. Furthermore, where environmental costs appear very large, a significant share of the total may consist of

| Case study | Finding |
|---|---|
| Amoco Oil | Nearly 22% of operating costs (excluding feedstock) were considered environmental at the Yorktown refinery. |
| Ciba-Geigy | The environmental component was estimated at over 19% of manufacturing costs (excluding raw materials) for one chemical additive. |
| Dow Chemical | Between 3.2% and 3.8% of the manufacturing cost for a polymer-based product was considered environmental. |
| Du Pont | Over 19% of manufacturing cost was identified as environmental for one agricultural pesticide. |
| S.C. Johnson Wax | Environmental costs identified for one consumer product were approximately 2.4% of the net sales. |

**Figure 1:** Aggregate Environmental Costs from Selected Case Studies

fixed expenses that have little net impact on cash flow. The important message is that no single number can adequately reflect the importance of environmental costs or their potential relevance to operational and strategic decision-making.

Furthermore, these estimates do not allow any meaningful comparisons of environmental costs across the case studies. A close inspection of Figure 1 shows that each company has its own way of expressing aggregate environmental costs (e.g. as a percentage of product manufacturing costs, as a portion of operating costs or as a share of net sales). Indeed, appropriate measures are industry-specific and often vary within a firm. In the petroleum industry, for instance, it makes sense to evaluate refinery operations exclusive of the changing cost of crude oil and other feedstocks. For a consumer products firm such as S.C. Johnson—where distribution, marketing and sales dominate product costs—net sales are a more meaningful gauge of the importance of environmental costs. In the Dow case, managers excluded the cost of operating labour devoted to environmental activities, arguing that attention to the environment is part of every employee's job. In the Amoco case, the portion of maintenance devoted to environmental activities was estimated and included in the total environmental cost number.

Clearly, there is no universal way to define environmental costs. Firms must tailor their own definitions to suit their intended uses—whether they be cost control, product pricing, capital budgeting, staff incentives or other uses.

## Breaking Down Environmental Costs

From a strategic perspective, firms must mind all costs. But, in the short term, managers generally focus on the most easily controlled costs. For these reasons, companies routinely distinguish between variable and fixed costs. Figure 2 shows the relative fixed and variable portions of environmental and non-environmental manufacturing costs of Du Pont's agriculture pesticide. Roughly four-fifths of the environmental cost of the product are fixed, suggesting fewer opportunities to reduce environmental costs quickly.

The conventional accounting distinction between fixed and variable costs can lead to some confusion about the 'controllability' of environmental costs. Over the long run, all costs become variable. Many costs are recorded as 'variable' or 'fixed', depending on whether they vary with production volume. Off-site waste disposal is a classic example of a variable cost because it rises and falls with the quantity of

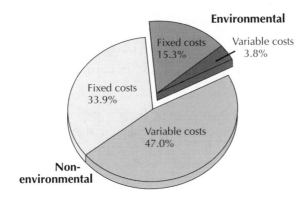

**Figure 2:** Environmental Costs for a Du Pont Agricultural Pesticide

production. But many types of environmental costs can vary with factors other than production. For example, those associated with emissions monitoring and reporting in a plant would change very little with the amount of product manufactured (ignoring the issue of regulatory thresholds), but could be lowered with a change of technology or materials.

What are the components of aggregate environmental cost within firms? As the breakdown of environmental costs at Amoco's Yorktown refinery in Figure 3 illustrates, environmental costs show up in maintenance, administration, and stipulations on product quality (e.g. volatility limits on gasoline). The more obvious costs of waste treatment and disposal, though significant, are dwarfed by the sum of environmental costs extracted from cost categories that are not exclusively environmental.

Other cases exhibit similarly diffuse environmental costs. For the consumer product at S.C. Johnson, the direct environmental costs of manufacturing were less than 0.3% of the manufacturing cost. But the environmental costs in marketing, research and development, and product registration proved significant.

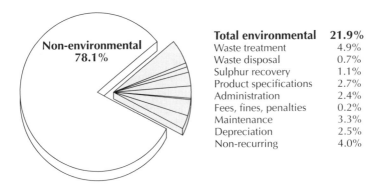

| Total environmental | 21.9% |
|---|---|
| Waste treatment | 4.9% |
| Waste disposal | 0.7% |
| Sulphur recovery | 1.1% |
| Product specifications | 2.7% |
| Administration | 2.4% |
| Fees, fines, penalties | 0.2% |
| Maintenance | 3.3% |
| Depreciation | 2.5% |
| Non-recurring | 4.0% |

**Figure 3:** Environmental Costs at Amoco's Yorktown Refinery as a Percentage of Operating Costs (Excluding Crude)

As the cases reveal, many environmental activities are simply not tracked and recorded as 'environmental'. One example is the time spent by operating personnel on environmental training, monitoring and compliance. The Ciba-Geigy, Dow and S.C. Johnson case studies speak to the difficulty of quantifying the environmental component of such ostensibly non-environmental accounts.

## Organisational Dimensions of Environmental Costs

The cases contain many examples of environmental costs that are allocated across departments and products, blurring the links between costs and their underlying drivers. As just one example, the time and resources that corporate environmental professionals at S.C. Johnson devote to issues arising from the use of solvents in certain products is allocated across all products on the basis of sales dollars instead of traced to specific products. Similar allocation practices were observed in all of the cases.

In larger, more complex corporations, decisions made in one part of the organisation frequently create environmental costs elsewhere in the company. The corporate hierarchy of divisions, business units and plants imposes cascading overheads on managers. The resulting disconnection between those who contribute to environmental costs and those ultimately held accountable hinders environmental management, even in smaller, less complex organisations. The practical challenge for accounting systems is to hold decision-makers responsible for their actions, regardless of where in the corporation the consequences materialise.

What difference does it make if maintenance on environmental equipment is tallied as an 'environmental' rather than 'maintenance' expenditure? In one sense, none. As firms assess their overall performance, whether a particular expense item carries a green flag in the accounting system is secondary to responsible management. Yet tracing costs back to specific decisions helps managers better understand cause and effect. Decisions about which products to manufacture and what technology to employ can increase or ease the environmental management burden on the firm. But if those costs show up only in the general maintenance budget, managers will systematically underestimate the benefits of cleaner production. This commingling of environmental costs in non-environmental accounts, combined with the gaps between costs and the activities that give rise to them, conceal the true magnitude of environmental costs. No wonder managers find it difficult to appreciate the full impact of environmental concerns on their business.

These case studies also demonstrate that firms can overcome these obstacles and better manage environmental costs. Specifically, the managerial accounting system can help decision-makers co-ordinate environmental activities and other activities within inherently complex organisations. The system can be used to monitor relevant environmental cost information and to communicate environmental objectives through charges and rewards to motivate employees.

Many managers may consider the practice of rolling environmental expenses into overhead accounts as a practical necessity and fear that implementing more rigorous accounting for environmental costs would be prohibitively expensive. They might be right, but the answer depends on whether better accounting for environmental costs will lead to more sound decisions and, ultimately, a more competitive company. To answer this question requires an even more detailed look at how such information can be put to use.

## ∎ Business Uses of Environmental Cost Information

Environmental accounting is much more than green 'bean counting'. However interesting the numbers, firms will not go through the trouble and expense of developing them unless they can be used to enhance productivity and profitability. Drawing on examples from the case studies, this section demonstrates several of the many ways in which information on environmental costs can influence business decision-making. In traditional business uses, environmental costs figure in the selection of product mix, the evaluation of alternative manufacturing inputs, the comparison of costs across facilities, and product pricing. Also explored here are the ways that better cost information supports more cost-effective environmental management, whether in evaluating waste management options, prioritising environmental initiatives or assessing opportunities for pollution prevention. Although these examples reflect private-sector experience, the uses have broad applicability in public-sector decision-making. Other uses of environmental cost information, including the capital budgeting process and product design, have been described elsewhere (White, Becker and Savage 1993; GEMI 1994a; Fiskel and Wapman 1994; Savage and White 1994).

### Product Mix Decisions

In choosing what to produce, all manufacturers must carefully weigh customer demand, capacity constraints, product profitability and other factors. Yet, because measures of profitability depend on projected revenues and projected costs, and because environmental costs are so frequently misallocated, products with relatively higher environmental costs are often subsidised by those with lower environmental costs. With only this distorted picture of profitability to work with, firms undervalue the direct economic benefits of products with lower environmental costs.

To illustrate, the principal environmental issue facing Spectrum Glass, a speciality sheet glass manufacturer, springs from the use of cadmium oxide, a colorant for which no adequate substitutes are available. Cadmium oxide is used in just one product line: 'ruby-red' glass. Although the manufacture of ruby-red produces more hazardous waste than that of other glasses, the resulting environmental costs are allocated across all products. At present, making ruby-red glass appears profitable. But if pending regulations on the release of cadmium take effect, costs will rise so sharply that the firm will drop the product from their portfolio. In this case, the prospective hike in environmental costs is easily attributable to the product responsible.

This is not always the case. The Amoco case study provides an especially rich illustration of how hard it can be to trace environmental costs to products. Petroleum refining is a highly integrated manufacturing process that produces a spectrum of hydrocarbon products. Any process change can affect the yield of all of the products. At Amoco, the product mix is determined above the refinery level, based on such factors as regional and seasonal demand and product margins. Once the product mix has been selected, the refinery uses a computer model to translate production targets into operating parameters for the individual units. This model helps management meet production objectives at lowest cost. As a practical matter, the model ignores all costs that are expected to remain fixed, including many environmental costs. But since some of the components of environmental costs do change with product mix, omitting them could bias the results toward mixes with relatively higher environmental costs. Conversely, incorporating variable environmental costs into decisions on product mix and refinery-operating parameters can lower overall costs. (Whether this effect is small or large could not be determined during the course of the case study.) To be sure, few industries are forced to contend with the extreme process complexity found in petroleum refining, but those that do should examine the implications of a changing product mix on environmental costs.

## Choosing Manufacturing Inputs

The choice of which materials to use in a manufacturing process often depends on cost as well as on safety, reliability, performance and other criteria. Consider a hypothetical choice between two chemical substitutes. The cheaper material triggers additional compliance costs, such as those for permit preparation, fees, monitoring and disposal. In contrast, using the more expensive material entails lower environmental costs. The additional environmental costs of the less expensive material may not be fully appreciated by the purchasing department, and production managers may be charged for only part of these higher environmental costs. For this reason, understanding how environmental costs depend on choice of materials can lead to more informed selection of raw materials, intermediates and other process inputs.

At Cascade Cabinet, a speciality cabinet manufacturer, a decision to switch from nitro-cellulose lacquer to a slightly more expensive conversion varnish saved the company significant sums. The lacquer, a hazardous material and source of volatile organic chemicals (VOCs), had been used to coat cabinet components. Residual dust from the lacquer, touched off by a welding spark, caused a serious fire and one million dollars in damage. Substituting the conversion varnish not only reduced the risk of another explosion, it also lowered the cost of insurance, waste disposal and air permit fees. Fuelled by these tangible savings, Cascade has asked one of its suppliers to develop a high-quality, water-based stain to replace the solvent-based product now in use.

Although Cascade Cabinet is a relatively small firm, the idea that different process inputs can lower environmental costs is also valid for larger organisations. For example, the world market for crude oil offers various grades, each containing different types and amounts of impurities. Removing these impurities carries environmental costs at the various stages of refining. Amoco is currently exploring how the choice of crude oil influences environmental costs. By developing a better understanding of how environmental costs vary with different varieties of crude, the company will be able to select those whose use lowers overall costs. This demonstrates how important the general link between process inputs and environmental costs is to the search for more cost-effective substitutes.

## Assessing Pollution Prevention Projects

Over the past three decades, efforts to limit industrial pollution have focused largely on end-of-pipe pollution control. More recently, the emphasis has moved upstream, toward process-oriented preventative solutions that reduce the amount of pollutants generated. Unfortunately, under typical management accounting practices, many costs avoided by such changes are not credited to the successful manager, so pollution prevention projects often compete on an unequal footing with projects reliant on the existing pollution control and waste disposal infrastructure. A study of 29 chemical plants found that source reduction activities were most likely where some sort of environmental accounting was being practised (INFORM 1992). Motivated by similar discoveries, others are providing tools and training to aid business in this effort (NEWMOA 1994).

Recognising the importance of this information, the Washington State pollution prevention planning rules require firms to scrutinise their operations and costs closely (see Case Study 6, page 187). Through this planning process, Eldec, an electronics manufacturer that sells to the military and civilian aerospace industry, sought to identify its most significant sources of hazardous waste and discovered that its generation of hazardous waste, and the corresponding costs, were driven primarily by the frequency of preventative maintenance and cleaning. By making minor procedural changes that did not reduce product quality, reliability or production, Eldec cut its use of hazardous cleaning solvents, its maintenance costs and its hazardous waste generation.

Also searching for ways to reduce hazardous wastes, Cascade Cabinet managed to save $100,000 annually on wood scrap. This unused material is now chipped and sold to a particle-board manufacturer, and sales bring in three times the disposal cost. The investment in a large grinder plus one additional hour of work each morning was offset by direct savings within just one year.

The case study firms in Washington State had taken steps to prevent pollution before the planning requirements were introduced. But even though the pollution prevention legislation cannot be credited for all of the newly uncovered opportunities, all four firms contend that preparing the required plans had been worthwhile. This finding is being corroborated in New Jersey, where similar, but even more demanding planning requirements apply (NJ DEPE 1995). In particular, quantifying environmental savings more precisely justified some projects that might otherwise have been rejected and strengthened the hand of in-house pollution prevention advocates. Washington State Department of Ecology publications document many examples of firms besides Eldec and Cascade that have undertaken pollution prevention projects as a consequence of the planning process (Washington State Department of Ecology 1993).

### Evaluating Waste Management Options

Firms must routinely choose among various waste management methods. Because of persistent pressures to reduce operating expenses, costs naturally figure into these decisions. However, the case studies reveal that quantifying the true costs of managing wastes or the potential savings from cleaner production is rarely simple.

The Du Pont case study demonstrates how a better understanding of environmental costs can make waste management more cost-effective. One of Du Pont's large wastewater streams has traditionally been managed by deep-well injection. As Figure 4 shows, if managers are charged the full absorption cost (both variable and fixed costs), deep-well injection—at 0.09 cents per pound—appears less costly than the alternatives. But look again. Much of the cost for deep-well injection is for pumping, while most of the cost for bio-treatment is fixed and does not change regardless of how wastes are managed. Once Du Pont analysed costs, it concluded that a switch to bio-treatment would save 0.04 cents per pound in out-of-pocket expenditures. While the facility has not yet phased out deep-well injection altogether, management now gives preference to biological treatment.

At Amoco's Yorktown refinery, management decided several years ago to treat wastewater sludge and other wastes in a process unit called the 'coker'. In a coker, residual fuel oils and heavy petroleum intermediates are thermally 'cracked' to produce solid coke and higher-value products such as gasoline. When the original decision was made, little information was available on the cost of coking wastes, though the benefits of

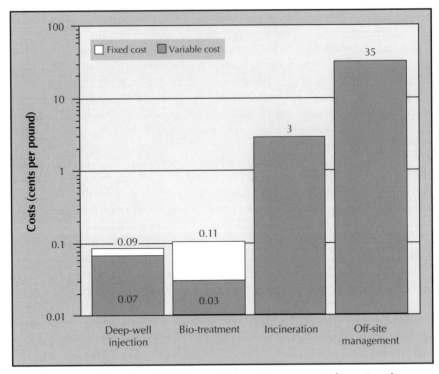

**Figure 4:** Fixed and Variable Costs of Waste Management Alternatives from the Du Pont Case Study

avoided costs for off-site sludge disposal were apparent. Later, a more detailed analysis revealed significant cost implications. Some had been anticipated, such as a slight reduction in the yield of higher-value products. Others, including higher maintenance costs, were not. Of course, even the best accounting information could not have predicted these unexpected costs. Yet, by tracking them after the decision had been implemented, Amoco could respond effectively. In retrospect, the company still backs the original decision, but its deepening financial understanding helps it to better manage the costs of waste coking and to respond more quickly to changing circumstances.

### Comparing Environmental Costs across Facilities

Environmental cost information can be influential as management compares operations within a firm. For example, firms can use such information to compare environmental performance across facilities and identify the best performers. The Ciba-Geigy case considered how environmental costs might be used in deciding where specific products should be manufactured. Many fixed costs of shared environmental facilities, such as existing incinerators and waste-water treatment plants, continue regardless of where the product is manufactured. The study identifies a subset of costs that would change if the product were shifted from one of the company's facilities to another. Internationally especially, these costs vary with both environmental regulations and such internal factors as the availability of trained personnel and waste management capacity.

Similarly, the case study based on a single Amoco refinery suggests how looking closer at environmental costs could influence the allocation of production among multiple refineries. For example, an earlier investment in technology for refining high-sulphur crudes permits the Yorktown refinery to meet new regulatory requirements for low-sulphur diesel fuel. Even though this investment raises the facility's cost burden incrementally, the refinery now has a comparative cost advantage in meeting this new product requirement.

### Prioritising Environmental Initiatives

Environmental cost information readily lends itself to the development of internal benchmarks of comparative environmental performance. For instance, a computer-chip manufacturer might want to determine the environmental cost per chip across its manufacturing sites or along the production chain. The Du Pont case study describes how the company prioritised more than 700 projects to reduce emissions by estimating the cost per pound of emission eliminated. With so many options, this cost-effectiveness measure helped managers choose among the vast array of options. Du Pont developed a methodology for rank-ordering projects that considered cost–benefit ratios, regulatory requirements and competitive factors. These projects each carry a 'technology indicator' that permits Du Pont to leverage solutions across the corporation. Regulations still force some relatively costly expenditures, and project scheduling and other considerations can also influence the decision of which projects to undertake. Still, this scheme provides a consistent and formal way to evaluate trade-offs among competing proposals.

### Pricing Products

From the 'polluter pays' perspective, it makes sense to include all attendant environmental costs in a product's price. If an entire industry adopted this approach uniformly, customers could make more informed decisions on what products to purchase. While this notion is at present far from reality, a better understanding of private environmental costs did affect product pricing in at least one of the case studies.

Dow Chemical faced new, tougher limits under the 1990 Clean Air Act Amendments in addition to their corporate commitment to EPA's 33/50 Programme, a voluntary campaign to lower emissions of seventeen toxic chemicals. These twin obligations implied significant new environmental costs in the manufacture of a polymer-based coating material and a large emitter of VOCs at the site. The company faced a stark choice: invest in pollution control or shut down the operation. Dow weighed the costs of additional investment against those of discontinuing operations. They also met with their industrial customers to explain that making the investment would guarantee a reliable continuing supply, but at higher cost. Thus, environmental cost information generated for internal decision-making purposes also helped justify the potential price increase. The new information convinced customers that Dow was making a genuine investment in the product's long-term viability, so, in essence, they accepted a price premium.

## ▌ A Practical Guide for Getting Started

Trying environmental accounting does not entail a complete overhaul of current accounting and information systems. Even modest incremental changes in how data are collected, conveyed and analysed can yield important environmental and economic benefits.

Moreover, as the case studies confirm, a well-designed and executed environmental accounting pilot project can uncover significant opportunities for improvement and suggest broader changes in company policy and practice.

There is no single approach for getting started. The strategic objectives of the firm, the complexity of its environmental challenges, and its size and corporate culture together dictate the right approach. Even so, the case studies presented at the end of this chapter offer general lessons about how to assess and manage environmental costs. More specifically, this guide outlines how firms might adapt environmental accounting to their own needs and capabilities and proposes mechanisms for integrating environmental costs into business decision-making.

Companies can have more than one reason for undertaking environmental accounting. An informal survey of some two dozen corporate environmental professionals listed general managerial control as the most common reason for monitoring environmental capital and operating costs (Nagle 1994). But respondents also cited other objectives, such as responding to investor interest, tracking the impact of government regulations, orienting waste minimisation efforts, reporting to the public and better understanding costs. This multiplicity of uses bolsters the case study findings.

Firms seeking a keener appreciation of their environmental costs may find some managers resistant. In part, their scepticism reflects a mistaken belief that environmental costs are already well known. At the outset of the case studies, a number of individuals indicated that they already knew where most environmental costs originate. But, more often than not, they were seeing only part of a much larger, more complicated picture. If the focus on sources and magnitudes of costs is limited, opportunities for improving environmental and economic performance will be missed.

Viewing environmental expenditures as an inevitable cost of doing business also impedes better management. Obviously, any company engaged in environmentally risky activities must take precautionary measures, and these actions will carry costs. But to assume that these are inherent and inescapable forestalls the search for cleaner materials, technologies and practices—a self-fulfilling prophecy and a self-limiting business outlook.

## Conducting a Pilot Project

This section offers some direction for managers interested in applying environmental accounting to their own operations.

*Step 1: Defining Boundaries.* In developing an environmental accounting pilot project, how broadly should the scope be defined? Is it practical to concentrate on an entire division? On a few product lines? On a single facility? The answer depends largely on the initial objectives. Some case study firms set out with clear objectives articulated at the corporate level for implementation throughout the organisation. Du Pont's corporate-wide objective of cost-effective waste reduction, for instance, cuts across its many business units and sites.

In other cases, external factors were the driving force for environmental accounting. The case studies of the Washington State firms demonstrate how mandatory pollution prevention planning can spur companies to re-evaluate environmental costs. In contrast, firms approaching environmental accounting voluntarily can start with a small number of processes, facilities or product lines. Consider Amoco's focus on a single refinery or the product-oriented approach represented by the Ciba-Geigy, Dow and S.C. Johnson cases. The key here is making the pilot project somehow representative of other company activities, so that the findings can be extended to the company as a whole.

***Step 2: Anticipating Resources.*** How much time and effort will the pilot project take? No company should expect to unravel its environmental costs all at once. As the case studies show, it takes time to sift through accounting data and to relate them to environmental and other business issues. A certain amount of detective work is required. So is time to understand how information flows through the firm, how this pattern influences behaviour, and whether and how to implement change.

The resources necessary for the pilot project depend on the scope of the analysis and the complexity of the business. Realistically, a team of three to five people working part-time over three to four months can identify the most important aspects of environmental accounting for review. The team should be free to devote attention at regular intervals. For more extensive reviews in large and complex organisations, the practice of uncovering costs can be integrated into ongoing business processes. For example, over the last four years, Du Pont has incorporated a systematic search for costs into corporate environmental planning. There is no substitute for learning as you go.

As in any new initiative, senior management support is essential before embarking on a pilot project. Winning this commitment requires being clear about what can be accomplished and what is expected of participants. From a practical perspective, high-level commitment signals that the undertaking is worth the time and energy of others with valuable information and insights. This initial backing also assures that the project's recommendations will receive serious consideration.

***Step 3: Selecting Participants.*** Who should be involved in the process? Finding the right mix of talents and experience requires thinking not only about the pilot, but also about the potential for sharing the findings within the organisation. Even though the outcome of the pilot project cannot be foreseen, involving a cross-section of business functions and responsibilities will strengthen the project and lend its conclusions greater credibility.

No single person, department or level in the company commands all the information and knowledge needed to analyse environmental costs in detail. Indeed, a variety of individuals and departments must be engaged, preferably as a multidisciplinary team. This team could include members of senior management with profit and loss responsibility, corporate risk managers, managerial accountants, product managers, operating engineers, designers, environmental specialists and others.

Involving individuals from outside the project's immediate focus can also be valuable. The case study companies worked in partnership with WRI and the academic investigators. In most cases, a senior manager responsible for environment, finance or corporate planning catalysed the company's efforts. But much of the information-gathering was carried out by managers, accountants, engineers and operators on-site at plants. The internal search for environmental costs can benefit from a fresh perspective on routine activities that those most deeply involved may not be able to recognise.

Critical to success is a dialogue across traditional boundaries. Sharing data and cross-fertilising expertise are valuable in their own right. In addition, the pilot project affords the chance jointly to explore the environmental dimensions of production, R&D, marketing and other compartmentalised functions. When Eldec began preparing its pollution prevention plan, the accounting department was cautious about participating. Later, the company acknowledged that interaction between environmental and accounting personnel is an important benefit. Such crossover can be fostered in the context of other ongoing activities. Involving managerial accountants in environmental compliance audits, for example, both acquaints accountants with general environmental issues and broadens the environmental staffs' view of business issues.

*Step 4: Gathering Information.* Once the scope of the analysis is defined and the team assembled, the next step is to begin collecting information. Where should the project team begin? Considering the major environmental issues facing the product, process or site is one way to start searching for environmental costs. For example, in both the Ciba-Geigy and Dow case studies, specific chemicals posed potentially significant environmental and occupational concerns, so a large portion of total environmental costs was driven by the handling, reclamation, recycling or treatment of these materials. In the pilot project at Amoco's Yorktown refinery, interviewing managers about their perceptions of current and prospective environmental challenges helped steer the team to costs with environmental relevance.

The search for environmental costs is not just a paper exercise. Environmental accounting requires an understanding of the overall business landscape, the company's core activities and capabilities, and the nature of environmental challenges. While this obviously involves an analysis of general ledgers, the accounting system is only one of many sources of information. Some others are listed in Figure 5. Still more are sure to come to light during the pilot study, supplementing standard information in the accounting system and helping the team estimate costs that are not currently recorded.

This process of gathering information is inherently open-ended and iterative, as the decisions on waste coking at the Yorktown refinery show (see Case Study 1). In that case, the company's understanding of costs evolved gradually over time. As Figure 6

| Environmental costs | Information sources |
| --- | --- |
| Permitting fees and fines | Regulatory documents<br>Management estimates |
| Maintaining environmental equipment | Maintenance logs<br>Service contracts |
| Non-product output | Emissions estimates<br>Production logs |
| Process penalties/shutdowns | Operating records |
| Depreciation | Capital asset ledger |
| Monitoring | Engineering estimates<br>Management estimates |
| Environmental auditing | Management estimates |
| Training | Personnel, EHS records<br>Management estimates |

**Figure 5:** Sample Sources for Environmental Cost Information

| Costs | 1987 Decision to coke waste | 1991 Coker upgrade | 1993 Coker upgrade |
|---|---|---|---|
| Capital | $ | $ | $ |
| Operating | $ | $ | $ |
| Coker outages | – | $ | $ |
| Maintenance | ✓ | ✓ | $ |
| Process penalty | ✓ | ✓ | $ |
| Permitting | ✓ | ✓ | ✓ |
| Product certification | ✓ | ✓ | ✓ |
| Compliance, record-keeping | – | ✓ | ✓ |
| Public, government relations | – | – | – |
| Fines, penalties | – | – | – |
| Future liability | – | – | – |

Key:
$  Costs considered and quantified
✓  Cost considered but not quantified
–  Costs judged insignificant or not recognised

**Figure 6:** Evolution of Cost Considerations in the Coking of Waste

shows, without a deliberate search it literally took years to appreciate all the cost ramifications, even without fully quantifying them. Whether inadvertent or intentional, this process of discovery is a prerequisite to understanding and managing costs.

As the pilot project progresses, the team will discover costs that are motivated as much by 'environmental improvement' as by 'yield enhancement' or other descriptors of improved productivity. A closed-loop solvent recycling system, while promising major environmental benefits, might be justified purely on the basis of reductions in solvent costs. How then is the company to score the costs associated with the environmental improvement? The adoption of more integrated, process-oriented investments for environmental protection makes answering this question more difficult. As the project team makes judgements on these issues, it must keep broader objectives in sight. Certainly, the case studies show that such difficulties do not pose significant obstacles, as long as assumptions are clearly identified and definitions uniformly applied.

Similar questions arise in the case of raw materials costs. Is the entire purchase cost 'environmental'? Not if most raw materials are incorporated in products rather than discharged as waste. At one extreme, roughly 99% of the crude oil entering a refinery leaves as product. This 1% differential between inputs and products can still cost the company a great deal, but the entire cost of crude oil would be a gross overstatement of the environmental cost. In contrast, in some very low-yield manufacturing operations with large volumes of waste—in the Dow case, roughly five pounds for each pound of product—the value of materials destined to become waste might be a meaningful component of environmental costs.

Clearly, companies do not need to think about environmental costs to understand the basic importance of productivity. Simply defining such non-product output as an 'environmental' cost will not change the basic drive to squeeze more product from a given amount of inputs. But recognising these dual benefits can tip the balance in business decisions. This is the essence of total cost assessment, a technique for incorporating pollution prevention benefits in capital budgeting and investment analysis (White, Becker and Savage 1993). When Heath Tecna, a manufacturer of composite materials for the aerospace industry, applied this technique, it found that it was spending about $85,000 a year to dispose of hazardous wastes derived from materials that had cost nearly $5 million. Recognising the disposal and materials savings helped convince Heath Tecna to invest in sophisticated technology for more efficient pattern layout and cutting.

Of course, even the best environmental cost information available will leave important gaps unfilled. For example, the case studies uncovered no direct evidence of accounting for the future environmental costs of current actions—an important concern of corporate risk managers and legal staff. Liabilities for current and future actions and intangible costs are extremely difficult to predict, let alone quantify. Where monetary estimates are possible, firms fear triggering financial disclosure requirements by coming up with a dollar figure. Even so, managers need to recognise what current activities could cost later. Although firms will find quantification difficult, these costs fall within the private cost domain and are integral to sound decision-making.

Future environmental liabilities also have positive counterparts. Unquantifiable environmental factors, including consumer perceptions of companies and products, can affect the bottom line through increased sales. These revenues could more than balance the readily quantified costs of labour, plant and equipment. Success in preventing pollution, for example, can translate into improved community relations, greater worker satisfaction, lower public opposition to possible expansions, and even faster permit approval—which can all reduce the firm's overall costs. Ignoring such intangible benefits, however hard to estimate, means making less than fully informed decisions.

***Step 5: Results of the Pilot Project.*** As these case studies demonstrate, the specific findings of the pilot project can be as varied as the businesses analysed. Still, the results of the pilot projects are likely to lead in three general directions. First, the scrutiny of environmental costs is likely to produce insights of direct relevance to the project itself. For example, identifying lower-cost waste management alternatives or more effective raw materials can prompt managers to alter practices. Second, the conclusions can highlight the need for change in company-wide environmental, accounting or managerial practices. Discovering systemic biases in the allocation of environmental costs or gaps in accountability may require broad remedies. Finally, if the pilot study's findings are compelling, the company can extend its recommendations to other portions of the business or even to the whole firm. Although simply repeating the pilot project for every facility and product would be unnecessary, expanding the project's scope to cover groups of facilities or a division may uncover opportunities for higher-level improvements in the accounting system.

In any event, to benefit fully from environmental accounting, companies must move from study to action. After reviewing the results of the pilot project, most firms will spot opportunities for improvement. Whether it is changing the flow of information, revising cost allocation schemes or devising new incentives for encouraging more effective management, these changes are all driven by a single objective: integrating environmental considerations through better business decisions and management.

## Managing Environmental Costs

There can be no universal prescription for how best to gather, evaluate and act on environmental cost information. Still, from the exposure to these case study firms, it is possible to distil five core recommendations for businesses:

- Inform decision-makers of the environmental costs they generate.
- Increase accountability of managers for environmental costs and benefits.
- Develop indicators that anticipate future costs and other measures of performance.
- Create incentives to address the causes of current and future costs.
- Incorporate environmental accounting into ongoing business processes.

This final section outlines some practical implications of these overarching business needs.

### Informing Decision-Makers

Information on environmental costs is meaningful only if it influences behaviour. At a minimum, it must be communicated to appropriate decision-makers. Regular summaries of environmental cost information can help key managers identify trends and promote greater awareness of where costs originate within the firm. This is not a question of providing more information, but more useful information.

As this project makes clear, many environmental costs are obscured by conventional accounting systems, so their magnitude and importance are underestimated. If their extent is communicated, the strategic significance of environmental management will be more apparent. Presenting aggregate cost numbers can be an effective way to direct attention to environmental issues. Yet, since costs can be managed only through the activities that create them, and since internal accounting systems typically order cost information by product, new ways of segmenting costs may be required.

One alternative emerged in the S.C. Johnson case study. A number of the company's popular household products can be provided in aqueous (water-based) or organic solvent-based formulations. Consumers prefer the solvent-based formulations, which are faster-acting but no more effective than the aqueous alternative. For either formulation, the manufacturing costs are quite minor in relation to the total product cost. But the organic solvent in the pesticide product investigated contributes to local environmental issues at the manufacturing site, ultimately adding to environmental costs. Even small cost differentials can spell the difference between profit and loss for highly competitive products.

The S.C. Johnson case indicated that product profitability statements that segment costs by aqueous and solvent-based products could be useful. In this hypothetical example depicted in Figure 7, environmental costs, signified by the shaded coins, are essentially equal across product lines A, B and C, where aqueous and solvent-based products are grouped together. But when the products are separated by formulation, it becomes clear that solvent-based products can account for disproportionately greater environmental costs. Such analysis can stimulate consumer-education efforts on the efficacy and environmental merits of water-based products.

There are many other formats for examining environmental costs. One could compare environmental costs across manufacturing operations both within and between firms, as alluded to earlier. Such analyses readily lend themselves to benchmarking and identifying best-of-class performance. Companies may discover other dimensions along which to disaggregate existing data.

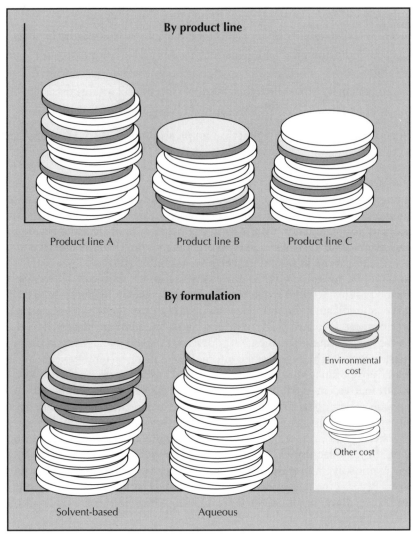

**Figure 7:** Hypothetical Environmental Costs by Product and Formulation

## Making Managers Accountable

While information on environmental costs is a powerful tool for decision-makers, it does not assure accountability. Assigning costs to the managers who control them is the objective of most firms. Unfortunately, bundled overheads that are opaque to managers and insensitive to their actions make doing so difficult. So why try? Because the company's objective is not just to cut expenses, but also to manage the activities that create them, and cost information provides a mechanism to connect responsible parties with environmental impacts.

Some managers may suspect that tracing costs to responsible products and processes more accurately will work to their disadvantage. No managers welcome significant increases in the costs they are paid to control, especially if the renegotiation of allocation schemes directly affects their performance and their pay. Indeed, any change in accounting practices creates internal winners and losers. But the conspicuous complaints of some must simply be weighed against the overall benefits to the firm of more rigorously assigning costs. Addressing transitional difficulties, including the timing of change and the reward systems for making good use of new information, should be a part of any recommended changes in the accounting system.

Many firms levy internal charges for use of on-site environmental facilities (e.g. for waste-water treatment, incineration, disposal) to bring the 'polluter pays' principle inside the plant gate. Besides charging direct operating costs to specific processes or product lines, firms can charge on a 'cost-plus' basis to lower waste generation even further. Dow, for example, has considered increasing internal rates above the estimated operating and maintenance costs to match off-site commercial rates (Kirschner 1994).

The practice of charging full costs or surcharges for waste management can have short-term consequences. For example, if on-site or off-site treatment costs were raised, managers might make changes in materials or operations that create higher out-of-pocket costs. In the extreme, a firm might discontinue products that are still profitable, but which appear too costly. But, despite this potential pitfall, such charges can also signal companies to move toward cleaner and greener processes and products over the long run.

## Managerial Incentives and Environmental Performance

Companies routinely evaluate managers' ability to meet specific objectives within budget. If predetermined business goals are exceeded, annual bonuses and other rewards may follow. While this approach might be applied to environmental cost targets, inappropriate measures can encourage self-defeating or myopic behaviour. For example, a heavy emphasis on reducing aggregate environmental costs could reward managers who cut back on environmentally related maintenance or R&D—activities essential to long-term environmental management.

An alternative is to connect incentives to the underlying activities that give rise to environmental costs. Such programmes have been used for decades to drive progress on targets for productivity, safety and quality. Amoco's gain-sharing plan, for instance, rewards employees for various measures of facility performance. The Yorktown refinery's waste generation metric for the gain-sharing programme excludes remediation costs associated with past contamination so that the current workforce is not penalised for decisions over which they had no control.

Including environmental metrics when managers are evaluated can rivet their attention and provide an incentive for reducing costs. Incentives schemes must also reflect the company's technological, economic and organisational realities. Recognising the inherently integrated character of petroleum-refining, Amoco based its waste metric on the facility's overall performance. Tying the metric to waste generation at the process level instead might have invited suboptimal behaviour. Specifically, penalising the tank-farm managers for the solid residues that settle out during storage would reward them for passing these impurities on to downstream process units. True, waste generation by the tank-farm would appear to fall, but downstream operating and maintenance problems and overall costs could significantly increase.

## Identifying Indicators for Environmental Costs

A common theme that permeates the cases is that future environmental costs, though inherently uncertain, are inevitable. Laws change; regulatory requirements evolve and expand. Actions that are perfectly legal today can create financial liabilities tomorrow. The challenge for managers is to avoid incurring future costs as they meet present demands.

One thing is clear: simply tracking historical costs is not good enough—it is blind to future changes in the rules of the game. Supplementing cost accounting with other indicators of environmental performance can help identify potential vulnerabilities before they become major cost factors. At Dow, a key solvent used in the manufacture of a polymer-based coating material represented only 1% of manufacturing costs, but release of this volatile organic air pollutant threatened to force the company to abandon an established product line. While the cost appeared modest, a changing regulatory climate and corporate commitments made use of this material increasingly costly. Attention was drawn to this challenge by tracking quantities of material rather than dollars.

Materials accounting—the systematic assessment of inputs, use and release—for particular environmentally significant chemicals offers a proxy for potential future environmental costs. Reducing the release of chemicals into the environment may reduce future liabilities (all other things being equal). That said, since the environmental impacts of chemicals vary widely, lowering total emissions does not necessarily reduce environmental consequences or liability exposure. Weighing the quantitative releases on the basis of existing scientific or regulatory estimates is a more meaningful way of comparing potential future impacts. As one example, a kind of toxicity-weighted pollution unit can be calculated using available environmental standards and toxicological information (Grimstead *et al.* 1994). However imprecise, such approaches enable managers to develop cost-effective ways of reducing environmental risks. An added advantage is that this method does not require monetary estimates of future liabilities, thus escaping some significant legal issues (Environmental Law Institute 1993). There is a great need for more work on practical indicators of environmental performance—to improve internal management as well as to report progress to outside stakeholders.

## Integrating Environmental Accounting into Business Processes

Identifying, understanding and managing environmental costs can be time-consuming and labour-intensive (Walley and Whitehead 1994). Yet this task is too important to be relegated to a one-time exercise. Rather than introducing another layer of internal reporting requirements or creating new accounting responsibilities for environmental units, firms should take advantage of their existing business processes.

Strategic planning offers a particularly promising opportunity for bringing environmental issues to bear on key company decisions. Corporate-wide environmental planning efforts, such as Du Pont's Corporate Environmental Plan, won't work without a realistic picture of the sources and magnitudes of environmental costs. Many other routine activities, such as capital budgeting and new product development, can be strengthened by explicitly incorporating environmental cost information. Firms undergoing major re-organisation or 're-engineering' can introduce some elements of environmental accounting. Both the internal auditing of management systems and environmental auditing can help firms gather better environmental cost information and field better practices.

Incorporating environmental accounting into these processes offers two practical advantages over stand-alone environmental accounting efforts. For one, expertise and information essential to environmental accounting can be tapped or acquired at little additional cost. More important, linking environmental accounting to these other business processes helps infuse environmental thinking into core decision-making.

As a business tool, environmental accounting can soften the economic impact of environmental compliance. It can help management better understand what gives rise to costs and how they can be managed. It can also help managers assess the business and environmental implications of different alternatives. Environmental accounting can track crucial indicators over time, using information that is routinely recorded but rarely exploited, thus enhancing a firm's self-knowledge and its environmental accountability.

As firms come to terms with current environmental costs, they will appreciate that the boundary between private and social costs is porous and moving. Other environmental costs, now borne by society, will exert a growing influence over the decisions made within companies. The emerging shift toward greater self-regulation, market-based mechanisms and green consumerism will inevitably drive companies to further internalise environmental externalities. Firms must begin to anticipate these broader environmental consequences and bring them to bear as they design new products, develop new technology and make investments for the future.

## ▮ Case Study 1: Amoco Yorktown Refinery

*Miriam Heller, David Shields and Beth Beloff*

### Highlights

Amoco Oil Company's Yorktown refinery, like the refining industry as a whole, faces significant environmental costs, both as a source of pollution and as producer of petroleum products. This case study underscores the importance of identifying and tracking environmental costs to understand better how much is being spent and why.

Accounting for environmental costs in a highly integrated refinery is especially complex. Early on, the investigators explored the possibility of tracing environmental costs to process units. But it became clear that assigning costs at this level could lead to decisions with adverse economic and environmental consequences for both the refinery and the company because the effects of process unit decisions are felt both upstream and downstream. For example, the decision to treat waste-water treatment sludge in a coker, while reducing off-site disposal costs, reduced yields of intermediate products and resulted in costly unanticipated shutdowns for cleaning and maintenance.

This study was valuable in revealing just how significant environmental costs are at Yorktown. The initial informal estimate of annual environmental costs was 3% of non-crude operating costs. Yet, after only six months, they were found to be approximately 22% of non-crude operating costs. And even this understates the total, which would also include estimates of future environmental liabilities.

This study has also provided insights into where environmental costs arise. For example, maintenance costs are estimated to be over 15% of total environmental costs. This percentage is much higher than the cost of waste-water treatment, which had originally been considered one of the most significant environmental costs at the refinery. This case study also considered the relevance of environmental cost information in complex product mix decisions.

## ▌ Case Study 2: Ciba-Geigy

*Ajay Maindiratta and Rebecca Todd*

### Highlights

This case study looks at an industrial chemical additive used to increase the shelf-life and stability of a wide range of products. It is produced by Ciba-Geigy, a Swiss-based diversified company which is a major producer of pharmaceuticals, speciality chemicals and agricultural products. The manufacturing process requires the use of several compounds that have significant environmental, health or safety implications. One, a highly reactive substance, is a key building block in production of the intermediate; two solvents generate VOC emissions and contribute to waste-water effluents.

Ciba-Geigy has a single, fully integrated general ledger system in use at all sites throughout the world. The company's accounting system is quite comprehensive in recording initial data. From the standpoint of environmental costs, most traditional cost elements that can be quantified are currently recorded in some part of the accounting system. The authors conclude that a number of environmental costs are not variable, but fixed. A further important distinction is between out-of-pocket and historical costs; only the former can be reduced. However, for planning and control purposes, particularly in the case of capital investment, historical cost items may signal important trends.

Many important environmental cost items cannot currently be determined from the accounts, since the system does not explicitly separate them. To support decision-making and direct attention, the authors suggest an 'environmental profiling' of each product be done, with not only traditionally identified cost elements, such as waste-water treatment and solvent-recovery charges, but also such less traditional costs as public relations for products with hazardous materials or effluent.

The case study also identified environmental costs that could be expected to vary by manufacturing site in different countries. The focus in any comparisons must be on 'escapability' from the corporation's point of view, rather than simply the individual product. For example, if a certain level of cost will continue whether or not the product is manufactured, then the cost is not relevant for international manufacturing cost comparisons.

## ▌ Case Study 3: Dow Chemical

*Ajay Maindiratta and Rebecca Todd*

### Highlights

This case study looks at a polymer-based coating material made in a number of grades by Dow Chemical, the second largest US chemical company. In manufacturing this product, significant quantities of two VOCs are released. Corporate commitments to reduce wastes, together with pending regulations under the Clean Air Act, require major changes to curtail emissions if the plant is to remain open.

Dow Chemical was interested in the case study first as a general review of its environmental cost accounting practices. Second, plant and product managers wished to consider how accounting practices related to the pending decision to upgrade or abandon this unit. Third, management wanted to review the current costing practice with a view to product portfolio and pricing decisions.

The company has a single, fully integrated general ledger system designed to provide a traditional 'full product costing' recharge of the costs of manufacturing services, facilities and corporate overhead to manufacturing plants and divisions. The system is relatively comprehensive in recording initial data. From the standpoint of environmental costs, most traditional costs that can be quantified are currently recorded in some part of the system.

In analysing the standard cost sheets for manufacturing the high-grade product, the authors found that 3.2% of the total cost fell in the conventional 'environmental' category: waste disposal and treatment. The environment-related components of such items as operating labour, managers' salaries and planning efforts are not distinguished as environmental. Nor are the environmentally relevant portions of capital investment, represented by depreciation allocations, as provided by the manufacturing cost-sheets.

The authors categorised costs according to their economic nature: variable, fixed, direct, indirect, out-of-pocket and historical. Those with potential environmental significance are predominantly fixed. Only out-of-pocket charges can be escaped (that is, controlled), and only variable costs can be controlled in the short run. If the plant were closed, the only costs the company as a whole would escape are the out-of-pocket direct and indirect (recharged) variable costs.

## ▌ Case Study 4: Du Pont

*David Shields, Miriam Heller, Devaun Kite and Beth Beloff*

### Highlights

This case study of E.I. Du Pont de Nemours and Company, the largest chemical producer in the United States, focuses on an agricultural pesticide manufactured at the LaPorte, Texas, facility. The company's current efforts to develop an environmental cost accounting system at LaPorte are presented, followed by an estimate of the environmental costs of producing this pesticide. For a company that spends more than $1 billion a year on environmental protection, the importance of understanding environmental costs better cannot be overstated.

Baseline determination of environmental costs at LaPorte involves a two-stage process. First, all costs labelled as environmental are isolated. Many of these obvious costs relate to waste management involving incinerators, bio-treatment, steam strippers, deep wells and off-site waste handling. Second, environmental costs hidden within other costs are identified. One example is the time spent by non-environmental management on recurring environmental activities. After environmental costs are identified, they are separated into fixed and variable components.

For the agricultural pesticide, over 19% of manufacturing costs are deemed 'environmental'. Almost one-third of fixed manufacturing costs were determined to be environmental, compared with about 7% of the variable costs. But, the authors explain, some costs classified as fixed have a variable component. For this reason, the volume sensitivity of environmental costs may be understated.

The case describes how a better understanding of the nature of environmental costs can influence waste management decisions at the facility level. When comparing the relative costs associated with waste management options, the company discovered that costs, as given by the accounting system, can be potentially misleading. By focusing on incremental costs, Du Pont was able to realise a real cost saving in the transition from deep-well injection to biological treatment of some process wastes.

This case study also shows how a company, armed with environmental cost accounting information, can make decisions that have a positive economic as well as environmental effect. For example, by comparing the projected reductions in waste against the estimated cost for several hundred corporate waste reduction initiatives, Du Pont was able to evaluate trade-offs between competing initiatives, while taking into account the company-wide leveraging of technological solutions.

## ▌Case Study 5: S.C. Johnson Wax

*Ajay Maindiratta and Rebecca Todd*

### Highlights

This case study looks at the Insect Control business of the North American Consumer Products division of S.C. Johnson Wax, one of the leading providers of chemical speciality products for the home and workplace. Most of the environmental challenges in this business arise in product registration, marketing and post-consumer product management (such as recycling of aerosol cans). Proliferating state regulations on pesticide labelling and use represent a major environmental issue for the company. Regulation affects lead times for product registration, lead times for developing formulations, and the incentive to develop new active ingredients given diseconomies of scale.

Some executives expressed concern that effective cost management may suffer because the demand placed on resources by environmental issues is not being adequately and comprehensively reflected and communicated in the organisation's management accounting systems. SCJ's current system is a full-cost one, and thus every historical cost of doing business appears somewhere in the line items of the product profitability statements.

Yet the environmental cost analysis done for one household pesticide product found that environment-related expenses for marketing and R&D could be a significant fraction of the total operating expenses. Waste-processing and the other costs of the factory's environmental departments are negligible—a mere 0.25% of the manufacturing cost-of-sales. But about 17% of marketing administration can be directly attributed to the environment (registration fees and mill taxes) per existing accounting systems. In addition, about 21% of personnel expenses (including wages and salaries) can be attributed to the environment.

The authors recommend that S.C. Johnson construct statements that segment the firm along lines that are more natural from the perspective of key environmental concerns, as a supplement to the profitability statements now used. They also encourage the company to develop better communication and strategic planning tools and quantify the impact of the scenarios in dollars and cents.

## ▮ Case Study 6: Accounting for Pollution Prevention in Washington State

*Christopher H. Stinson*

### Highlights

This collection of case studies looks at four relatively small firms (annual sales of $12–140 million) in Washington, a state that requires companies to tabulate environmental costs and benefits as they prepare corporate pollution prevention plans. Quantifying the environmental savings associated with pollution prevention options helped companies identify cost-effective projects that might otherwise have been overlooked.

At Heath Tecna, a manufacturer of composite materials for the aerospace industry, the value of lost raw material far exceeds the cost of disposal. Taken together, these costs supported a decision to invest in technology for increased materials efficiency. Management does not charge individual production units for off-site disposal costs, so that budget constraints do not create a short-term incentive for inappropriate waste disposal.

At Cascade Cabinet, a manufacturer of wood cabinets for kitchens and bathrooms, the major environmental issues include solvents, VOCs and wood waste. The company switched from a lacquer to varnish on environmental, safety and cost grounds, and it is converting wood scraps into a saleable product, eliminating disposal costs.

Spectrum Glass manufactures about 30% of the world's stained sheet glass for art and architectural applications. A key environmental issue is the use of cadmium oxide, a highly toxic and hazardous chemical, as a colorant for ruby-red glass. Increasingly restrictive regulations on the manufacture of this chemical could lead to the disappearance of ruby-red glass in the near future.

Eldec Corporation designs and builds electronic equipment for aerospace applications. In preparing the pollution prevention plan, the accounting department was originally concerned that accounting systems might have to be modified to comply with the regulations. However, after preparing the plan, the company felt they benefited from the process. Since nearly half the company's revenues come from military contracts, current Department of Defense accounting standards may discourage revised accounting for environmental costs.

# 10 Environmental Cost Accounting for Chemical and Oil Companies

## A Benchmarking Study[1]

David Shields, Beth Beloff
and Miriam Heller

## ■ Introduction

IN 1994, the Institute for Corporate Environmental Management (ICEM) at the University of Houston and Pilko & Associates initiated the Environmental Cost Accounting Benchmarking Project to develop an understanding of environmental cost accounting through a benchmarking study of corporate practices. The project had the following objectives:

- To collect information on what companies in the chemical and refining industries are doing in accounting for environmental costs and how environmental accounting can provide decision-makers with the information they need
- To develop case studies from each participating company on how they are grappling with the issues of environmental management and cost accounting for environmental activities
- To establish baseline environmental cost information as identified by all participating companies
- To develop a framework for understanding the nature and uses of environmental accounting

---

1. This chapter is an edited version of a longer report of the same title published by the US Environmental Protection Agency in June 1997 (BCSD and University of Houston 1997). The report was prepared by the Institute for Corporate Environmental Management at the University of Houston for the EPA's Environmental Accounting Project. It was also sponsored by the Business Council for Sustainable Development–Gulf of Mexico. The full report can be read on the Project's website at *http://www.epa.gov/opptintr/acctg* or can be obtained free of charge from EPA's Pollution Prevention Information Clearinghouse by telephoning +1 202 260 1023.

At the same time, member companies of the Business Council for Sustainable Development–Gulf of Mexico (BCSD-GM) were developing projects to demonstrate the business opportunities inherent in incorporating sustainable development approaches in business decisions. BCSD-GM recognises the value of developing a tool to assist in identifying economic opportunities in reduction of environmental impacts and encouraged its US and Mexican companies to participate in the Environmental Cost Accounting Benchmarking Project. In addition, several companies outside of the BCSD membership were approached.

Four companies eventually participated in the project fully:

- **CIBA-Geigy**, a Swiss-based producer of human and animal healthcare products, agrochemicals and industrial chemicals with nineteen operating units in the US, Canada and Mexico
- **Grupo Primex**, a Mexican producer of plasticisers and PVC resins and compounds with a single plant employing 540 staff at the time of the study.
- **International Refineries**, a producer, refiner and marketer of oil, gas, petroleum products and chemicals with thirteen operating locations in the US.
- **Specialty Refiners**, a refiner and marketer of motor oils and speciality oils and chemicals with 120 operating locations in the US

A fifth company, Celanese Mexicana, a producer of chemicals, fibres and packaging, also participated in the early stages of the project.

## ▌ The Benchmarking Process

The benchmarking was based on Pilko's Co-operative Benchmarking$_{SM}$ process,[2] which has four parts:

- Defining project objectives from factors and issues identified as important by participants (see ranking in Fig. 1)
- Data collection through questionnaires
- Benchmarking sessions (two of five days' total duration)
- Synthesis

Figure 2 captures the flow of activities characterising the entire effort.

## ▌ Environmental Accounting System Development Framework

The framework shown in Figure 3 illustrates the relationship between different management processes involved in environmental cost accounting. This framework was the basis for the questionnaire developed in conjunction with the Environmental Cost Accounting Benchmarking Project. Each of the elements in the framework was the subject of one or more discussion sessions during the benchmarking sessions.

The framework reflects the inter-relationship between corporate culture (attitude toward environmental stewardship) and the availability of resources for development

---

2. Co-operative Benchmarking$_{SM}$ is a service mark of Pilko & Associates, Inc.

| Factor | Rank |
|---|---|
| Environmental cost monitoring | 1 |
| Full-cost accounting | 2 |
| Environmental accounting information systems | 2 |
| Capital versus operating | 3 |
| Management control | 3 |
| Environmental cost allocations | 3 |
| ECA information for decision support | 3 |
| Budgeting process | 3 |
| Future environmental liabilities | 4 |
| Compliance versus voluntary costs | 5 |
| Reporting systems; accounting systems | 5 |
| Remedial/clean-up costs | 5 |
| Other metrics | 6 |
| Life-cycle cost accounting | 6 |
| EC information for external-focused issues | 7 |
| Environmental cost drivers | 7 |
| Public financial disclosure statements | 7 |
| NAFTA | 7 |

**Figure 1:** Ranking of Factors and Issues

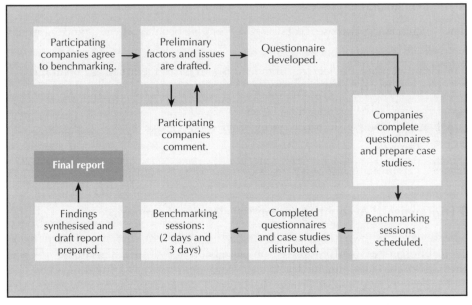

**Figure 2:** Environmental Cost Accounting Co-operative Benchmarking_SM Process

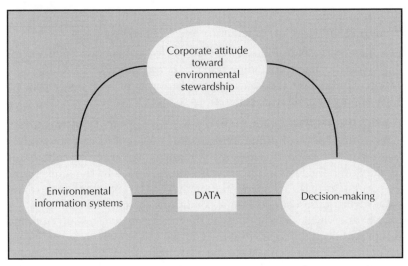

**Figure 3:** Conceptual-Level Framework for Environmental Cost Accounting

of an environmental accounting system. Only companies with a clear desire to integrate environmental decision-making into the normal business context are likely to make such an investment.

The environmental information system will generate data that would otherwise be unavailable to managers. These data may be cost-based or may consist of non-financial measures. Non-financial measures can be used either to generate financial measures with the addition of other data, or they may be used directly, as real-time data.

In either case, the data provide evidence relating to the decision to be made. Ultimately, the creation of better information systems generates better data that support better decision-making. This, in turn, may have a feedback effect on the organisational learning and culture of the company. This culture change brings about conditions that could cause future changes in the organisation's information systems as well as the kinds of decisions that management will make.

## ▮ Attitude toward Environmental Stewardship

During this project it became clear that, for the five participating companies, an organisation's attitude toward environmental stewardship is directly related to the company culture and, in fact, is a subset of the company culture. An example for assessing how these elements function together was offered by one partner company. Ciba-Geigy has published its environmental policy: 'Vision 2000', which describes a balance between the economic, social and environmental responsibilities that will ensure the prosperity of the business beyond 2000.

The Vision 2000 policy reflects the company culture at Ciba-Geigy, defining the nature of the organisation and describing the core values that determine the actions of the company:

- **Economic success.** This is ultimately the measure by which companies and their shareholders evaluate their success.

- **Social responsibility.** Ciba-Geigy accepts responsibility for the effects of its actions on employees, the community and other stakeholders.
- **Environmental responsibility.** Ciba-Geigy strives to maintain a policy of sustainable development, including resource conservation and pollution prevention. The goal is to conduct business in a way that will not impede the ability of present and future generations to meet their needs.

This attitude toward environmental stewardship helps to define the organisation through its culture because the culture represents the set of values, norms and procedures that serve as the foundation for any environmental project. The culture is affected by the dominant individuals in the organisation, as well as by the industry and, perhaps, the nationality of the company. Nationality can be important due to differences in laws and regulations, socio-economic conditions and cultural traditions.

The five partners made presentations on the topic of attitude toward environmental stewardship. Although the topics discussed during this session varied with each partner, most of the discussion centred around a few main areas. These areas include: the environmental organisation and associated roles and responsibilities; environmental management systems, policies and procedures; environmental standards, compliance efforts and voluntary programmes; and organisational changes.

### Structure of the Environmental Organisations

All five partners have some form of environmental organisation. The roles and responsibilities of these environmental units vary among the partners. Some companies organise their environmental group in a decentralised manner. In these cases, significant resources are placed at operating facilities while the corporate group provides an oversight and limited support role.

Other partners allocate a greater share of resources to a centralised organisation. The centralised environmental group provides proportionately more top-down support to the facilities. The amount and allocation of environmental resources provide an initial indicator of the company's attitude toward environmental stewardship. However, differences in organisational demands for environmental services may also explain the different organisational structures and resource allocations.

### US/Mexico Environmental Climate

Each partner has developed environmental management systems, policies and procedures that reflect their national cultures, as well as their corporate cultures. All partners, whether Mexican or US-based, are concerned with compliance with applicable environmental rules and regulations.

The regulatory requirements in the US and Mexico are generally similar, but there are also a number of differences. Many of the regulations created in the US are subsequently issued in Mexico after a time-lag of several years. Consequently, it may be easier for Mexican companies than for US companies to predict accurately what future regulations will be.

The Mexican companies in this study claim that they face a more regulated environment than US companies in the use of water. Water usage is taxed at a higher rate than in the US, and is taxed twice: once when it is extracted from groundwater wells, and a second time either as the cost of treating the water to standards or the costs of re-injecting water to water quality standards. This double taxation results in very conservative water usage policies by Mexican partners.

The extent to which the partners participate in voluntary environmental programmes varies with the nationality of the company. The US companies tend to have more voluntary programme participation than do the Mexican companies. However, there are fewer voluntary programmes in Mexico than in the US. The direction of causation is not clear: do voluntary programmes generate interest among companies, or do companies provide the impetus for voluntary programmes?

All the partners have undergone or are in the process of undergoing significant environmental organisational changes. This phenomenon is not limited to the current wave of 're-engineering' that is occurring in many US industries. Additional competitive pressures, brought on by dynamics such as the globalisation of industry and political policy changes such as NAFTA, have caused the Mexican partners to undergo significant organisational changes as well. (Partner companies ranked NAFTA-driven environmental changes as relatively uninteresting during the planning session.)

The following sketches out relevant aspects of the corporate culture of the participating companies:

## Ciba-Geigy

Within Ciba-Geigy, compliance is the responsibility of the line organisation. If the line organisation requests assistance in determining compliance levels, or for related support, several groups are available to provide it. These include Environmental Affairs, Regulatory Affairs, Legal Department and Regulatory Networks. If local standards are below Ciba-Geigy's internal standards, they will comply with the higher standards.

Ciba-Geigy participates in a number of voluntary programmes including ICC Charter for Sustainable Development, Responsible Care, ISO 14000 (site by site), EPA 33/50, Environmental Leadership and Green Lights. In general, all employees are responsible for both regulatory compliance and company internal standards. Environmental performance goals are part of the organisation's objectives. Consequently, there is a corresponding impact on compensation. Adherence to these environmental regulations and standards is everyone's responsibility day in and day out.

## Grupo Primex

Within Grupo Primex, compliance issues are a top priority. Compliance with effluent discharge limitations is currently under evaluation. The company plans to start up a waste-water treatment system to address chemical oxygen demand, suspended solids and oil and grease exceedences.

In addition to its membership in the BCSD–Gulf of Mexico, Grupo Primex participates in the Responsibilidad Integral: El Compromiso de la Industria Quimica (Integral Responsibility: The Commitment of the Chemical Industry). This voluntary programme is analogous to the Responsible Care programme in the US. The programme is aggressive, as companies have agreed to full implementation in five years.

Grupo Primex re-organised and downsized during 1993. The company closed a plant and is now conducting all operations at a single complex. Upon completion of the re-organisation, Grupo Primex began a programme of paying employees who submit ideas that improve environmental performance, but only after the idea is implemented.

## International Refineries

Within International Refineries, the operating unit is responsible for conformance with the environmental policy. International Refineries employs a decentralised staff, placing resources at operating locations. A lean and flat organisation provides preventative support to the operating organisation for ongoing efforts.

International Refineries' environmental policy is signed by the CEO and was last updated in 1990. The management systems emphasise preventative programmes, and senior management supports efforts by the corporate environmental group to improve performance. As demonstrated in the case studies included in this report, International Refineries management sees value in better understanding its environmental costs.

International Refineries participates in a number of industry and government voluntary programmes. These include Responsible Care, American Petroleum Institute (API)'s STEP, EPA 33/50, Green Lights, Natural Gas Star and Energy Star. Incentives have been created for improvements in environmental performance. Individual performance review has an environmental element, and senior management performance review also includes an informal environmental element.

International Refineries has significantly changed its organisation and asset structure from 1990 to 1995. Assets in the exploration and production and retail marketing areas have been rationalised. The upstream operations have been consolidated and new business units were created for the downstream and chemical operations. These changes have led to an approximate 30% decrease in headcount. During this period, the Health, Safety and Environmental (HSE) group has increased headcount and created a stable structure.

## Specialty Refiners

Specialty Refiners conducts a compliance auditing programme to evaluate the effectiveness of compliance efforts at the company's facilities. The company also requires annual assurance letters from facility management that certify compliance with regulations. It also participates in a number of voluntary programmes including: EPA 33/50, API STEP, Green Lights, Waste Wise and ISO 9000.

Specialty Refiners has been a leader in developing a used-oil stewardship programme in support of its car service business. Major efforts under the auspices of the API included the formation of a used-oil coalition which consisted of 12–15 oil companies. Another effort was in the area of developing model laws and regulation for the used oil, mainly to prevent it from becoming a hazardous waste.

Specialty Refiners re-organised during 1995, effectively moving corporate resources out to operating facilities. The corporate group is currently responsible for regulatory matters and environmental policy. There are also two shared services groups: remediation and occupational health, both located within the marketing division. All other functions are within the operating divisions or at the facility. There are approximately 100 full-time equivalents involved in the Environmental, Safety and Health (ESH) group.

## ▌ Environmental Cost Accounting System Development

In the sessions, each of the partner companies discussed their current system for generating environmental cost information. A variety of methods was used. Some systems were wholly integrated, or layered, onto the general ledger. Other systems were in the fledgling stage, existing as ad hoc systems that access information from other established systems to fulfil decision-making needs as they arise. Each system is described more fully in the following section.

## Ciba-Geigy

Ciba-Geigy has identified two uses for environmental cost accounting. The first use is to assign the proper current cost to the correct product, for profit determination and inventory valuation. The second use is to provide management with support for controlling and managing costs. The company recognises the impact of misallocation of overhead costs, which often include environmental costs, to product. The system implemented by Ciba-Geigy addresses most issues regarding environmental costs by instituting an improved cost accounting system that accurately associates each cost with each product.

Ciba-Geigy generates environmental cost information as part of the general ledger system. Each product has associated with it a detailed product cost sheet which defines all cost elements associated with the product. Given for each cost element is the unit of measure, consumption per 100 pounds of product, unit price and cost per 100 pounds of product (see Fig. 4 for a fictitious cost sheet). The cost elements include raw material, direct labour, equipment, repair and maintenance, electricity, steam, analysis and quality control, waste-water, incinerated wastes broken down into solid, organic and aqueous types, general facilities, services and overhead (general FS&O), and unit overhead.

| Cost element | Unit of measurement | Consumption per 100 lbs | Unit price | Cost per 100 lbs |
|---|---|---|---|---|
| Intermediate material A | lb | 75.3604 | $6.0702 | $457.45 |
| Raw material B | lb | 125.2400 | $1.5239 | $190.85 |
| Raw material C | lb | 92.8863 | $1.7625 | $163.71 |
| Direct labour | hr | 1.3100 | $26.3200 | $34.48 |
| Equipment X | hr | 0.4400 | $95.3818 | $41.97 |
| Equipment Y | hr | 0.1364 | $170.6994 | $23.28 |
| Equipment Z | hr | 0.4000 | $47.9629 | $19.19 |
| Repair/maintenance equipment X | hr | 0.4400 | $45.2140 | $19.89 |
| Repair/maintenance equipment Y | hr | 0.1364 | $50.3554 | $6.87 |
| Repair/maintenance equipment Z | hr | 0.4000 | $16.9008 | $6.76 |
| Electricity | kWh | 104.9500 | $0.0467 | $4.90 |
| Steam | ml | 4.8000 | $4.2672 | $20.48 |
| Auditing and quality control | hr | 0.1400 | $53.3962 | $7.48 |
| Effluent: flow | million gallons | 6.0000 | $0.3339 | $2.00 |
| Effluent: TOC | lb | 35.2182 | $0.4519 | $15.92 |
| Incineration: solids | lb | 23.8282 | $0.4519 | $10.77 |
| Incineration: organic liquids | lb | 85.2605 | $0.4108 | $35.03 |
| Incineration: aqueous | lb | 65.4289 | $0.7204 | $47.13 |
| General facilities, services and overhead | | | | $21.02 |
| Unit overhead | | | | $42.33 |
| **Total** | | | | **$1,171.50** |

**Figure 4:** Environmental Cost Accounting: Cost Sheet for Product A

*Source: Adapted from Ciba-Geigy 1996*

While the cost sheets enable more accurate product costing, controlling and managing environmental costs are dealt with using transfer pricing mechanisms. Ciba-Geigy has considered several methods for determining the appropriate rate for the transfer price. One method, using the ratio of the total cost to total capacity, is easy to implement, provides a predetermined rate for the budget cycle, and it adequately reflects a lower cost for lower usage. However, since it is not normalised to reflect actual usage rather than total capacity, there may be unallocated costs.

Another method, which defines the rate as the ratio of total cost to actual usage, shows the real cost of services but results in a rate that varies during the budget cycle. This method may not provide incentive to minimise wastes, since the environmental cost will depend on the interactive effects of all users of the service. For example, if all users reduce their use of the service by one-half, none will realise a reduced rate for usage, as the corresponding unit rate will double.

Moreover, Ciba-Geigy recognises limitations of purely financial information. The company has developed another management tool for computerised reporting known as SEEP (Safety, Energy and Environmental Protection). The system requires all major production sites to input to SEEP data on safety, resource use and environmental releases, excluding those from the raw materials production and consumer use stages of the product life-cycle. Cost data are also manually entered by site. Aggregate and analytic forms of these data form the foundation of Ciba's corporate environmental report. Also, these data fit directly into a limited life-cycle assessment methodology currently under development to improve the cost-effectiveness of environmental decision.

**System Users and System Maintenance.** The system is used by line employees to manage ongoing environmental costs (as with other costs). Cost information is available throughout the organisation and each operating unit knows product costs. Corporate Environmental Affairs is responsible for remediation costs, but these are not handled by the cost accounting system. They are dealt with in a fashion similar to the way capital investment projects are handled. The maintenance of the systems rests with corporate, division and plant accounting functions.

**System Strengths and Weaknesses.** The system clearly defines responsibilities and accountabilities. The objective is to try to account for environmental costs at the location and level where individuals are responsible for them and can make decisions to manage them.

A primary weakness of the system is that environmental costs are part of many cost centres. This leads to environmental costs being general and summary in nature. While charge-back mechanisms should facilitate improvement, it is difficult explicitly to define environmental costs and accumulate them into meaningful categories. For instance, aggregating environmental cost information over multiple cost centres, as is required to be consistent with annual SEEP reporting needs, is very difficult.

### Grupo Primex

In 1996, Grupo Primex began environmental cost monitoring. Goals for the environmental accounting system were established. First, the environmental cost accounting system was to be a subsystem from its existing management information system. Second, the system needed to assist the company in determining the impact of an environmental cost on net profit and conversion (direct labour and overhead) cost. Finally, the system would have to enable performance measurement, at all levels in the company, based on ecological management ratios. The ecological management ratio is an index that measures pounds of waste produced per pounds of product, i.e. it is a waste-focused yield measurement. The index is not yet widely used in the plants.

Grupo Primex focuses on environmental costs in certain income statement categories: variable costs, maintenance and fixed expenses. Variable costs may fall into a general regulatory category, which includes environmental audits, risk assessment, penalties, remediation or research and development. Other variable costs are tracked by media. Variable costs associated with water may relate to duty payments for environmental permit exceedences not associated with audits, compliance fees for particular discharge conditions, utilities, electricity, water analysis, and labour costs. Water has double costs, since there is a fee to extract it from the ground as well as to dispose of it. Variable costs for air include monitoring and duty payments (fees and taxes). Currently, Mexico has nothing comparable to the Clean Air Act Amendments of 1990. Variable costs for hazardous materials may be associated with analysis, duty payments, treatment and disposal, transportation, and labour cost. Finally, costs for soil analyses and duty payments are variable.

Measurement instrumentation enables Grupo Primex to track environmental costs at the unit level. For instance, the costs of the waste-water treatment plant are allocated to each product according to loading and volume.

Mexican companies report usage and emissions to the government to determine duty payments. Payment is made when emissions are reported. Water is reported quarterly; air emissions are reported once per year. Soil is reported periodically. Funds are deposited in an escrow account, based on predicted duty payments. To facilitate reporting, Grupo Primex's general ledger tracks metric tonnes of production with costs. The financial and environmental management systems are reviewed on a monthly basis.

Maintenance costs include all expenses incurred as a result of preventative and corrective maintenance for ecological assets, e.g. the waste-water treatment plant. Fixed expenses are integrated by salaries and wages paid to ecology department personnel.

**System Strengths and Weaknesses.** Grupo Primex has begun separating environmental costs from operating costs. The system is free-standing and does not directly access data from any other system. However, the company has made strides in procuring measurement equipment that enables real-time data collection. Environmental costs can be related to financial information using the management information system.

One principal problem with the environmental accounting system is implementation. People resist what they perceive to be forced compliance or behavioural change. The company is trying to provide incentives through monetary recognition of individual initiatives once implemented.

Another issue to be resolved is the company's operating definition of environmental costs. This definition process is not complete, so not all potentially interesting information is being collected. For example, the current system does not attempt to track less tangible costs.

Depreciation costs of environmental equipment and production assets are commingled, i.e. environmentally related depreciation is not separated from depreciation of non-environmental assets. This distinction is more important in Mexico than in the US, because, in Mexico, if a capital investment is for environmental assets, it can be fully depreciated in the first year, resulting in significant tax savings.

### International Refineries

Four overarching goals led International Refineries to the decision to extend their accounting system to incorporate environmental costs. First, the company would like to improve reporting capabilities for governmental, public and industry concerns. Second, the system would provide additional detail to management about Health, Safety and Environment (HSE) expenditures. Third, such a system would provide a standardised approach to capturing HSE data through divisions. Finally, International Refineries opted to 'enhance' the system rather than risk disruption and change through the construction of an entirely new system.

Specific requirements for the system were internally generated. The system would be simple, but would have the capability (1) to track and define all expenditures related to health, safety or environmental projects by category, media, location and line of business; (2) to track penalties and fines imposed by regulatory agencies; and (3) to spot errors. System development entailed three people working for seven months: two accountants and one environmental specialist.

The enhanced Health, Safety and Environmental (HSE) Accounting System builds on the existing accounting system by defining HSE product codes. These HSE product codes are teamed (via prefix) with existing accounting, location and expense codes. There are eight categories of HSE product code: safety, air, water, solid and hazardous waste, remediation, spill clean-up, medical services and other (e.g. maintenance). Other categories correspond to those defined by the US Census Bureau and the American Petroleum Institute (API) in their annual surveys.

HSE product codes for safety include compliance fines, monitoring, asbestos, safety equipment and safety supplies. HSE product codes for air include compliance fines, emissions testing and stack sampling. Solid waste product codes include compliance fines, hazardous waste, non-hazardous waste, and waste testing and analysis. Remediation codes include soil remediation, compliance fines, testing and analysis and site assessment. A last category (other) includes miscellaneous environmental expenses, e.g. bird cones for the stacks. About thirty expense codes are explicitly associated with HSE product codes. Examples of environmentally connected expense codes include environmental professional fees, public awareness programmes, external analytical laboratory fees, and waste transportation.

Field personnel approve an invoice and select the correct product and expense codes. Accounting 'load' forms are filled in using specific accounting codes assigned by location, product and expense codes. The accounting department pays invoices and enters coding into the general ledger system. Therefore, any expense can be identified and selected according to any of these three codes. Utility costs can be listed in this format and one can even specify the utility according to whether it is electric, water, gas, etc.

Although environmental costs have been tracked to some degree for the last several years, only data after 1993 are accurate and comparable.

*System Strengths and Weaknesses.* The system has many strengths, such as its simplicity and the nature of its extension to the existing accounting system. Data are continuously and instantaneously available, so expenditures can be instantly tracked by category. By including environmental costs into the cost base of project accounting, individuals who are in the best position to manage environmental costs can give them full consideration. In turn, these individuals can be held responsible for their decisions. Extraordinary activities can be captured because the system is expandable through the simple addition of new account categories.

The ease with which new accounts can be added can also be risky, since it facilitates inappropriate changes as well as appropriate ones. New account classifications may be added, resulting in overlapping account codes. This inconsistency means that time-series comparison of environmental costs will not be accurate. The system will reject codes of impossible formats, but no other checks are performed as the data are entered. Thus, classification accuracy will depend on the account assignment, as determined by an accounting clerk.

Another weakness of the system is its lack of distinction of recurring and non-recurring costs. Distinguishing between these costs could produce cost time-series comparisons of higher fidelity, since non-recurring costs could be screened out.

Future improvements in the system include developing a procedure routinely to review product codes to delete redundant codes and ensure that new product codes are not included in the wrong categories.

The account classifications do not facilitate tracing costs arising from incidents, such as spills, back to a specific unit, since not all units have separate cost centres (e.g. the waste-water treatment plant). These systems are viewed as general overhead costs, without any charge-back mechanism. The system, therefore, does not currently support transfer pricing to all units. Additional metering capabilities, e.g. tracking volumes sent to flares, would provide measures of activity that could be used as a basis for transfer pricing.

International Refineries is aware that excessive reliance on profit-maximisation strategies using internal charge-backs will inevitably cause internal arguments, and may not be good in the long run, since cost cutting may short-change safety. On the other hand, International Refineries believes charge-back could contribute to improved operations management. A transfer pricing mechanism would help track line leaks or determine which unit ought to recycle.

The system has some limitations regarding labour costs. It does not capture personnel time by project, and includes labour only if it is associated directly with an 'environmental unit', e.g. the desulphurisation unit, or the waste-water treatment plant.

## Specialty Refiners

Specialty Refiners' primary goal for an environmental cost accounting system is to provide 'point-of-generation' environmental expenses assigned to product lines. Spills, disposal costs, over/under-treating for lube oils, permitting, waste-water treatment, etc. would all be assigned relatively accurate costs. The cost information would be distributed to appropriate field personnel to demonstrate the cost of specific actions as well as inaction. Given Specialty Refiners' trend toward decentralised ESH, environmental costs could serve as a unifying metric and communication tool.

Specialty Refiners has begun examining environmental costs, although a separate environmental cost system is not in place. Environmental cost data are generated and analysed on an ad hoc basis. Environmental cost data are generated using the mainframe general ledger and the Maintenance Management System. Labour costs are not included since salary information is stored on the payroll system. Capital contractor work requests are maintained and accessed on a database. All of these data are used to synthesise environmental costs which are stored on a manually generated spreadsheet. The spreadsheets are generated by the accounting department or by management on an as-needed basis. The level of detail depends on need, and is usually based on a particular commodity, chemical or contractor.

Specialty Refiners can modify its general ledger in a fashion similar to that of International Refineries. Any environmental area can be set up as a cost centre at the business manager's discretion. One refinery is tracking costs using the system. This capability permits detailed information to be made available from high-level categories, e.g. materials and supplies, labour, contract services, etc. This capability is rarely utilised and depends on a facility's desire to capture information. Although cost centre codes are easy to set up, gaining approval for them may be politically difficult: cost codes require authorisation of each division and each division's controller.

Capital project costs can be retrieved from invoices. There are tax benefits to capturing details on capital costs. These costs have been identified and broken down for several annual reports for the last twenty years. In determining which costs fall under the category 'environmental', there is not much discretion since Specialty Refiners uses the percentages recommended by the American Petroleum Institute.

Work order and material detail information can be obtained from the Maintenance Management System. Material detail is available only if a requisition is written or a warehouse has entered information. The discretionary use of credit cards for expenses under $5,000, instituted three to four years ago, confounds the ability to track all costs accurately. These expenses translate into approximately $25,000 per supervisor per month, based on EHS personnel estimates. Labour costs can only be obtained using the payroll system and are based on time-cards.

Specialty Refiners maintains a capital projects database which monitors project information, including status, contract work orders (CWO), authorisation requests and authorisation for expenditures. This system generates and tracks CWOs, excluding vendors that are often charged on credit cards, for expense work. Contractor cost detail can be generated by the desired category.

**Users of the System.** The system's use depends on each facility's demand for environmental cost information. There is typically a monthly review of information by a management team. The information is analysed with both an environmental as well as non-environmental perspective. The information is used by waste management teams consisting of two or three individuals focused on a specific problem, and by project managers to justify environmental expenditures. Finally, the environmental department uses the various systems mentioned above to generate and report required regulatory data.

**System Maintenance.** Accounting is charged with maintaining the general ledger system. Maintenance is responsible for the Maintenance Management System. Interfaces are the responsibility of the Corporate Information Technology Group.

**System Strengths and Weaknesses.** The systems in place that are used to generate environmental cost data have many advantages. The general ledger is a very reliable system and the accounting staff are well trained in its application. Cost data from the general ledger are available on a monthly basis, so analyses can be fairly current. Capital projects are easier to monitor because they are well defined and have a paper trail associated with them, except for the smaller credit card charges. The Maintenance Management System can be used to respond to custom requests: for example, volatile organic compound (VOC) monitoring information was successfully generated using this system.

The systems used to provide environmental cost information are not very effective. They are primarily high-level systems directed toward meeting government requirements, not business requirements. There is only limited user training, despite difficulties in manipulating the data, the complexity of the cost structure and the inflexibility in report generation. The system can allocate costs, but the process is labour-intensive.

The system allows environmental costs to be accumulated primarily at the cost centre level, so detailed cost analysis is difficult. Specialty Refiners is converting a new accounting system (SAP AG), which should allow for more flexible cost monitoring, not only by cost centre, but also by product line and product code. The allocation mechanism of the system will be improved and environment costs will be integrated into the facility cost structure. The allocation mechanism will be facilitated by the installation of better measurement equipment. For instance, Specialty Refiners is in the process of installing total organic carbon (TOC) analysers on effluent streams to allocate back to 'pieces' of the plant. Such information would also be useful for identifying alarm conditions or for source reduction.

Given the shortcomings associated with the high-level nature of the system, it is not surprising that Specialty Refiners identified 'point-of-generation' environmental expense assignment to product lines as a future system need. A system that could determine the cost of spills, disposal costs, over/under-treating for lube oils or permitting, etc. would enable field personnel to understand and take control of specific action or inaction costs to the company.

## ▌ Environmental Costs

As part of the initial questionnaire, the participating companies were asked to complete cost matrices, whose results are summarised in Figure 5. This was requested for three reasons.

First, the degree of effort required to complete the task reflects the quality of the current environmental cost accounting system, if any. It is possible to evaluate the quality of the systems as they currently stand by reviewing the information provided by each company.

Second, the costs are organised by the type of activity that generated the costs. By comparing the magnitude of the costs for each activity, the important activities could be identified. As it turned out, item 3, 'Environmental aspects of ongoing operating', and item 4, 'Remediation function', contained the most dollars for all participants. This was not surprising. Many costs associated with the ongoing operations are tracked as separate expense elements in traditional cost accounting systems. As such, they are easily discernible as 'environmental'. Similarly, costs associated with remediation are easily identified as 'environmental' and more easily quantified (if not accurately), since US Securities and Exchange Commission (SEC) regulations governing financial accounting require they be reported.

**Sales (worldwide) '94**

A  International Refineries        $3.4 billion
B  Ciba-Geigy        ($15.3 billion) $4.6 billion
C  Grupo Primex        $184 million
D  Specialty Refiners        $1.51 billion
E  Celanese Mexicana did not provide any cost table information

| 1 | 2 | 3 | 4* | 5 | 6 |
|---|---|---|---|---|---|
| Environmental function | Total cost ($) (most recent fiscal year) | Source of information in column 2 | % of total cost allocated to product as part of cost of goods sold (COGS) | Source of information in column 4 | Anticipated annual percentage change over next three years |
| **1.Centralised environmental activities** | | | | | |
| Permitting | B. $1,000K | | 0% | | 5% |
| Training | A. 1,292K | | 0% | | 8% |
| Environmental compliance auditing | C. 173K | | 0% | | 15% |
| Environmental management systems audit | D. 561K | | 0% | | -33% |
| Other | | | | | |
| **2. Regulatory affairs** | | | | | |
| Lobbying | B. 1,300K | | 0% | | 5% |
| Political action committee | A. – | | n/a | | – |
| Regulatory agency involvement | C. – | | 0% | | – |
| Involvement in legislative process | D. – | | – | | – |
| Other | | | | | |
| **3. Environmental aspects of ongoing operations** | | | | | |
| Annual depreciation of environment assets | B. 61,500K | | 100% | | 5% |
| Stripping column | A. 23,076K | | 0% | | 5% |
| Incinerator | C. ~630K | | 17% | | 0% |
| Waste-water treatment system | D. ~14,500K | | 100% | | – |
| Labour for operating the environmental assets | | | | | |
| Maintenance of environmental assets | | | | | |
| Operations and maintenance | | | | | |
| Other | | | | | |
| **4. Remediation function** | | | | | |
| Labour | B. 60,000K | | 0% | | Significantly less |
| Contractors/studies | A. 25,800K | | 0% | | 10% |
| Disposal | C. 128K | | 0% | | 0% |
| Annual depreciation of assets | D. Not broken down | | – | | – |
| Operation and maintenance | | | | | |
| Other field work/implementation | | | | | |

**5. Environmental aspects of community/external affairs**

| | | | |
|---|---|---|---|
| Public relations | n/a | – | |
| Community affairs | n/a | n/a | |
| Philanthropy | 0% | – | |
| Industry groups | – | – | |
| Other | | | |

B. (in #3)
A. Don't track
C. 5K
D. Not broken down

**6. Environmental aspects of product and process development**

| | | | |
|---|---|---|---|
| R&D: design for environment | 30% | 10% | |
| Product engineering | 0% | n/a | |
| Process engineering | 0% | 0% | |
| Pilot plant | – | – | |
| Marketing | | | |
| Procurement | | | |
| Other | | | |

B. 2,600K
A. Don't track
C. 30,172K
D. Not broken down

**7. Other (please specify)**
Marine Preservation Association
National Polystyrene Recycling Corp.

| | | |
|---|---|---|
| A. 2,464K | 95% | 0% |

\* Including cost of feedstock, labour and other refining/manufacturing costs

What percentage of cost of goods sold (COGS) relates to feedstock? _____ %

Provide capital expenditures for the last three years for environment-related plant and equipment:

| $1994 | $1993 | $1992 |
|---|---|---|
| A. 22 million | A. 26 million | A. 41 million |
| B. 15 million | B. 30 million | B. 49 million |
| C. 2 million | C. – | C. 1 million |
| D. 2.289 million | D. 26.263 million | D. 5.059 million |

**Figure 5:** Comparative Environmental Costs

Third, by comparing the costs for each activity across companies, it is possible to determine whether companies are having similar experiences in dealing with environmental issues, and whether their accounting systems are having similar success in capturing cost data.

The general results indicate that the participants varied greatly in their reported costs, although two variables—company size and type of business (refining or chemicals)—can explain most of that variation. The level of detail in the breakdowns varied by company. Grupo Primex and Specialty Refiners provided detailed costs for sub-categories in each relevant item, while Ciba-Geigy and International Refineries provided global totals. However, even Grupo Primex and Specialty Refiners acknowledged that it was very difficult to complete the questionnaire accurately, despite having environmental cost accounting systems in place. One problem may be that the environmental cost accounting systems tend to be aggregated at the plant level (especially Grupo Primex and Specialty Refiners), not at the corporate level.

There was general agreement that most costs will not increase drastically over the next few years. It was pointed out that some of the costs are discretionary, in that the timing of implementation is partially driven by what the company can afford in the given period.

There was surprisingly little reported investment in 'Centralised environmental activities' (item 1) or 'Regulatory affairs' (item 2), given that all of the US companies report a centralised function, including active interaction with federal and state agencies.

In fact, the presentation of the aggregate results to the participants was met with a surprisingly cool reception. It became clear that the participants do not feel that environmental costs *by themselves* are very interesting or important. However, the participants were able to identify and categorise a number of uses for environmental cost information. Although there appears to be some degree of overlap between them, three basic types of uses were identified:

1. **Decision-making.** Environmental cost information can illuminate issues such as determining the level of value added, use of risk-based versus necessity acquisitions, capital budgeting including lower hurdle rates for environmental projects, product costing and discretionary versus regulatory investments.

2. **Baseline cost information.** By developing baseline environmental cost information, managers can improve resource allocation decisions, support lobbying efforts and improve cost control. Comparing baseline costs to current costs may help identify cost reduction opportunities, estimate future project costs, serve as a basis for budgeting and capture cost avoidance.

3. **Management incentives.** Participants agreed that management's incentives for environmental stewardship were sometimes complex, and that environmental cost information might be useful for improving the management performance evaluation process by incorporating environmental costs into financial performance measures. These costs could also be used for public relations purposes, by providing evidence of the efforts made by the company in environmental stewardship. For example: several partners discussed community relations problems that may lead to future costs if not resolved, i.e. the legal release of irritating gases; and the 'environmental justice' issue, in

which the community that developed around the existing plant now claims environmental damage from the plant's proximity to the community. There was consensus that, even if company activities were above legal reproach, building better community relations can help avoid lawsuits, denied permits and similar costs. Partners are often willing to invest in good community relations to avoid these future costs.

Finally, the participants agreed on the difference between production economics, politics and the emotions of the press and of private citizens. These differences make 'rational management techniques' somewhat risky, in that public opinion is often far removed from what is right or fair. In discussing the use of environmental costs for decision-making, the participants differentiated between capital expenditures and routine operating and maintenance costs. All participants felt that environmental capital expenditures were special, because they are often mandated by government.

If technologies are mandated, the task is to implement them on a timely basis, even when mandated technologies represent an excessively expensive, suboptimal solution. Particularly, mandated technologies tend to represent end-of-pipe solutions, which may be less efficient than pollution prevention approaches, which depend on re-designing the waste-generating processes. The companies would often rather be given a mandated outcome than a mandated method or technology for reaching that outcome, but the time-horizon for meeting these targets is often too short to arrive at the best solution.

In addition to selected environmental cost information, all of the participating companies use non-financial input and output measures, which provide more detailed information more directly than anything that could be provided by the cost system. Non-financial indicators, such as number of incidents reported and TRI (Toxics Release Inventory) statistics, are often more useful for operating managers. Cost information is most useful as a way of determining the overall economic effects of current methods and of alternatives.

## ▌Use of Environmental Cost Accounting in Managerial Decision-Making

### Decisions Using Cost Data

A number of specific decisions were identified that could directly make use of environmental cost information. The list is not necessarily exhaustive. In many of these decision contexts, environmental cost information would be treated as just another cost of doing business, such as in product pricing or product mix. In other situations, the environmental cost accounting information may have a unique role in the decision process. This might be the case in waste management decisions, pollution prevention alternatives or market-based environmental options. In all cases, identifying and quantifying environmental costs, whether currently captured by the accounting system or as part of future liabilities or intangible costs and benefits, will enrich the quality of the following decisions.

1. **Internal/external benchmarking.** How are we doing against competitors? How are individual plants doing, on a comparative basis?

2. **Product pricing.** Better environmental cost accounting can lead to better understanding of what a particular product costs to produce. For products with price flexibility (differentiated products), this may be reflected in price adjustments.

3. **Product mix.** Better environmental cost accounting can be beneficial even with commodity products, for which the price is market-driven. The company may choose to adjust its product mix to maximise overall profitability.

4. **Waste management decisions.** Better understanding of environmental cost structures will lead engineers and managers to make more cost-effective choices in treating and disposing of waste.

5. **Pollution prevention alternatives.** A better understanding of current environmental costs, as well as that of prospective alternatives, will result in better capital expenditure decisions.

6. **Materials/supplier selection.** Companies committed to environmentally responsible manufacturing understand that a 'cradle-to-grave' mentality is necessary. Through better sourcing of materials, companies can push environmental responsibility up the supply chain. They may partner with suppliers to make pollution prevention options more cost-effective. Also, can significant environmental costs be avoided through outsourcing?

7. **Facility location/layout.** Companies may find that their by-products can be used as inputs to other companies. Such companies often co-locate with these partners.

8. **Outbound logistics.** These issues pertain to finished product, by-products and waste. Packaging of finished product has significant environmental implications, if the packaging must be destroyed to use the product. Is additional cost of design and materials worth the investment, if the environmental liability might be reduced? For by-products and waste, off-site disposal raises the risk of future liabilities for activities that are currently legal, and off-site transport moves the material beyond the control of the company.

9. **Market-based environmental options.** An active market in $SO_2$ and other pollution allowances is developing. Understanding the cost of reducing these emissions is key to establishing values for these allowances.

10. **International environmental standards.** ISO 14000 was finalised during 1997. It requires that environmental standards be documented and followed, though environmental accounting is not explicitly mandated. Certification may be required to maintain the customer base, especially in Europe.

11. **Public relations/lobbying.** Understanding the cost of this activity, and of the costs of not participating in this activity, will help to rationalise the level of investment to be made here.

12. **Training.** The best level of training (from a cost–benefit point of view) is easier to determine if the benefits are quantifiable.

## Decisions Using Non-Financial Data

As previously mentioned, one crucial issue is the degree to which financial indicators, such as environmental cost accounting information, are used, as compared with non-financial indicators, such as the number of incidents or TRI release information. The value of non-financial indicators is linked to the specificity and timeliness of those indicators. The value of financial information is that the net economic effect of the interplay of complex physical systems can be captured on a summary basis.

Participants seemed to agree that, at the plant level, non-financial indicators are essential, and cost accounting measures are secondary. Perhaps this is because the decisions that must be made at the operations level tend to be relatively straightforward, technical and immediate. Plant operations must be monitored and adjusted more quickly than cost information can be provided. The value of cost information seems to be limited to decisions requiring a common denominator.

The higher the management level, the more likely that financial information will be used, as costs can be seen as a shorthand for multiple factors. It is also likely that, the higher the management level, the more the manager's performance evaluation will be based on financial measurements. Thus, the high-level manager will be more interested in summary financial measurements of environmental activities than will plant operations. This is because the high-level manager is responsible for more diverse activities, and for activities with longer time-spans. Of course, high-level management will be interested in aggregate non-financial data as well.

Ciba-Geigy presented its use of life-cycle analysis as a method for evaluating relative environmental impacts of processes and products. This is found in case studies later in this chapter.

## Company Use of Environmental Cost Information for Decision-Making

Four companies made presentations on this topic: Ciba-Geigy, International Refineries, Grupo Primex and Specialty Refiners. Two primary types of decisions were addressed: capital acquisitions and operating decisions.

Environmental cost information was used primarily for capital acquisition decisions. The companies indicated that the processes used for capital acquisitions of environmental items tended to be made using the same process as that for non-environmental items. However, as indicated below, in individual company coverage, the environmental items tended to be given priority, especially if the environmental investments were mandated by government regulation. Desirable environmental projects tended to be given priority in an informal rather than a formal way, in that most companies did not use a lower hurdle rate for environmental investments. This was true of both US and Mexican companies.

## Identifiability of Environmental Portion of Pollution Prevention Projects

The preference for environmental capital items is reflected in an issue also discussed in the 'costs' section: that end-of-pipe types of acquisitions are clearly identifiable as environmental expenditures, but improved technology (which may also result in cleaner processes, thus addressing environmental issues at least as well, but less directly) are often considered operating improvements, not environmental. Because integrative solutions, such as improved technology, are clearly the superior approach to environmental stewardship (pollution prevention being superior to post-production clean-up), the preference for clearly environmental capital expenditures over non-environmental may inhibit companies' ability to achieve greater environmental responsibility through pollution prevention.

Perhaps one solution is to identify differential economic benefits to investments that can be legitimately classified as environmental, whether they represent pollution prevention or end-of-pipe investments. The issue of determining what portion of the investment is environmental has already been tackled by the American Petroleum Institute (API). One company, Specialty Refiners, systematically allocates a portion of their investment in pollution prevention projects as being environmental, using API guidelines. It was pointed out that, in some jurisdictions, environmental projects are currently given special tax status, such as shorter depreciation periods (US) or immediate write-off of cost (Mexico). These tax concessions may be justified by the environmental benefit to citizens within these jurisdictions.

The following describes how each of the participating companies uses environmental cost information:

## Ciba-Geigy

*Capital Expenditure Decisions.* Ciba-Geigy tracks environmental costs separately as certain cost elements on product cost sheets. However, environmental capital expenditures are seen as part of normal business expenditures; they are viewed as a cost of doing business. The biggest difference between environmental and non-environmental expenditures relates to government mandates, whether the project is required, and whether the government regulation stipulates technologies or performance. Performance-based regulation is generally preferred, because companies can identify and implement the best alternative solution.

All potential capital projects are evaluated based on traditional capital budgeting criteria, and are all subject to the same approval levels, hurdle rates and other financial tests, except in the case of mandated environmental projects. If the project is governmentally mandated, hurdle rates are meaningless; the project must be done.

*Operating Decisions.* Ciba-Geigy tracks environmental remediation costs separately. While significant differences exist, the internal approval process is similar to the capital expenditure approval process.

## Grupo Primex

Grupo Primex tracks environmental costs relating to both capital expenditures and ongoing operations and maintenance. However, like the other partner companies, decision-making with environmental costs focuses primarily on capital expenditures.

*Capital Expenditure Decisions.* The process for developing cost estimates is the same for any kind of capital project, whether environmental or not. The Environmental Manager decides whether a project is to be classified as environmental. However, environmental capital expenditures are tracked independently of other capital expenditures. Once needs or opportunity areas are identified, preliminary engineering studies identify critical parameters, leading to an estimate of the total capital expenditure required. If the project is required to meet regulations, the project will be approved in virtually all cases, subject to final approval by the General Director. If the project is not required to meet regulations, the payback period and internal rate of return are determined. If the payback period is less than three years and the internal rate of return exceeds the hurdle rate, if sufficient cash is available, and other aspects of the project meet with the approval of senior management, then the project is undertaken. If the project does not meet the financial criteria, the project is put on hold until regulations change, unless there are significant intangible benefits associated with the project.

*Operating Decisions.* Grupo Primex captures the ongoing cost of environmental activities, but does not exhaustively differentiate between environmental and other operating and maintenance activities. The company views environmental activities as a part of doing business, and manages these activities accordingly.

## International Refineries

The system has helped International Refineries better understand where they are spending money. Two areas with unexpectedly high costs proved to be waste hauling and operations analysis. With International Refineries' system, the cost per pound of waste can be estimated. They can break down waste-hauling costs by vendor as well. Despite what might be expected, it is not uncommon for the same vendor to charge different prices for the same services that were negotiated separately. A side-project developing around this system involves co-ordinating purchases and looking at volume discounts. A refinery manager could also use this type of information to improve selection or negotiations with catalyst suppliers, waste disposers, etc.

Improved understanding of environmental costs improves the budgeting process. The focus on expense codes is particularly useful, since budgets are constructed around expense codes. As a result of the Enhanced HSE accounting systems, there is a better understanding of remediation costs, as well as the costs of different types of remediation.

Accounting for environmental costs has also proved useful as a communication tool. International Refineries is now able to track environmental, health and safety (HSE) operating expenditures, HSE capital expenses as well as environmental accruals. Superfund expenses are tracked and are included depending on the information displayed. HSE expenditures are also tracked according to business sector, i.e. upstream, downstream and chemical.

*Capital Expenditure Decisions.* International Refineries evaluates environmental capital expenditures in the same way as non-environmental capital expenditures. However, the priority of the capital project is directly linked to whether the project is required to comply with government regulations.

HSE capital and operating budgets are developed by business units with input from on-site HSE personnel. Included are corporate HSE costs that are allocated to business units, based on a rough estimate of percentage of time spent by corporate HSE staff on each business unit.

International Refineries has developed a Capital Budgeting Guidelines hierarchy of projects, ranging from A to G. Category A consists of environmental projects, of which A1 projects are required for compliance with regulations in the current year, and A2 projects will be required in the future. B projects are health and safety projects, with B1 referring to legal mandates effective in the current year, and B2 projects referring to future legal mandates. C projects are expenditures required to avoid the loss of an asset, or to meet the terms of an existing contract. D projects are discretionary projects with a current year positive net cash flow; E projects with high returns where delay in investment will significantly reduce future returns; and F projects are those with returns that exceed the hurdle rate for the year. G projects are projects with undefined economics. These G projects are virtually never funded.

**Operating Decisions.** International Refineries has a relatively well-developed environmental cost accounting system, but does not base ongoing operating and maintenance decisions on costs specifically identified as environmental. However, the company uses this system to track all expenditures of HSE functions by category, media, location and line of business. It functions well in spotting irregularities. Examples of uses of the International Refineries information include:

- Budgeting (by expense code)
- Identifying opportunities for volume discounts/leverage with vendors
- Making business units responsible for their own costs

### Specialty Refiners

Specialty Refiners uses environmental cost information extensively for both capital expenditure and operating decisions.

**Capital Expenditure Decisions.** Specialty Refiners' capital budgeting process is similar to that of the other companies, consisting of a multi-step process. The initial project is often conceived at the plant level, where the problem is most often identified, but it may originate at corporate or at the tech centre, depending on the specific problem. An initial set of alternatives is evaluated, and the leading contender(s) is/are further refined. Process design specifications are generated, including a detailed safety and environmental checklist, which must be approved by corporate HSE before the project can proceed. A capital budget proposal is generated, consisting of both the technical and economic evaluations of the alternative. The approval process will be affected by the size of the project, the economics and the degree of linkage to other projects in the current capital budget. As a part of the authorisation for expenditure, the cost is apportioned between the environmental part and the non-environmental proportion, using API's guidelines and Specialty Refiners' experience. These environmental proportions are rough, usually 50% to 100%, but may be as small as 25%.

During the construction of the project, concurrent cost audits are regularly performed, providing a basis for evaluating both the budget and the actual performance. However, upon completion of the project, post-implementation audits are rarely, if ever, performed.

**Operating Decisions.** Specialty Refiners takes environmental costs into account when making operating decisions. For example, product-mix decisions are affected by potentially high environmental costs of new products. Additional environmental responsibility brings the need for better environmental performance, which could translate into either more environmental staff, or a larger non-environmental staff who take responsibility for some environmental activities. Finally, environmental costs can affect product pricing for speciality products, for which there is pricing flexibility. One difficulty in using environmental costs for operating decisions is the accuracy and timeliness of cost assignments to specific products.

In addition, Specialty Refiners is aware of how current practices have a real, future cost but of uncertain magnitude and timing. For example, environmental liabilities and future disposal costs are real, but difficult to estimate. Environmental regulations are likely to change, and may bring about new retroactive liability. Consumer confidence and the support of the community are essential, but can be influenced either positively or negatively by public relations, quite apart from real environmental issues. These issues affect operating decisions, but are difficult to quantify accurately.

# 11 Applying Environmental Accounting to Electroplating Operations

An In-Depth Analysis[1]

Mark Haveman and Terry Foecke,
Waste Reduction Institute for Training and
Applications Research (WRITAR)

## ▌ Introduction

THIS CHAPTER contains the findings and results of an eighteen-month investigation into the application of environmental accounting practices at 24 electroplating facilities. The purpose of this research was to conduct a detailed examination of the mechanics of implementing environmental accounting practices within a specific industry.

## ▌ Background on Research Strategy and Implementation

The investigation centred on hexavalent chrome processes and zinc cyanide processes— both of which are under regulatory pressure and are common targets for change, and therefore likely opportunities for using environmental accounting analysis. Project researchers used an environmental cost category template developed by the Tellus Institute (as adapted and reported in NEWMOA 1994) as the analytical framework for the investigation (See Fig. 1). In examining the various labour, materials and overhead cost elements, four questions comprised the focus of this portion of the research.

---

1. This chapter is an edited version of a longer report of the same title published by the US Environmental Protection Agency in May 1997 (WRITAR 1997). The authors prepared the original document for the EPA's Environmental Accounting Project. The full report can be read on the Project's website at *http://www.epa.gov/opptintr/acctg* or can be obtained free of charge from EPA's Pollution Prevention Information Clearinghouse by telephoning +1 202 260 1023.

| **Materials** | **Regulatory compliance** |
|---|---|
| Direct product materials | Monitoring |
| Catalysts and solvents | Manifesting |
| Wasted raw materials | Reporting |
| Transport | Notification |
| Storage | Record-keeping |
| | Training (right-to-know, safety, etc.) |
| **Waste management (materials and labour)** | Training materials |
| Pre-treatment | Inspections |
| On-site handling | Protective equipment |
| Storage | Labelling |
| Hauling | Penalties/fines |
| Insurance | Lab fees |
| Disposal | Insurance |
| | R&D to comply with regulations |
| **Utilities** | Handling (raw materials and waste) |
| Electricity | Closure and post-closure care |
| Steam | |
| Cooling and process water | |
| Refrigeration | |
| Fuel (gas or oil) | **Revenues** |
| Plant air and insert gas | Sale of product |
| Sewage | Marketable by-product |
| | Manufacturing throughput change |
| **Direct labour** | Change in sales from: |
| Operating labour and supervision | Increased market share |
| Manufacturing clerical labour | Improved corporate image |
| Inspection (QA and QC) | |
| Worker productivity changes | |
| | |
| **Indirect labour** | |
| Maintenance (materials and labour) | |
| Miscellaneous (housekeeping) | |
| Medical surveillance | |

**Figure 1:** Potential Operating Costs Included in an Environmental Accounting Analysis

1. What is the applicability of these costs to electroplating?
2. How are they currently allocated and how likely are they to be allocated to processes or parts?
3. What is the significance of various types of cost to electroplating?
4. If an assignment is made, would it be accurate?

Reviewing for applicability involved identifying which cost elements were relevant to electroplating operations and obtaining a better understanding of how they were relevant. Reviewing for allocation involved exploring how the facility currently accounted for these costs and assessing their ability to assign or allocate those costs to responsible products or processes to support decision-making. This review enabled researchers to gain an understanding of how environmental costs are currently handled, how they might be managed differently, and how to make the transition from a conventional cost analysis to an analysis based on environmental accounting.

The final two questions were investigated to understand better the practical implications of using environmental accounting in electroplating operations. WRITAR considered cost significance to be an important qualifying issue in adopting environmental accounting methods. It was assumed that establishing a threshold of significance helped ensure that the probable benefits of using this cost information for decision-making purposes would be greater than the costs of gathering and analysing this information. For the purposes of this report, 0.5% of gross annual revenues was selected as a 'significance threshold'. This was based on previous WRITAR experiences in working with metal finishers which indicated that cost issues of this magnitude are most likely to get the attention and interest of facility management. 'Environmental management' costs in aggregate well exceeded the significance threshold; 6%–12% was a common range found in the course of the research.[2] Within this collection of costs, particular emphasis was placed on reviewing and analysing those elements exceeding the 0.5% criteria.

Allocation accuracy was chosen as another critical issue to help ensure against situations in which the value of environmental accounting analysis might be rendered moot by inappropriate relationships between various activities and elements of cost. It is evident that facilities must often estimate, use best professional judgement, or a combination of both practices to assign many environmental costs to processes. The research placed an emphasis on identifying circumstances in which the accuracy of allocation efforts would probably be low because it was presumed that inaccurate allocations could potentially mislead or misdirect facility decision-making.

Throughout this report, qualifiers are used to describe facility conditions and practices. There follow some definitions used by the project researchers during the course of the investigation and found in this report:

- **Allocated by product:** cost is used as a line item for calculating a unit price for finishing a specific part or type of part.
- **Allocated by process:** cost is used to calculate the contribution of a unit operation to facility costs and/or unit price for a specific part or type of part.
- **'Accurate':** data from purchase records or similar; auditable
- **'Mostly accurate':** data from purchase records or similar, subdivided based on number of operations, number of tanks, number of parts produced, etc. and verifiable by production records.
- **'Somewhat accurate':** based on data from production records and logs; values are obtained by multiplying volumes by cost data and should balance over long periods of time with purchase records.

## ▌ Analysis of Environmental Costs in Electroplating Operations

The research focused exclusively on operating costs—both direct and indirect. Contingent costs (costs based on probabilities of events occurring in the future) and image costs (intangible costs associated with company goodwill of improving environmental performance) were not included as part of the investigation. The rationale for their omission is that these costs are highly context- and situation-specific, while the purpose of this report was to generate information generally applicable to all electroplating facilities.

The following synopses of operating costs is based on the cost categories found in Figure 1. During the analysis, special attention was given to cost elements that other

---

2. This range was based on a consideration of all labour, materials and other expenditures directly associated with environmental management activities and keeping a facility in compliance.

environmental accounting research studies had found to be frequently under-recognised in operations in order to shed additional light and information on identifying, quantifying and allocating these costs.

There follow synopses of each cost area.

## Materials

Key wasted materials costs in electroplating are:

- Process chemistry
- Water
- Addition agents
- Miscellaneous chemistries (e.g. strippers, chromates)
- Cleaners
- Acids

***Direct Product Materials: Applicability and Significance.*** Costs for coating materials (zinc, nickel, copper, gold, silver, etc.) appear to vary greatly, both comparatively and with respect to their significance in the overall facility cost structure. Precious metals and some other materials (such as electroless nickel, electroless copper and some alloys) are expensive to deposit. Significance is also affected by the scope of processes within a shop: if a facility focuses on a particular type of plating process, plating materials associated with this process will be a large item for the facility even if the materials are comparatively inexpensive. Hidden costs in terms of materials waste can be traced to two sources: a differential between the amount of coating materials purchased and the amount of metal that actually goes out on parts, and another differential between the amount of metal that goes out on parts and the amount that is actually needed to meet the desired production requirements. This, of course, has subsequent implications for waste-water and disposal costs.

Costs for addition agents (brighteners; pH adjustment; consumable components such as cyanide) demonstrate enormous variation by process, and are dependent not just on the market price of metals (as in the cost of coating materials), but also on the cost of formulation, size of market (small niches may feature high materials prices enabling manufacturers to recoup development costs), age of product (new materials cost more) and function of the process solution (run to depletion [such as electroless nickel] versus maintain for many years [such as nickel sulphate]). The research suggests that by far the most ubiquitous and expensive (cost per unit volume as well as total cost) are brighteners. Research also suggested that cost understanding is often poor in this area as standard operating practices for additions are often highly subjective. As a result, a scan for price and use volumes is highly recommended in reviewing plating cost structures.

As with other direct materials inputs, costs for acceptable water vary greatly, and will depend on the type of process solution and products. Water inputs can be characterised most readily in reference to either (1) current water quality (usually most important when using groundwater rather than drinking water); and/or (2) market niche (electronics requiring extremely pure water, most others increasingly requiring at least supplemental treatment up to and including de-ionisation). Also influencing the variability and relative significance of water costs are noted trends toward generally higher water costs and, in some cases, restricted availability (South San Francisco Bay Area; Phoenix; Los Angeles County). Water costs are also linked to pre-treatment costs, since capital costs for equipment are directly linked to hydraulic loadings of the system. This, in turn, establishes a relationship with other cost issues such as indirect labour since the greater the hydraulic loading, the larger the system, and the more maintenance is required. Its significance is enhanced by the fundamental relationship between water use and quality/process control.

***Direct Product Materials: Assignment and Allocation.*** Costs for coating materials are assigned at the product level, especially among users of precious metals, and more of this type of assignment could probably be done. However, it does not appear valid to use the experience of the users of precious metals as a guide. Those companies are able to allocate costs at the product level because extremely accurate records are kept of additions, precious metal concentration in the process solution, surface area of parts, and thickness of precious metal coatings.

Base metals were only found to be allocated to the product level when it was either a high-volume material or when it was expensive for other reasons. In these circumstances, allocation was found to be nearly always done using a gross multiplier during cost estimation (e.g. 1,000 square feet surface area to be plated to a thickness of 0.0005 inches gets a 'factor' of 20% added to the price). This factor moves with the thickness and sometimes the surface area. Greater allocation to the product level would be technically possible, but would probably be thwarted by the amount of record-keeping required that is not useful for any other purpose. Therefore, most allocations are likely to be restricted to the process level. It is believed that this allocation would be accurate in nearly all cases.

Research suggests that costs for addition agents could probably be assigned but restricted to the process level. It is believed that this allocation would be accurate in nearly all cases. Costs for water could be allocated, but only with an investment in monitoring and control equipment. Such investments have proven to be unlikely unless linked to other facility needs. However, even if assigned, costs for water would only be somewhat accurate, with an estimated variance of ±20% as this is the estimated range for water flow through the most common valves in use. Even in high-cost situations, water was found to be included with other general overhead charges and allocated in whatever way the facility chose to allocate general overhead.

***Indirect Materials Use: Applicability and Significance.*** In finishing operations, a wide variety of indirect materials is used in surface preparation, in the finishing process itself, and in rework operations. As with direct materials, there is a significant amount of variability concerning the absolute and relative significance of indirect materials depending on the type of finishing operation. As alternative cleaners replace traditional organic solvents, the significance increases in two ways. First, they are more expensive to purchase; research found situations where purchase price of new cleaners exceeded old solvents by a factor of twelve. Second, they can be more expensive to treat, or at least this is the presumption.

Costs for acids are not particularly significant except in easily discerned cases such as electronics; plating on plastics; and etching/deburring as pre-plate operations. Costs for filters and miscellaneous chemistries tend to be small in most shops but under-recognised and are therefore a good addition to environmental accounting.

***Indirect Materials Use: Assignment and Allocation.*** Costs for acids, filters, cleaning materials and miscellaneous chemistries (strippers, chromates) tended not to be allocated outside general overhead. These materials demonstrate dispersed use patterns, episodic ordering, and a lack of other reasons to keep the required records (mostly use logs), thereby inhibiting process or product assignment. Evidence from some facilities suggests that, even if the allocation were attempted, it was unlikely to be accurate because of a reliance on individual effort and a perception of low value ascribed to the effort. Some evidence was found of allocation of cleaners to product and in other situations by process, since it involves a discrete unit operation. However, no approaches were found in the research that break that cost down between solvent actually used to clean parts and solvent lost to evaporation. Therefore, costs are likely to be completely accurate only where complete elimination is contemplated. Where more efficient operations (increased freeboard; reduced withdrawal rate) are proposed, costs are likely to be only mostly accurate.

By definition, process-level allocation of filters is possible given that each filter is 'assigned' to a particular process. Product-level allocation is impossible.

***Materials Storage.*** A final category of cost analysis pertaining to materials is storage. These costs are not very applicable to P2 (pollution prevention) analysis for electroplating operations. Most P2 options related to materials storage involve replacing materials and equipment that still require storage. Research suggests that storage related to input materials is not usually a significant cost, and not often connected with other costs likely to be analysed.

### Waste Management

Key waste management costs in electroplating are:

- Handling
- Waste-water treatment: labour, chemicals, energy
- Storage
- Insurance
- Disposal
- Transportation

*Applicability and significance.* Waste handling, storage, insurance and waste-water treatment appear to have considerable potential to elucidate a P2 project analysis. Typically, handling and storage costs are subsumed together in a facility waste management line, while insurance costs are often found in a general operations line. Insurance pertaining specifically to waste management operations is exceedingly rare. Insurance costs (overall) are clearly significant in most operations.

Moving materials from place to place, especially as segregation increases, appears to require considerable labour and equipment expense. Options short of elimination (segregation to facilitate recycling, for example) appear to increase this cost burden.

Storage costs can be significant in larger operations where waste treatment areas can be 25% the size of production areas. In smaller operations, these costs are marginally significant and it is probably best to think of them in combination with other related waste management activities.

Much of the cost analysis has its greatest relevance when a facility is being designed. Capital costs related to buying tanks, building storage sheds, diking warehouse areas, etc. are very significant. Since sludge in many cases can stay on-site for only ninety days, the relevance to pollution prevention project analysis is typically limited to those circumstances when a massive overhaul could possibly prevent the need to build more storage. The same logic applies to insurance costs since systems/options can be assembled that could, if done well, change risk categories. Research suggests that a pollution prevention option that reduced the *need* for storage would not reduce the *costs* of storage, or at least not very often. Likewise, an option that reduces the *need* for storage seldom reduces the *costs* of insurance.

The primary and priority cost centre, operation and maintenance of a conventional waste-water treatment system, combines a wide variety of labour, chemical and utility costs. The significance of cost burden for facilities is high and made higher by the need to reduce chrome or oxidise cyanide if those processes exist in a facility. Costs may also be unnecessarily inflated by less than optimal management practices. Similarly, the use of recovery technologies as 'pre-treatment' to the treatment system carries potentially significant costs and will generate wastes themselves. Caution must be exercised when reviewing cost implications of this technology use, since costs appear to be shifted as often as they are actually reduced.

*Assignment and allocation.* Allocation of waste storage and handling costs only to processes or products generating the need for handling appears straightforward, especially when considered as part of the overall waste management line. Typical assignment would be on a unit volume basis. Insurance, however, poses a more challenging problem, and allocation of these costs only to processes or products generating the need for some discrete component of insurance seems unlikely. Such an insurance line item would involve estimation and/or calculation of the cost per unit volume to insure activities that seem hardly to be recognised in policies as risk vectors.

The accuracy of waste management costs is difficult to ascertain. It is doubtful that these costs could be readily audited, since waste management relies heavily on a labour component that would in turn rely on individuals differentiating their activities. Accuracy even of the overall waste management line is also somewhat suspect, since records seem to rely on time assigned to duties, rather than tracking of actual time spent.

Waste-water treatment system costs typically appear as an overhead charge and are most frequently allocated across all processes based on square footage of production or some other related measure. Occasionally facilities were found in which veteran estimators knew that certain processes created larger treatment burden than others and made some attempt to differentiate charges. This practice, however, more often occurred in response to pricing sensitivity rather than to seek an understanding of the 'true costs'. In other words, more refined allocation measures were employed to figure out a way to pass the cost on rather than to reduce it. In other, typically larger, facilities the treatment system was a cost centre with charge-backs to departments based on volumes and/or incidences such as bath dumps. However, charge-backs appear to be based on rather crude calculations.

## Utilities

Key utility costs in electroplating are:

- Energy
- Sewage
- Process air

*Energy: Applicability and Significance.* Electricity in general as a cost to electroplating facilities is extremely significant. However, its applicability as an area of cost analysis for pollution prevention purposes is inherently limited. By far the majority of electricity is used for application of coatings in heating and electrodeposition, which have physical limits to their efficiency and therefore rarely form the focus of pollution prevention analyses. Steam and natural gas are used in some facilities for process heating, and, for facilities with large boilers or ovens, this can be a significant cost. It may also, in some cases, be part of the need for an air permit. Natural gas may also be used for firing bake or dry ovens. These costs are nearly always subsumed in a 'general operations' line.

*Energy: Assignment and Allocation.* Allocation of these costs to only processes or products that generate the need for some discrete component of power or fuel use seems unlikely. This cost seems nearly impossible to allocate: it would involve estimation and/or calculation of the cost per unit volume of something that varies constantly; at a minimum, significant investment in monitoring equipment would be required. Likewise, assignment of steam costs only to processes or products generating the need for some discrete component of steam use seems unlikely.

An interesting sidelight discovered in the course of the research is the use of electrical consumption as part of a ratio to adjust measures of P2 progress. The theory is that use of electricity declines when work activity declines, so an electricity–water use ratio could be derived that would shed some light on whether water use declined as a result of P2 activity or a decline in business activity. Unfortunately, this would still be very hard to use at anything other than the facility aggregate level.

*Sewage: Applicability and Significance.* Sewage costs should be applicable to the analysis of any facility and any specific pollution prevention project dealing with water use and water waste streams. It is worthwhile to note that sewer charges can and are computed in many different ways, many of which have no direct relationship to the amount of water actually being discharged from the facility. Some facilities have a standard sewer charge based on inputs into the facility rather than outputs. Others feature a cost computation based on broad categories rather than actual amounts; for example, the size of pipe combined with the amount of water purchased is determined to be equal to the amount of water sewered with a multiplier applied to monthly activity. Sewer costs can be especially significant if special surcharges have been applied because of local conditions and requirements. These costs are nearly always subsumed as part of other categories.

*Sewage: Assignment and Allocation.* It appears possible to assign these costs to processes or products that generate the need for some discrete portion of sewer use. However, as with other utilities, this cost seems nearly impossible to allocate accurately as it would involve estimation and/or calculation of the cost per unit volume of something that varies continually. As with other utilities, an increase in accuracy would require significant investments in monitoring equipment.

The quality of cost analysis and assignment in sewage could be improved, however, by separating process-related from non-process-related sewage. The research found many instances of large volumes of non-contact cooling water, water from cooling towers, domestic sewage and even storm-water being combined with process water, thus raising volumes and charges or, at a minimum, obscuring industrial use information that would be the focus of a P2 project analysis.

*Process Air: Applicability and Significance.* Costs for clean process air, especially compressed air, may be applicable to a P2 project analysis, although the issue here is ensuring that this cost is accounted for in the economic analysis of the P2 option. Clean compressed air is an important component in several 'keystone' source-reduction concepts for electroplating facilities, e.g. dry air blow-off of process solution to reduce water use; clean air to increase agitation in water rinses and reduce need for water use; warm air mixed with water rinses to create 'fog' rinses and reduce need for water rinse. A facility implementing the above source-reduction option could easily outstrip existing compressed-air capacity, especially since clean and/or dry air is quite expensive to generate. These costs are often not significant, but may become significant with substantial source-reduction activity. Several facilities contacted in the research reported costs for air outstripping costs for some acids. These costs are always subsumed in a general operating category.

*Process Air: Assignment and Allocation.* Assignment of these costs by unit operation are possible, since most applications are designed with specific flows, both required and controlled. However, two complicating factors exist. First, the cost of producing clean compressed air is difficult to calculate as no generally accepted default values were found to exist during the project research. Second, this allocation would only be somewhat accurate because of the wide variation in use possible and the amount of access by operators to the system.

## Direct Labour: Clerical and Inspection

*Applicability and Significance.* Because pollution prevention projects in electroplating typically have strong process control, record-keeping, monitoring and analysis themes associated with them, manufacturing clerical labour expenses will tend to be applicable. P2 project analysis may also require specification and contract review, another purpose for which manufacturing clerical labour would be used. These can be significant costs in many operations and are related to similar activities undertaken to improve quality and/or productivity that may be part of a P2 project analysis.

Some P2 projects may also affect the need for parts inspection, especially in cases where overall process quality is improved or a particular finish is modified. Therefore, these costs may be applicable to P2 project analysis, albeit marginally. Research found that inspection costs are part of overhead in about 50% of facilities investigated, and part of a separate quality management and documentation system in the other 50% of facilities. In the latter case, inspection costs are sometimes identifiable as separate costs, but not at the product or process level. In a few cases of high-volume or sophisticated products, inspection costs may be both specific and allocated. Inspection activities tend not to be significant costs. In addition, the 'quality revolution' has caused most inspection activities to be combined with the production duties of operators, which further reduces the significance of the costs, since inspection is almost always done in the 'downtime' that occurs while waiting for parts to be processed.

***Assignment and Allocation.*** Manufacturing clerical costs could be assigned, but only in settings where workers in these labour categories already use activity coding for other purposes, or are specialists who could respond accurately to interview-based analysis. Allocation of these costs seems to be only somewhat accurate, apparently because a 'life-cycle' of record-keeping was found in project research: intensive in the early stages of project analysis; decreasing to a somewhat lower level during decision-making and implementation; and possibly disappearing entirely at a later point. However, several cases were described that incorporated the new records and analysis into quality monitoring systems, thus permanently increasing these costs.

Assignment of inspection costs seems to be unlikely and, if attempted, inaccurate.

## Indirect Labour

***Applicability and Significance.*** Applicable indirect labour elements in electroplating include maintenance and medical surveillance. Maintenance costs do appear quite applicable to a P2 project analysis and can come in many forms. Process solution maintenance tends to be done by production operators, with significant exceptions in the electronics products and aerospace industry sectors. Mechanical maintenance is usually done by specialists, either in-house or contracted. Maintenance carried out by operators is often done during downtime while waiting for parts to process, and may or may not be recorded separately, while maintenance carried out by contractors, in-house specialists, and for special projects (process solution changeover on a weekend, for example) are often recorded and tracked through a system of work orders or job sheets. These costs are often significant, especially in larger facilities.

Medical surveillance costs may be applicable to a P2 project analysis, especially if the project contemplates eliminating particular substances or unit operations that cause the need for monitoring. These activities are nearly exclusively performed by clinics or outside contractors and are accessible as separate line items only in rare cases. In most facilities, these are treated either as 'general operating' costs or even as part of 'employee benefits'. These costs are apparently significant in only a small group of facilities, but the cost significance issue may be rendered moot. Even if the P2 option reduces or eliminates the need, monitoring may still be either required, prudent or very desirable.

***Assignment and Allocation.*** Maintenance carried out by contractors, in-house specialists, and for special projects could probably be allocated accurately to the process or unit operation level. However, maintenance carried out by operators is probably impossible to assign separately unless supported by a work order/paper trail and distinguished from production activities.

Assignment of medical surveillance costs seems likely to be possible and will probably be quite accurate, inasmuch as they are driven by specific monitoring and not general screening.

## Regulatory Compliance

*Applicability and Significance.* As with other industries, electroplating features an extensive array of cost items pertaining to the regulatory compliance and the environmental management function. For purposes of analysis, it is useful to segregate these compliance costs into two categories: those driven by a facility's use of particular substances and those driven by the facility's generalised need to respond to environmental regulations.

Substance-driven costs would include:

- Monitoring
- Protective equipment
- Lab fees
- Manifesting
- Labelling
- R&D to comply with regulations

These costs are usually applicable to a P2 project analysis and typically subsumed in a 'general operating' category. Of this list, lab fees and monitoring can often be significant costs.

Regulatory response driven costs would include:

- Reporting
- Closure and post-closure care
- Handling
- Record-keeping
- Personal injury
- Fines and penalties
- Inspections
- Notification
- Insurance
- Training and materials

These costs too are usually applicable to a P2 project analysis and usually subsumed in a 'general operating' category. With the exception of closure, which can be associated with very high costs, the individual cost elements may not be very great. However, taken together, their sum can be quite significant.

*Assignment and Allocation.* The purpose of distinguishing these compliance cost categories lies in their ability to be assigned to processes or products. The former set, use-driven costs, can probably be assigned quite readily and the allocations would probably be quite accurate. Costs associated with generalised compliance response prove far more problematic. They can probably be assigned only on a volume basis (e.g. 10% of sludge is nickel, so 10% of cost pool goes to nickel processes). Such assignments would only be somewhat accurate.

## Incremental Revenues

*Applicability and Significance.* Revenue accruing from the production and sale from a marketable by-product is common in electroplating and applicable to P2 project analysis. Most electroplating facilities selling by-products seem to include these revenues as part of a general income line. These revenues tend not to be very large components of gross revenues and, furthermore, vary considerably because of their exposure to price swings in commodity markets. As a result, reliance on these revenues for cost–benefit analysis when doing project analysis is fairly rare.

Sales increases as a result of image improvements might be possible, but the research found no instance of a facility willing to consider this approach to valuing the results of P2 projects.

*Assignment and Allocation.* By-product revenues can be allocated accurately, but generally cannot be projected for use in capital budgeting efforts.

## ▌ Issues in Implementing Environmental Accounting in Electroplating Operations

Electroplating firms contacted and visited for this project have recognised that waste management and regulatory compliance involve more costs than those in evidence as the total payments made to treatment and disposal firms. In addition, these firms are seeing that more benefits accrue to their facility—and financial statements—from pollution prevention approaches than has usually been portrayed in the past.

Environmental accounting has a role in facilitating this understanding, supporting better project analysis and prompting more implementation. A generally accepted environmental accounting implementation strategy is comprised of the following steps:

- Identify environmental management costs
- Prioritise and select the costs to investigate in more detail
- Quantify or qualify the costs
- Allocate costs to products or processes responsible for their generation
- Integrate costs into facility decision-making

There follows a discussion of the relevant issues pertaining to each of these steps as they relate to finishing operations:

### Identifying Environmental Costs

As Figure 2 illustrates, environmental management costs in electroplating facilities might best be broken into two primary categories: (1) direct and indirect operating costs associated with environmental management and compliance; and (2) opportunity costs stemming from not using raw materials as efficiently or as productively as possible. The direct and indirect costs associated with environmental management issues vary in their relevance to finishing operations and in their 'ability' to be identified, depending on the types and nature of processes employed at the facility. Generally speaking, however, these direct and indirect costs can be quite readily identified by a facility. Those that require actual checks on a regular or semi-regular basis were quickly identified by

| Type | Environmental management costs | Lost material costs |
|---|---|---|
| Examples | ✓ Labour cost for environmental compliance<br>✓ Regulatory fees<br>✓ Purchased materials for water treatment | ✓ Wasted process solution<br>✓ Lost process chemistry |
| How identified | Cost accounting review<br>✓ General ledger review<br>✓ Payables review | Cost accounting review plus process review<br>✓ Materials balance<br>✓ Process benchmarking |

Note: Two other cost categories that may be part of environmental management costs are contingency costs and image costs. However, a consideration of these costs is not included in this chapter.

**Figure 2:** Environmental Cost Components

the facilities visited and leave a paper-trail in the system. Smaller, more miscellaneous expenditures such as regulatory fees were also identified with some limited prompting.

Environmental management costs that do not involve monthly cheque-writing (e.g. indirect labour, etc.) were also readily acknowledged by facilities and recognised as something that should be reduced. An important complicating issue here, however, is the degree to which facilities can link them to specific activity drivers (e.g. separating the indirect labour pool into its components of record-keeping, monitoring, notification, etc.). Although facilities recognise their existence, they proved to be far less willing or able to identify costs at this level of individual detail. From a standpoint of environmental accounting analysis, such an effort may not be necessary given the relative size of these items in relation to other cost elements.

The drivers of these direct and indirect costs, rather than the costs themselves, were most often under-recognised by a facility. This was especially true in target areas such as waste-water treatment and hazardous waste disposal. While facilities could generally develop cost totals for these areas, the understanding of the contribution and impact that episodic events—such as disposal of tank bottoms, rework of reject parts, and bath dumps—have on these items was under-recognised. Applying environmental accounting to these activities was found to generate interest in reducing the frequency of these events through better process management and control.

A larger and more significant set of environmental costs for electroplating is the opportunity costs of unoptimised materials use. These opportunity costs are the value of lost process materials—materials purchased but not sold, wasted, or not used as efficiently or as productively as could be. They are implicitly more difficult to identify for the following reasons:

- There are few readily available 'best industrial practice' benchmarks to indicate to a facility how efficient or productive their processes could be from the standpoint of materials use.

- Reviews of electroplating cost and pricing literature found no way to track lost material through explicit or implicit loss allowances. It may exist in some electroplating facilities for very specific reasons, but nothing was uncovered in this project research.

• Cost accounting systems do not capture such opportunity costs with a line item called 'wasted raw material'. In fact, the opposite may be true: cost accounting systems may artificially hide this cost of waste by building costing systems on standard loss allowances rather than actual losses.

Adding the opportunity costs to the direct and indirect environmental management costs gives a facility a more accurate measure of the facility's total cost of waste. They are also obviously inter-related: reduction in these opportunity costs will decrease primary environmental management costs such as waste-water treatment and hazardous waste disposal.

### Prioritising Costs to Investigate

For reasons described above, facility attention should first be focused on valuing the opportunity costs of materials losses before attempting to 'unpack' environmental overhead in deeper levels of environmental accounting analysis. The research showed that for electroplating facilities the significance of environmental management overhead items generally pales in comparison in amount and significance to these opportunity cost issues. For example, in one of the zinc shops visited:

| | |
|---|---|
| Apparent cost of F006 sludge based on payments to TSD (treatment, storage and disposal facility) and haulier | $525/ton |
| environmental accounting-generated share of overhead cost | $67/ton |
| Apparent cost of sludge adjusted for allocated overhead | $587/ton |
| Cost of lost process chemistry[3] | $306/ton |
| Cost of sludge adjusted for environmental management and opportunity costs | $893/ton |

In this example, the opportunity costs of lost material exceeded the overhead allocated costs by a factor of nearly five. In many of the shops visited, opportunity costs exceeded environmental management costs by a factor of ten.

The opportunity costs likely to generate motivating numbers for the facility to consider are:

• Value of lost process solution and chemistries

• Excess water use

Once this understanding is gained, it can be used as part of the analysis for other direct 'big-ticket' environmental accounting cost items such as waste-water treatment, sewage and waste disposal. For example, every gallon of excess water use creates an environmental management cost that is typically undervalued. Care should be taken always to divide total environmental management costs related to waste-water into waste-water volume, since environmental accounting analysis has shown that waste-water costs can be surprisingly high.

### Quantifying or Qualifying Costs

Efforts to quantify environmental management costs begin with the ability to access cost information. Assuring the accuracy of this quantification effort is equally important.

---

3. [(Sludge at 1% zinc metal/ton = 320 oz zinc metal/ton) ÷ (zinc cyanide plating solution at 4.7 oz/gal)] × (zinc cyanide plating solution at $4.50/gal) = $306.38.

Quantifying direct and indirect environmental management costs, as described earlier, can be quite straightforward for certain items (those with invoices) and less so for other elements (e.g. labour component, liability). The challenge found throughout project research was the ability to quantify these costs at a level of detail to help facilitate an environmental accounting analysis. Electroplating facilities visited and contacted for this project tended to aggregate many costs into very general categories, at least when preparing financial statements such as balance sheets and profit and loss statements. Moreover, 'mining' this information at a level of detail necessary to enable allocation is potentially an expensive thing for a facility to do and there often appears to be no other particular production or management reason for collecting and managing information at this level of detail. Without other opportunities for using this information, its collection and management is even less likely for a facility.

Labour costs are often major components of 'hidden' environmental management costs. Both direct and indirect labour are captured on the general ledger and subsequent statements as aggregated costs. Supporting documentation, usually time sheets, use codes, part numbers, job names or similar appellations are used to distinguish between work activities for production and support staff, and usually use considerably less detail for salaried personnel, e.g. a manager in charge of research and record-keeping for regulatory compliance. These can be used to help allocate costs to responsible products or processes. However, two issues arose that affected the use of labour costs in an environmental accounting analysis:

- **Minimal or no effect on cash flow.** In no facilities reviewed would a strategy offer the potential actually to reduce labour costs in such a way that cash flow is affected. Discussions with facility representatives on the labour valuation issue suggested a hesitancy to assign 'savings' to activities that did not directly affect payroll or payables.

- **Need to recognise labour trade-offs.** Facilities recognised that reducing necessary investments in environmental labour offered the potential to reallocate human resources to more productive activities. However, finishing facilities are quick to point out that whatever pollution prevention technique or technology is being evaluated is likely to also have its own increases in labour costs.

Environmentally preferable processes or recovery technologies typically have tighter operating windows requiring more careful monitoring practices and new standard operating procedures. Recovery technologies typically are quite sophisticated, demanding increased maintenance, monitoring, training and control efforts. The National Association of Metal Finishers has reported technology failure rates of 30%–40% for various types of recovery technology within the industry, and experts believe a portion of this failure may be due to labour issues and the sophistication of the technologies.

It is clear that environmental labour issues are important and relevant to an analysis, and decreases in regulatory-related labour activities of all types is certainly in the best interest of a facility. Many facilities also seem to benefit from quantifying these labour costs to obtain an understanding of the potential productivity gains that may be available. However, for actual economic evaluations of projects, many of these labour issues might be best factored into an analysis in a qualitative fashion.

Quantifying the opportunity cost component is an ongoing challenge of finishing operations because it requires an ability to assess 'how good the process could be'. Quantifying the opportunity cost component faces two key problems:

1. **Accuracy of cost information is frequently questionable.** The reality in electroplating is that only information based on an invoice or similar documentation can be trusted to be completely accurate. Data derived from internal records and calculations, or even from interviews of personnel, are at their most reliable still highly variable, simply because of the ever-changing nature of electroplating operations. Some facilities will add to that variability with spotty record-keeping, personal bias and lack of equipment for proper monitoring. The only exceptions seem to be records that are also required to be kept for other purposes, such as demonstration of compliance with customer specifications or documentation for quality management systems.

2. **Need for strong supporting production information systems.** Robust production management information is needed to be able to support the environmental accounting analysis and numbers. In parallel with the system of book-keeping and accounting is the record-keeping and analysis required to produce job cost estimates, schedule production and maintenance, do short- and long-range company planning, and track and assure quality. This system tends to be less accurate, mostly because it is heavily influenced by subjective responses, records kept by humans and the ebb and flow of customer demands. This is the very system that must produce documentation that would be used to cross-check estimates (acid purchased versus acid used), analyse options (length of time required to rinse 'thoroughly'), and gauge information needs for implementation (estimated downtime for conversion to non-cyanide zinc).

## Allocating Costs to Products or Processes Responsible for their Generation

The question of whether to allocate to processes or products depends on the type of project being investigated. Some pollution prevention technologies—ion exchange, reverse osmosis, replacement of organic solvent cleaning, cyanide elimination, and others—are best supported by cost allocations to the process level. That is to say, as many costs as possible that are associated with a given process (the 'target' of a pollution prevention technology) should be disaggregated and made part of the pollution prevention project analysis. Cost of process solution, cost of utilities and cost of labour all become a legitimate part of the analysis. In that case, the technologies and allocation level 'match'; what is needed happens to be for the most part available.

For other projects, especially those having implications regarding water use, procedure modifications or process redesign, allocations to the process level are often insufficient. The success of these projects is often product-specific, meaning that the relative effect of a product on the costs of operating a process is at least as important as the costs of process chemistry and waste management costs associated with a process. In these cases, pollution prevention project analysis requires knowing not just how much a given process costs in terms of process operations and waste treatment, but also what percentage of those costs are caused by Part A, Part B and Part C. Cost allocation at the product level is quite rare.

For most facilities visited, *how* to allocate was typically a greater issue than on what level to allocate. As in quantifying costs, individual pieces of environmental accounting analysis vary in degrees of difficulty with respect to allocating to their sources. Three allocation options that facilities may consider are allocations based on actual contribution, allocations based on estimated calculations, and allocations based on professional judgement.

**Allocation Based on Actual Contribution.** Most of the 'hidden' environmental management costs are extremely difficult to allocate based on their actual contributions. The number of product and process variables affecting actual contribution are typically too complex and systemic for most finishing facilities to be able to (or want to) generate actual contribution numbers. This is true even for the high-profile environmental management costs. For example, in many shops, a large environmental management cost is the system used to satisfy the limits on concentrations of various materials in waste-water discharge permits. In order to allocate accurately to the process level the costs of creating and managing metal hydroxide sludge, a shop would have to account for:

- Variation in product mix
- Potential variation in operating procedures
- Variation in chemical concentrations
- Potentially numerous sources for each constituent of concern

A review of the literature found company examples that had documented in excruciating detail the relative contributions of different parts and solutions to different waste-water streams. Site-specific data were in fact generated, but so many variables were controlled as to render general transferability of this data to other facilities completely moot. The research required to allocate large costs such as water, sludge generation and concentrated process solution disposal appears to be too complicated and too expensive to be conducted by any but the most sophisticated electroplating operations.

**Allocations Based on Estimated Calculations.** Given these inherent difficulties, facilities have responded by allocating these costs based not on actual but estimated contribution. Estimations are typically generated in one of two ways: using standard surcharges or 'kickers' which the facility uses to price products; or using activity-based measures.

Standard surcharges are commonly found in electroplating. A finishing operation might add 15% to the cost of the chrome-finished products to cover treatment costs, but a 30% charge to cover cyanide processes. These percentages might also be used for cost allocation purposes. This approach is well accepted in the industry. WRITAR found evidence in the research of using 'default values' based on sources such as the plating literature, discussion with peers and personal or organisational experience. However, as with labour costs, these surcharges are 'loaded' with extraneous costs not pertinent to a pollution prevention project analysis, typically assembled to recover production costs, not to reveal activity costs; and therefore yield high levels of inaccuracy.

An alternative way of allocating is through engineering estimates. Environmental management activity is first subdivided into separate categories of activity (e.g. cyanide destruction, chrome reduction, etc.) These activities are then assigned a cost, in this case usually on a 'per-gallons-treated' basis by calculating the predicted costs of treatment for a specific concentration, flow, method, etc. These controlled variables are chosen to be as nearly representative of actual facility conditions as possible, and the cost is 'built' by analysing each step of the process. Engineering estimates have higher levels of accuracy, and the more closely that the controlled variables reflect reality in the facility, the more accurate the estimates will be. Developing engineering estimates, however, is a sophisticated, technically demanding effort, and may be outside the scope of non-process experts.

Allocations based on an appropriate production factor (square feet processed, hours of operation, etc.) is perhaps the simplest type of estimated calculation and a possible short-cut to full engineering estimates. This allocation approach can be complemented by electroplating literature which will contain published industry averages such as 'dollars per gallon treated'. Users, however, must recognise that these estimates are averages and can vary significantly based on the production factors cited earlier (product configuration, concentration, flow, other materials in waste stream, etc.) As an example, in the hexavalent chrome facilities visited for this project, the dollars-per-gallon chrome reduction estimates varied by a factor of 22. As a result, production factor allocations are probably best used when high levels of accuracy are not determined to be critical.

**Allocations Based on Professional Judgement.** Another response to the difficulty of deriving actual values from records in electroplating is deriving relative contributions of the different components of an aggregated environmental cost using the best judgement of facility staff. Because real numbers are not generated, this approach is not used for capital budgeting. However, it could be used for targeting opportunities and prioritising improvement activities.

To evaluate the efficacy of this approach, WRITAR had facility staff generate contribution estimates based on professional judgement. The costs were then allocated accordingly. These cost allocations were then cross-checked through WRITAR's assessment procedure.[4]

The best 'match' between 'estimated' allocation and assessment-based allocation was obtained when the following constraints were met:

- Production records were already gathered and analysed for other purposes.
- Process analysis and control was documented and cross-checked by supervisor.
- Approval was required for discharge to treatment.
- Waste treatment and environmental management staff were familiar with production operations.

However, in only three facilities did staff allocation estimates come within 25% of what were eventually accepted as somewhat accurate allocations (see definition on page 214). In several facilities variations were as high as 400%–500%.

---

4. WRITAR's assessment procedure entailed an allocation based on engineering estimates and materials balances. Materials use amounts were generated from a variety of production records and cross-checked. Electroplating engineering handbooks were then used to calculate the amount of process consumption that would be necessary to accomplish the facility's production volumes and specs. The balance would be considered lost process material requiring treatment.

Many interpretations are possible of the basis for these inaccuracies. The most likely, based on our research, is that many electroplating operations rely on cost structures that are very general unless allocation is required by some outside force (customer, shareholder, banker, etc.). This leads to working with and accepting information that is perfectly acceptable for day-to-day operations, but which can falter when brought to bear on a more rigorous and detailed effort such as improvement targeting. As a result, allocation based on professional judgement should be interpreted with caution.

### Integrating Costs into Decision-Making

Environmental accounting has potential application in several areas of facility decision-making, including capital budgeting and targeting of improvement actions. The following section describes some of the research findings and conclusions about environmental accounting use in these two decision situations for electroplating shops.

*Capital Budgeting Decisions.* In the sample of shops visited, conventional costs were routinely captured in project analysis. In six of the shops in our sample, parts of environmental accounting addressing direct and indirect costs associated with environmental management had been used to justify pollution prevention projects requiring capital investments. The parts of environmental accounting selected focused on large environmental costs that could be fairly easily allocated because they were either comprised of the costs to manage a single waste or type of waste that would be eliminated (e.g. switching from organic solvents for degreasing to aqueous alkaline cleaners) or because the linkage to environmental management costs was completely obvious (e.g. eliminating cyanide inputs from the flow-through portion of the pre-treatment system which reduced disposal costs and eliminated the need for a unit operation). Many shops have been using parts of environmental accounting without knowing it or calling it that, much in the way that shops perform 'pollution prevention' without calling it that in the normal course of improving operations.

The idea of consistently applying a full environmental accounting framework including contingent and image costs did not seem likely to any representatives in the sample. There is a very strong sense of individuals using only what is (1) of high potential in terms of cost reduction; and (2) solid in terms of accuracy. Although facilities found indirect costs (those not directly affecting payroll and payables) useful to consider and quantify, metal finishers demonstrated hesitancy to factor them formally into an actual economic evaluation of a project. Electroplating facilities were strongly oriented toward using conservative cost estimates in decision-making.

A broader issue concerns the type of capital projects best supported by environmental accounting analysis. The optimal situation occurs in capital projects in which entire operations such as cyanide destruction, chrome reduction or solvent cleaning can be eliminated. This avoids the need to engage in difficult allocation mathematics. However, research found that, in most cost areas, process substitution results in incremental reductions rather than complete elimination. For most of the facilities, some cyanide destruction and chrome reduction capacity needed to remain because of other processes in the facility. While theoretically estimated savings based on volume reduction might be generated, the problems of data quality and tracking and allocation challenges arose. No shop in the zinc cyanide sample could approach being able to figure out a percent contribution by source for cyanide destruction. Difficulties of this nature are found in other environmantal accounting cost categories including waste management and regulatory compliance costs. As a general rule, the best candidates for a more fully developed environmental accounting analysis are those that are focused on a single waste stream and address complete, rather than partial, elimination of environmental cost categories.

Perhaps the most useful application of environmental accounting in the capital budgeting area is to help the facility make better choices on the types and sizes of capital project needed in the facility. The need to invest in process optimisation before making capital budgeting decisions has long been recognised in electroplating. Perhaps the best example of this principle can be found in efforts to create a closed-loop waste-water system. According to *Plating and Surface Finishing* (October 1993), a pursuit of zero water discharge without upfront process optimisation can cost a facility 2–5 times *more* than conventional end-of-pipe treatment. Using environmental accounting to examine costs of loss material in a facility and identify where process improvement is prudent can provide valuable input into facility capital-planning.

***Targeting and Prioritising Improvement Opportunities.*** Environmental accounting also has value in targeting and prioritising improvement opportunities for facility operations. However, the key to targeting was again found to be rooted in understanding and quantifying the cost of the lost material portion of an environmental accounting analysis. Moreover, to obtain the best results, environmental accounting is best used in conjunction with traditional targeting methods such as reject rate analysis to get the best environmental as well as financial benefit.

Targeting improvement areas based solely on relative contributions to facility environmental management overhead and ignoring lost material issues may misdirect facility attention from where the greatest gains could be realised. WRITAR compared the results of targeting based solely on contributions to environmental overhead with those based on an assessment approach based on materials accounting and lost process solution. The best 'match' between allocated-overhead-based targeting and assessment-based targeting was obtained when the following constraints were met:

- Limited variation of products
- Limited use of multiple-use process solutions (other than rinses)
- Limited total number of processes

The project lacked sufficient resources to define 'limited' in much detail, but this much is clear: when the total of the number of products plus the number of multiple-use solutions plus the total number of processes (yielding total 'factors') exceeded twenty, the overhead-focused methodology actually became *worse* than a guess, since it actively misdirected the investigation because of its issue-specific approach. As an example, a manufacturer of nickel-plated hand-tools:

- Has **three** products
- Uses **two** different cleaners, **two** different acids  } Total 'factors' = 9
- Uses **two** electroplating solutions

In this extremely simple, probably anomalous, production situation, the overhead allocation method and the assessment method both pointed to the nickel-plating solution as being the most deserving of P2 attention with the goal of reducing environmental management costs. (In this case, the calculated approach focused on allocating water, sewer and sludge costs because of the effect of waste-water volume on total sludge volume.)

In a more typical (at least for this coating/product mix) production situation, a manufacturer of enclosures for the electronics industry which are zinc-plated and chromated:

- Has (on the one line chosen for analysis) **twelve** 'primary' products
- Uses **two** cleaners, **three** acids
- Uses **one** electroplating solution, **three** chromate solutions

} Total 'factors' = **21**

In this case, the target recommended by environmental overhead allocation (once again allocating water, sewer and sludge costs) was the chromate portion of the process. However, the assessment-based approach pointed to the zinc solution and related process control issues (because of the effect on reject rates and corresponding impacts on materials use and waste issues), with the chromates a close second. As the cases examined increased in 'complexity', at least as indicated by the 'factors' approach used above, the two approaches pointed less and less frequently in anything like the same direction.

Targeting based solely on the allocation of environmental overhead may not result in the best projects in terms of environmental and financial returns. Targeting based on an environmental accounting analysis which first examines loss process solution costs and then allocates environmental management cost components can be a powerful targeting tool.

### Applying Environmental Accounting to Hexavalent Chrome and Zinc Cyanide Substitution Projects.
To examine and illustrate some of the issues and challenges of environmental accounting analysis in pollution prevention projects, WRITAR investigated two plating processes that appeared to be best tailored to potential environmental accounting application: substitution of zinc cyanide processes and substitution of hexavalent chrome processes. Besides being regulatory targets and therefore the focus for existing change efforts, these two processes appeared to satisfy the condition of well-defined, technology-based changes that appear to be very amenable to environmental accounting analysis.

WRITAR visited eight shops using zinc cyanide processes and ten shops using hexavalent chrome processes for decorative plating. Working jointly with the facilities, WRITAR applied an environmental accounting analysis to the decision on whether to switch to appropriate alternatives.

With the exception of two cost categories, utilities and insurance, WRITAR found that applying environmental accounting methodologies offered the potential for elucidating and incorporating other costs that would help justify a process change. However, three issues arose which are likely to be found in other types of project evaluations:

1. **More cost savings are not necessarily the implementation key.** Because of the regulatory pressures, nearly all the firms were already quite aware of the general economics of alternative processes. That the economic justification could be 'beefed up' was generally not an issue. General resistance to change, inability to change due to perceived or codified requirements, labour skill-base and other factors rather than the numbers were stopping implementation. This issue is not a fault or limitation of environmental accounting; rather, it simply demonstrates that cost is just one of a multitude of potential issues affecting pollution prevention implementation.

2. **Environmental accounting analysis is better served by decision situations featuring complete, rather than incremental reductions.** As discussed earlier, allocation challenges created the major roadblock for environmental accounting analysis. In facilities where entire operations such as cyanide destruction or chrome reduction could be eliminated, environmental accounting offered tremendous potential. The corresponding environmental management and regulatory costs associated with cyanide destruction could be readily assembled and factored into the analysis. However, research found that, in most cost areas, process substitution resulted in incremental reductions rather than complete elimination. For most of the facilities, some cyanide destruction and chrome reduction capacity needed to remain because of other processes in the facility. While theoretically an estimated savings based on volume reduction might be generated, the problems of data quality and tracking described earlier immediately came to the surface. Difficulties of this nature were found in other environmental accounting cost categories, including waste management and regulatory compliance costs.

3. **Indirect labour costs exist for alternatives as well.** Indirect labour 'savings' associated with environmental management could be roughly approximated, but full costing demanded 'equal time' on the other side of the ledger and labour cost changes were approximated for alternative processes. Many of the environmental management labour savings were offset by increased labour costs associated with the alternative processes which generally demand more careful maintenance and control practices and embodied their own training issues.

## ▌ Guidance for Investigating Lost Material Costs in Electroplating

Because the environmental management cost structures of a facility are really a function of how, and how well, materials are used in operations, obtaining an understanding of existing process performance and materials use is a good starting point for environmental accounting in electroplating. Investigators should begin by exploring what types of effort have been made to:

- Improve process controls
- Reduce water use
- Understand and optimise materials use
- Implement and monitor standard procedures

to obtain a sense of where facilities have and have not invested in process improvement.

Once this understanding is gained, the economic justification for further activity can be made through an application of environmental accounting. The value of waste as lost process solution is the critical calculation worth performing, since it provides a sense of the opportunity costs available to the facility, identifies the best cost saving targets, and it can be very useful as part of the disposal costs portion of an environmental accounting framework. This can be done through the recommended steps described in the two appendices. These opportunity costs can then be added to the environmental accounting-generated environmental cost numbers (which capture the environmental management cost components) to get the true 'cost of wasting'.

Linking environmental management costs to ongoing improvement needs proved to be another useful way of generating facility interest in pollution prevention. Environmental accounting can be used to bolster the economic argument and identify the facility's 'true cost of quality'. The key to realising the value-added of environmental accounting in electroplating is to link this tool with materials balancing and materials accounting. The opportunity costs of wasted raw materials and materials losses in electroplating are not only the most significant costs to consider, they are the linchpin in understanding the reasons for the facilities' environmental management cost structures.

## ▮ Appendix A: Calculating Value of Process Solutions and Lost Process Solution

**Step 1** Determine total annual cost of material inputs.

**Step 2** Determine list of material inputs that individually are responsible for at least 5% of the total material input costs.

**Step 3** Determine value of individual process solutions on annual basis and in cost per litre of solution. Start with electroplating solutions, then do acids and cleaners. Occasional-use solutions (strippers; chromates; other special-use surface treatments) are optional at this time; focus on 'big tickets' first.

> *Step 3a* Determine 'formula' for each process solution.
> – List of constituents
> – Concentration of each constituent
>
> *Step 3b* Determine cost of each constituent.
> *Step 3c* Calculate value of solution.

**Step 4** Rank process solutions by cost.

**Step 5** Choose top-ranked electroplating process solution. Determine total metal purchased for that solution as kg/year. Assure that all sources are accounted for. For example, nickel metal in an electroplating process solution is derived from, at a minimum:

- Solid nickel anodes
- Nickel sulphate solution

and may have other sources.

**Step 6** Determine total metal plated from that solution as kg/year. This is done by determining total square decimetres of product processed through the solution, and multiplying by the average thickness of the plated coating. Convert to kg using weight-to-thickness tables.

**Step 7** Subtract total in Step 6 from total in Step 5. Difference is metal lost to dragout, although recognise that incorrect data can skew the analysis.

**Step 8** Repeat Steps 5–7 as necessary until 80% of costs due to purchase of electroplating process materials has been analysed.

## ▌Appendix B: Full Environmental Accounting Costing of Rejects

**Step 1** Determine list of most common reject situations.

**Step 2** Describe actions taken to correct rejects, starting with most common cause or problem.

- Inspect
- Sort and re-plate only bad parts
- Strip and re-plate all parts
- Purchase new parts and plate

**Step 3** Determine cost of each action.

- Cost to detect rejects (inspection; return from customer)
- Cost to correct reject (purchase new parts; strip and re-plate parts; spot re-plating)
- Cost of lost production time (rejected parts took the place of parts that could have been sold to generate income)
- Cost of treatment and disposal of materials associated with corrective action
- Overhead costs associated with corrective action

**Step 4** Repeat Steps 1–3 until 80% of reject volume has been analysed.

# 12 Reducing the Uncertainty in Environmental Investments

## Integrating Stakeholder Values into Corporate Decisions[1]

### Graham Earl, Tuula Moilanen and Roland Clift

## ▌ A New Agenda: Integrating Stakeholder Values in the Corporate Sector

The environment, and how governments and companies impact on it through their decisions, is a complex and often controversial and emotional subject. In part, this is explained by the potentially numerous and diverse stakeholder groups[2] affected by such decisions, many of whom on the surface seem to have contradictory and highly emotive environmental stances. On the other hand, the decision-makers[3] themselves may be hesitant to engage in debate with these same groups, driven by a fear that this may result in a stalemate of conflicting demands or, even worse, in a lengthy, acrimonious and confusing public debate.

Effective management of such debates is complicated by the fact that decisions that can impact on the final environmental performance of a project typically involve multiple objectives and therefore value trade-offs. Furthermore, it is likely that the decision process may hinge on intangible benefits, such as environmental image, and the

1.  This paper summarises part of the work carried out by Graham Earl for the degree of Doctor of Engineering at the University of Surrey, UK. Financial support from EPSRC and Paras Ltd is gratefully acknowledged. Special thanks are also extended to British American Tobacco, especially to Dr David Robinson and Mr Steven Hemsley, for their time and efforts and for providing the opportunity to apply this research, and also to Mr Graham Long (of WS Atkins and the University of Surrey) for some characteristically stimulating suggestions.

2.  Stakeholder groups are defined here as a more-or-less organised group of people who stand to be affected by the implications of a decision and who can directly or indirectly influence the consequences of the decision. For example, enraged customers may choose to boycott a certain product.

3.  Decision-makers are the stakeholder group charged with analysing and justifying a decision choice to all other stakeholders of the decision.

benefits and costs of the investment may evolve over a long time-horizon. From a company's perspective, all of these factors introduce uncertainty and financial risk into the decision process.

The predominant theme that emerges from the foregoing discussion is that stakeholders and their needs and desires are becoming increasingly important in shaping corporate environmental investment decisions. Indeed, one does not have to look far to see examples of this: Shell's initial foiled attempt to dispose of the Brent Spar was unusual in the amount of press coverage it generated, but is now regarded as an example rather than a unique event.

Put simply, the way in which stakeholders react to company decisions can have a dramatic effect on the success or failure of a company's investment decision, be it on sales revenue, market share or any other strategic indicator.[4] In the case of environmental investments, which have the potential to impact a widely disparate set of stakeholder groups, there is a clear incentive for decision-makers to consider their needs and priorities prior to making an actual investment choice.

Although businesses have always considered numerous factors in their decision process, such as legislation and changes in consumer tastes and behaviour (these factors are called business drivers), the environmental business driver is both relatively new and complicated by the fact that it can affect many aspects of a company, ranging from its operations through to its operating environment and equity value. Environmental issues, in turn, may be promoted by stakeholder groups that will be new to a company, and differ from traditional stakeholders, such as shareholders, in that their motivation is not purely financial. Consequently, a company may perceive its own objectives to be contradictory to those of environmentally oriented stakeholder groups.

This situation adds new and difficult challenges to a company decision-maker, and ones that are not readily resolved through traditional tools such as net present value (NPV), discounted cash flow (DCF) or payback analysis. Put simply, it is not enough to produce financial figures to justify environmental investment decisions if the analysis does not attempt to tackle the uncertainty inherent in these figures.

This chapter argues that the best way to reduce these uncertainties—and hence confidently to justify the final decision to an often intransigent and sceptical audience—is to find out what the priorities and values of the decision's stakeholders are and to incorporate these into the decision process.

## ▌ The Stakeholder Value Analysis Toolkit

The driving force behind this research has been the hypothesis that the uncertainties and risks[5] faced by decision-makers charged with making environmental investments can best be managed by having a knowledge of the values that stakeholders place on a decision, or, in other words, knowing the 'total value' of the decision. As the previous section has argued, the value of a decision to a company is driven by the expectations,

---

4. This no doubt explains why Shell is now committed to public consultation, under the heading 'The Way Forward', to try to define the eventual fate of the Brent Spar (Shell UK 1996).

5. The terms 'risk' and 'uncertainty' are often used interchangeably, but are distinctly different. Uncertainty relates to a situation in which the probability distribution and indeed even the type of event is not known. Risk relates to a situation in which the probability distribution of an event is known.

perceptions and priorities of the decision's stakeholders. From a decision-maker's perspective, this is problematic because not all of these values can be transposed into monetary terms, but will take different forms, some of which on the face of it seem intangible. It is with these challenges in mind that the work summarised here has been carried out to develop the Stakeholder Value Analysis (SVA) Toolkit, which is a methodology that can capture the total stakeholder value of environmental investment decisions.[6]

Figure 1 illustrates the SVA Toolkit structure, and its relationship to a stakeholder value analysis methodology, which aims to capture, analyse and maximise stakeholder project values of an investment decision and in so doing to minimise the risks and uncertainty of the investment decision. The basic elements of the SVA methodology are:

1. The inputs to the Toolkit models
2. The Toolkit's decision and analysis models, which are all interlinked and can be used in any combination or form, and include:
   - The Multi-Attribute Decision Environment (MADE), which is the central tool used for defining, valuing and processing less tangible stakeholder values
   - The Paras financial model, which has been developed by Paras Ltd and supports the financial quantification of a wider array of financial attributes
   - Sensitivity analysis and option creation models to support the analysis
3. Outputs from the models for use in risk and strategy management of the decision

The basic sequence of the methodology follows the heavy full lines in Figure 1, while feedback paths are shown by lighter dotted lines.

Reference to Figure 1 shows that the SVA methodology provides a structure with a number of different paths for the stakeholder input information to follow before it can be used to make an investment decision. The subsequent subsections aim to describe how this information is gathered and used within the Toolkit's models. This will start with descriptions of the two main components: the Paras and MADE models. Following from this, the synergistic value of the SVA Toolkit will be developed by showing how the models that comprise the Toolkit can be combined to produce quantitative metrics to support the management of a company's risk and its strategic objectives.

### The Paras Model
The Paras model was developed in response to industry's call for a practical methodology to support the holistic financial appraisal of environmental investments, particularly those that aim to employ 'Clean Technology' solutions. While the Paras methodology encompasses the principle of using an expanded cost and revenue inventory in the analysis, it differs from what have become known as total cost assessment (TCA) methods, the most notable examples of which have been developed by the Tellus Institute (1991), General Electric (1987) and George Beetle Company (1989).

---

6. The SVA Toolkit builds on the previously developed Multi-Attribute Decision Environment (MADE) model (Earl 1996a) and the Paras model for environmental investments (Earl 1996b; Moilanen and Martin 1996).

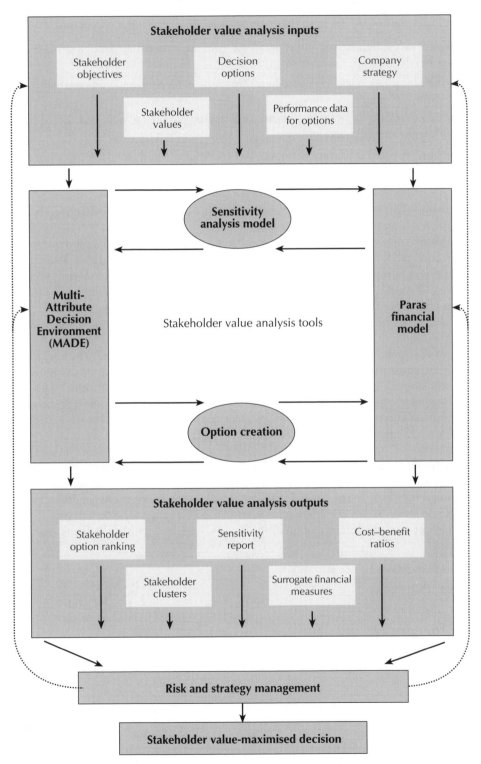

**Figure 1:** Stakeholder Value Analysis Toolkit

The Paras method differs from these methods in the following fundamental ways.

- The investment is put into a strategic context rather than seen as an isolated decision.
- The interdependencies between internal and external parameters are made explicit.
- The investment analysis is operationalised by providing suggested lines of thought through *aides-mémoire*.
- The expected monetary value (EMV) technique is built into the model to support not only the definition of financial risk but also the expansion of cost and benefit categories.

The principle features of the Paras model that support these facets are described below.

**Modular Approach.** Conceptually, the Paras financial model uses a modular structure. Each module is defined within this context as an operational area in a company within which investments can directly be made or where investments can have an impact. Figure 2 illustrates the structure of the model, the modules defined by it and their interdependencies.

To illustrate the 'module' concept, let us take an investment in a new product design as an example. It would be wrong to assume that the impacts of such an investment stop at the boundaries of the design module. First, a product may need to be researched and developed within the R&D function. Second, the new product may require different manufacturing processes and/or materials. Third, the product needs to be sold and this may require additional marketing input. Last but not least, consideration must be taken of the impact that the new product could have on the company's external and internal image.

**Holistic Approach.** The Paras model actively encourages and supports the quantification of those costs and benefits not typically included within traditional investment appraisal methods. In the case of environmental investments, these can have a considerable influence on the financial consequences of a project. Below are listed some of the costs and benefits whose quantification is supported within the Paras financial model.

- Hidden costs
- Liability costs
- Insurance costs
- Financing costs
- Environmental benefits/costs
- Public image benefits/costs[7]

---

7. The SVA Toolkit combines the MADE and Paras financial models to allow decision-makers to include the public image effects of an investment into the decision analysis without needing explicitly to assign a monetary value to this element.

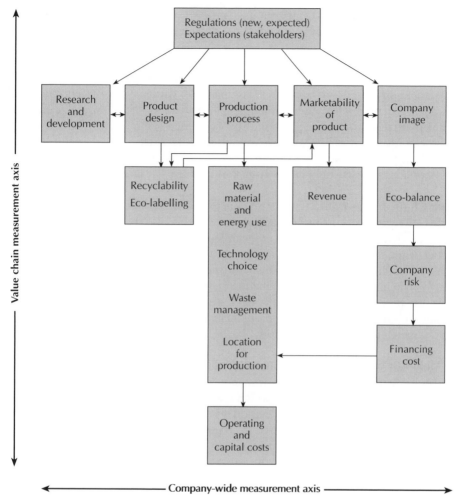

**Figure 2:** Structure of the Paras Financial Model

*Source: After Moilanen and Martin (1996)*

***Risk Evaluation.*** Traditional investment analysis techniques (e.g. net present value and internal rate of return) do not allow risk to be individually assigned to the cost and benefit streams in an investment appraisal. This has led to misapplication of the models through the use of various contingencies and abnormally high discount rates. For example, if the discount rate is raised to counter worries concerning environmental liabilities, this risk increment is wrongly assigned to all the parameters of the analysis.

The recommended technique is to use expected monetary values (EMVs) of projects, which is the technique adopted in the Paras financial model. In order to derive the EMV for a particular parameter, the following needs to be defined.

- The possible values for the variable parameters in a scenario
- Their relative likelihood of occurrence (expressed as probabilities summing to one)

The EMV is then calculated as the summed product of these values and probabilities. Each EMV value can be discounted in the normal way to give a risk-weighted internal rate of return (IRR) or net present value (NPV).

In the case of an environmentally driven investment, the EMV approach is particularly effective, since some parameters have high discounted cash flow effects but their probability of occurrence is low. For example, while an investment may carry with it the potential for an environmental liability (in the US this would be enforced by Superfund legislation), the probability of occurrence is often a function of the technology employed, and a critical factor for option differentiation. Although EMVs are a useful technique for measuring and managing investment risk, the technique on its own does not explicitly help decision-makers quantify the uncertainty of a decision.

**Value Chain Integration.** The Paras financial model has been designed to measure not only the values occurring 'across the company' but also those occurring along the full 'value chain'.[8] To illustrate this we can look at a simple example, set within the product design module (refer to Fig. 2). Here, as with design for the environment (DFE), the model would guide the decision team into considering the recyclability of a design, since this can have a direct cost implication on the company if, further down the value chain, the product is subject to 'take-back' legislation. Clearly, investment options that allow for recyclability and re-use will incur lesser costs on the company than those that do not share these characteristics.

### The Multi-Attribute Decision Environment Model

The Multi-Attribute Decision Environment (MADE) model provides a framework to capture and process stakeholder-wide views, opinions and knowledge (inputs) on a decision. Using a multi-attribute decision methodology, the financial and non-financial multivariate metrics are combined to produce output performance indicators that can usefully be employed by company decision-makers to assess the stakeholder-wide implications of possible decision options, and also to assess the implications of the decision on the company's longer-term strategic objectives. The steps followed by the MADE model are illustrated in Figure 3 and detailed in the subsequent sections.

**Stakeholder Identification.** The first step in the MADE process involves identifying the relevant stakeholders of the decision.[9] In this context, the definition of a stakeholder encompasses not only stakeholders external to the company, such as customers, but also includes internal stakeholders, such as employees and company owners.

The result of the identification process may be a long list of stakeholders. In the course of this research, decisions with implications on a regional scale have generated lists of 15–20 organisations. Winterfeld and Edwards (1987), on the other hand, quote lists of 40–50 organisations, which is not surprising given that their research dealt with wider national policy decisions.

---

8. The value chain describes the value of a material as it moves along its life-cycle.
9. Although shown here as a component of the MADE model, the stakeholder identification process step is equally applicable to the Paras model.

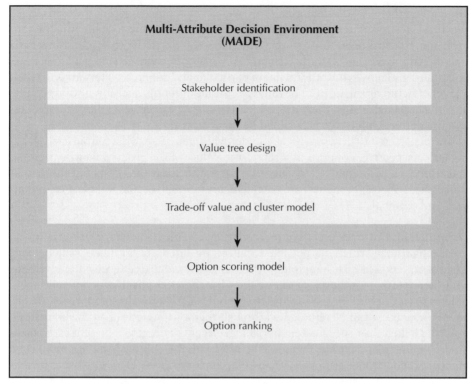

**Figure 3:** Multi-Attribute Decision Environment Flow Scheme

This research has found that it is relatively easy to classify these groups on the basis of prior knowledge about the views and the people they represent. Thus, although fifteen stakeholder groups may have been identified, this will not necessarily result in fifteen different value systems. It is therefore neither necessary nor advisable to interact with all groups. Conclusions drawn from applications of this research have shown that it is possible to limit the stakeholders to a short list of between five and ten groups without compromising the integrity of the analysis.

Clearly, the value of the SVA methodology hinges critically on the involvement of stakeholders in the process. Therefore, while it is desirable to limit the number of groups in the process from a logistic and practical viewpoint, this aim has to be reconciled with the very real danger of isolating a potent and influential group. This issue is discussed further below.

**Value Tree Design.** The next step in the process involves establishing the concerns of each stakeholder, the aim being to identify and structure the values (performance attributes) that they consider relevant for evaluation of the decision options.

The MADE methodology includes a generic questionnaire to use on a one-to-one basis with the leading stakeholder representatives. For each of the applications discussed later (see below), the approach has remained relatively stable apart from slight tailoring of the questionnaire to issues specific to the case.

Following directly from the interviews, the analyst can identify a hierarchy of values for each interview, referred to here as a value tree. A value tree's hierarchical levels follow directly from definition of an overall aim for the project at the top (termed the 'top-level aim'), to performance attributes that contribute to achieving the top-level aim, to third-level sub-criteria (which jointly contribute the second-level criteria directly above) and so forth. The process starts at the more general (and sometimes uncertain) and moves towards the more particular and concrete. The aim is to reach a level of detail at the bottom of the tree so that different decision options can be directly measured against the bottom-level criteria. A specific example of this process and the derived value tree is outlined below.

The final step is to generate a stakeholder-wide value tree, which, from each stakeholder's perspective, contains as a subset the values of the group. Likewise, from an operational view, Keeney (1976) suggests that the decision analyst should ensure that the set of attributes (branches) are complete, operational, decomposable, non-redundant and minimal.

***Definition and Clustering of Value Trade-Offs.*** Value trade-offs or weights provide the mechanism by which it is possible to reflect the different priorities of stakeholders. The basic premise is that important attributes in the value tree receive high weights, while unimportant attributes get low weights. It is therefore possible for different stakeholders to agree on a common value tree even though their value preferences are different, since these can be represented through their own weighting of the branches.

While stakeholder-specific weights offer an elegant method to represent each stakeholder's value tree, there are obvious difficulties with this approach. First, given a large number of performance attributes, it can be a complex and bewildering task to try simultaneously to assign actual numerical weights to all the performance attributes shown in the value tree.

Second, averaging of the weights defined for different stakeholders can seriously degrade the value of the analysis. For example, if one stakeholder group assigns a very high weight (95%) and another a very low weight (5%) to a performance attribute, the average of these two weights, 50%, is not necessarily representative for either or both stakeholder groups.

The MADE model employs two separate methodologies to deal with these challenges.

*Pairwise comparison model for weights and consistency definition.* The methodology used to support this task is based on the pairwise comparison technique developed by Saaty (1980). The MADE model has drawn on Saaty's pairwise method because it offers an intuitive and appealing means of eliciting the weights that stakeholders place on the branches (performance attributes) in a value tree.

The principal advantage of the pairwise comparison method over other weighting techniques is that it allows users methodically and systematically to determine their weights for a value tree hierarchy's performance attributes simply by comparing pairs of attributes one at a time.

The analysis of the pairwise preference data produces a distribution of criteria weights and also provides a measure of consistency (known as a consistency index). The method has also been found to be relatively simple to use, which is an essential ingredient if the weights of lay-persons are sought.[10]

---

10. It is fair to say that there are numerous opponents (e.g. Islei and Lockett 1988) and proponents (e.g. Harker 1989) of Saaty's pairwise technique as a method for weight definition. It is sufficient to say here that the discussion centres on technical issues: for example, which mathematical

*Stakeholder cluster method.* The underlying basis for the cluster model is that it is possible to find groups or 'clusters' of stakeholders who share broadly similar priority weights. The idea therefore is to try and classify stakeholders by their value systems, but without compromising their individuality.

Although the 'cluster' process may sound confrontational, in practice the research has found that group members generally find that the process leads to a fair and just representation of their priority weights. The possible limitations of stakeholder identification and clustering are discussed below.

**Option Scoring.** This step in the process involves scoring the performance of each option against each of the performance attributes defined at the lowest level of the value tree. This step of the process is generally much less controversial than the value judgements required in the weighting process. In fact, if the value tree has been carefully designed, the performance attributes being scored should be well suited to quantitative objective scoring of options by suitably qualified experts.

The MADE model uses an interval-scale 'word' model technique. A 'word' model uses an absolute scoring system, which is calibrated against word descriptions to ensure consistency. Figure 8 shows an example word model developed for the test case described below.

**Option Analysis.** This is the final stage of the analysis and involves the calculation of a 'total value' index for each option being assessed by the analysis. Starting at the bottom of the tree, it involves multiplying the performance scores for each option (defined using the word models) by the weights for the branches in the value tree and adding these together. These cumulative scores then define the performance scores for the next hierarchical level of the tree, which are multiplied by the branch weights and summed in the same way until a total value score is calculated at the top of the value tree. Calculated scores (ranging between 100 and 0) at the top level of the value tree are then representative of each option's performance in meeting the top-level aim for the project. This process can be repeated using weights specified by each stakeholder cluster to determine stakeholder-specific option rankings.

## Using the Toolkit for Risk and Strategy Management

Traditionally, the information used in capital investment appraisal to estimate profits falls into two categories: quantitative—or tangible—financial information and qualitative—or intangible—information. Criteria based on tangible information, for example NPV, have been developed and widely accepted. However, intangible information poses major problems in the decision process. Indeed, the intangible element often takes on greater significance when the investment has the potential to impact on the company's environmental performance and hence on a widely disparate group of stakeholders. Consequently, the level of uncertainty, or possibility that the project will not meet the company's profit targets, grows in direct relation to the importance of the intangible element.

procedure is best suited for extracting weights from a pairwise comparison matrix. Research by Golany (1993), which compares the Eigenvector approach proposed by Saaty and used by the MADE weighting model with five other methods, concludes that the Eigenvector approach is not dominated by any other of the methods investigated. Consequently, this research has used the Eigenvector method rather than other techniques for calculating weights from pairwise comparisons.

The SVA methodology working through the Paras and MADE models provides a process for addressing these uncertainties and consequently for quantifying and managing the risks. The mechanisms in place include:

- EMVs that encompass parameter-specific risk estimates rather than a discount rate premium or hurdle rate to reflect project risk
- Expanded cost and benefit inventory
- Explicit measurement of stakeholder expectations and value trade-offs between project parameters (financial and non-financial)

Although the Paras and MADE models have been designed so that they can be used as stand-alone models, the emphasis of the Toolkit and the SVA methodology is that both models are used during the decision process. The exact method of use is by nature problem-specific but in most cases will either involve the models in series, where the outputs of one model feed the next, or in parallel, where outputs of both models are used independently.

A typical example might involve using the Paras model to measure the financial performance of investment options in terms of EMVs. At this stage, the decision team can consider whether the financial analysis can conclusively differentiate the projects. If not (and this is more likely the greater the strategic significance of the investment), the financial performance information can be passed to the MADE model. The MADE model can then be used to construct a value tree and to define stakeholder priorities for the tree's performance attributes. The financial performance data can then be integrated, through the value tree and priority weights, with the other quantitative performance data to calculate a total stakeholder-wide value index for each investment option. (A test case application that integrates the financial assessment into a multi-attribute analysis is detailed below.)

The underlying premise is that the non-financial and financial metrics can be integrated as the weighting process underlines the strategic objectives of the company. Although the non-financial measures are not explicitly converted to financial values, the implicit assumption is that these will eventually translate themselves into financial benefit (or cost) for the company. The likelihood and importance of these is reflected through the priority weights.

Alternatively the MADE model can be used as a preliminary screening tool to reduce investment options to a list that meets minimum company performance criteria, as set by the company's environmental policy, for example. The Paras financial model can subsequently be used to dictate the final decision. The process would help ensure that the projects being considered would not undermine strategic performance targets set for the company.

Clearly, this is not an exhaustive list of how the tools can be used in combination. Part of the decision process is to decide on the objectives of the decision and it is this step that will have the greatest influence on how the components of the Toolkit are used.

For all applications, the SVA Toolkit is based on the premise that information flows can and should circulate round the models through feedback loops. It is this iterative process that helps to support the option creation stage of the methodology, whereby the initial list of options can be re-appraised in light of the analysis of stakeholder value trade-off and performance data. Indeed, the option creation step may be critical to the decision process if the list originally considered does not include an option that is acceptable to the stakeholder set. This step in the SVA process, therefore, supports the creation of a new options, possibly based on a combination of options that maximises total value of the decision, taking into account all stakeholder value trade-offs. While nothing guarantees that this can be done, detailed assessment of weights, followed by an attempt to design options that exploit them by recognising and complying with the most highly weighted values, is intuitively an attractive and promising approach.

## ◼ Application and Validation of the SVA Toolkit

Development of the SVA methodology has throughout been supported and validated by real-life industrial applications. So far, five separate applications of the methodology have been explored, two of which are summarised here by way of example. The first demonstrates the use of the Paras financial model and largely precedes the current SVA development work. The second involves the application of the SVA methodology and includes the integration of financial and non-financial metrics to produce a value metric to inform the investment decision. Further case studies will be reported in due course.

### Holistic Investment Appraisal of a Totally Chlorine-Free Pulp Manufacturing Facility

This test case demonstrates the use of the Paras financial model for carrying out a holistic financial investment appraisal. The financial assessment in hand was carried out by Paras and concerned a plant for the manufacture of totally chlorine-free (TCF) pulp for a Finnish paper manufacturer, Wisaforest Oy Ab. Figure 4 provides a résumé of this test case. Further details will be found in Moilanen and Martin's book on the Paras financial model (Moilanen and Martin 1996).

### Total Value Assessment of Waste Management Contractors

*Study Aims.* British American Tobacco wished to employ a waste management contractor to service the waste produced at their Southampton, UK, site. The waste management contract would involve:

- Collecting all the site's waste to a central processing area
- Sorting the waste into discrete waste streams
- Waste removal to recycling facilities or landfill
- Administering and managing the waste collection and disposal

On the basis of a tender document issued by British American Tobacco, three proposals for the contract had been received—from waste management contractors hereafter referred to as Companies A, B and C. At the opening stage of the decision analysis, the project manager revealed that British American Tobacco had a number of broad criteria against which they wished to judge these proposals.

## Background

**Company/business area**   Wisaforest Oy Ab, manufacture of pulp and sawn timber
**Investment**   Manufacture of Totally Chlorine-Free (TCF) pulp
**Driver**   Demand for TCF pulp from German buyers (60% of the market)

## Options

- Purchase TCF pulp from outside suppliers
- Develop a technology for their own production
- Wait for others to develop the technology and buy this when ready
- Start producing pulp bleached using chlorine dioxide and address other than the German-speaking countries

The second option was chosen based on the interpretation of corporate strategy objectives and the prevailing and anticipated market condition in the long term.

## Focus

- Demonstrate how a research and development project is not an isolated business area, but needs to be considered in the context of the whole plant
- Factor risk into the evaluation in a structured and objective way

## Original versus Model Calculation

Wisaforest calculated some financial indicators for the project based on its original expectation and figures. The analysis was developed using the Paras model to include a wider spectrum of revenues, costs and risks.

| Revenues | Original analysis | Paras analysis |
|---|---|---|
| Operating margin | ✓ | ✓ |
| Grant | | ✓ |
| Subsidised loan (interest differential) | | ✓ |
| **Costs** | | |
| Testing (excluding test pulp cost) | ✓ | ✓ |
| Revenue lost through sale of inferior test quality pulp | | ✓ |
| Interest on TEKES loan | | ✓ |
| Storage cost, incremental | ✓ | ✓ |
| Market study | | ✓ |
| R&D capital investment expenditure | ✓ | ✓ |
| New common facilities (12% of total) | | ✓ |
| New waste management costs | | ✓ |
| Future liability cost (using HAZOP [hazard and operability study]) | | |
| **Financial Indicators (IRR/EMV)** | 17.6%/ n/a | 14.9%/ 29.6 million Finnish Marka |

## Discussion and Conclusions

The revised investment appraisal brought out two main failings of the initial analysis. First, the investment team working from within the R&D function failed to incorporate the absolute costs and revenues accruing in other areas of the company as a direct result of the investment. This included such fundamental revenues as the loan from the Finnish research organisation (TEKES) and the cost implications of the pilot plant on the core production facility. The Paras approach, which supports and encourages a cross-disciplinary decision team and identification of cost and revenue parameters across the whole organisation, was able to redress these oversights in the original calculation. Second, the original analysis failed to analyse the expected monetary values of the identified parameters. The revised calculations using the EMV technique were able to incorporate risks into the calculations and, most notably, account for future liability costs, exchange rate and pulp price movements.

In this case, the effect of the Paras calculations has been to reduce the project IRR, which could be misconstrued as counter-intuitive. This effect, however, is predominately caused by the omission in the original calculations of the very significant additional common facility costs needed to accommodate the new pilot plant. The appeal of the project was not altered by the lower IRR calculated by the Paras analysis because the company recognised that, although the revised calculations better accounted for the costs of the project, the analysis did not explicitly account for financial benefit associated with improving the company's image. The Paras analysis was therefore seen as more representative of a worst-case scenario and not really comparable with the original analysis.

**Figure 4:** Financial Appraisal of a Pilot Plant for Non-Chlorine Bleaching of Pulp

- Cost per tonne of waste managed
- The environmental impact of the disposal options and the impact of this on British American Tobacco policy commitment and image
- The operational ability of the contractors

The first of these measures clearly falls into the tangible financial category, while the second two fall into the intangible category. Since the intangible element was likely to constitute a significant proportion of the project value, it was agree that the SVA methodology, especially the MADE process, would help not only define, measure and trade off the intangible attributes but also impose a structure and process that could be replicated in later projects as required.

*Financial Analysis of Tangible Values.* A financial analysis of the performance of the three contractors was carried out. This included an expanded cost and benefit inventory to cover the fixed costs of the contract, costs of waste disposal for different waste streams and revenues if applicable for recycling of materials. A coarse risk analysis was appended to the financial model to account for:

- Variations in waste volume
- Variations in transport and landfill costs
- Potential landfill taxes
- Variations in recycling revenues for different materials

The cost, benefit and probability estimates were combined to calculate total contract cost and expected cost per tonne removed for each bidding contractor.

*Stakeholder Value Analysis of Waste Management Contractors.* The SVA Toolkit's MADE model was applied to support British American Tobacco in structuring and systematising their decision process and to help account for the significant non-financial value attached to the decision and to integrate this with the financial assessment previously carried out. The MADE analysis was carried out over four phases described below.

*Phase 1: Value tree design.* The first step in the MADE process involved developing a hierarchical value tree for the waste contractor decision problem. The process employed to design the tree was based on a series of seven interviews with diverse departmental staff from British American Tobacco. The aim of these interviews was to define on an individual level what issues were at stake and what particular performance attributes were of concern to all the stakeholders of the decision. Figure 5 shows the company areas that were represented in these interviews.

The choice of area to include was largely intuitive and aimed to cover the company roles likely to be affected by the decision. The choice of individuals to interview was dictated by responsibility and or knowledge, so in most cases the most senior individual was chosen. Because 'Production' was considered to be the area most impacted by the decision, it was subdivided and represented by the three main disciplines spanning this area.

Information gathered through the interviews was used to define the full spectrum of performance attributes relevant across the whole company and to establish an overriding top-level goal for the investment decision.

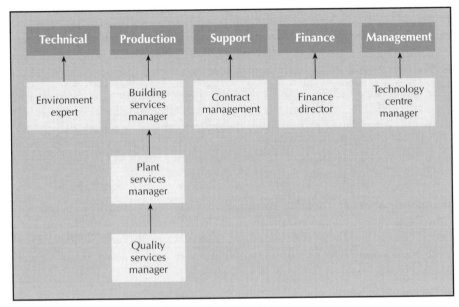

**Figure 5:** Interview Coverage

Figure 6 illustrates the stakeholder-wide value tree that resulted from this process. Stakeholders of the decision were also given the chance to comment on the design before it was 'frozen' in the form shown here. Reference to Figure 6 shows that financial performance is included within the seven main attributes important in deciding the waste contractor. The significance of the financial metric would be determined by the weights set by the decision team.

*Phase 2: Weight analysis.* The second phase of the MADE analysis involved establishing the priority weights for the value tree branches. The weights analysis was done on both individual and group bases. Initially, individuals were supplied with input sheets to fill in their pairwise data and return for analysis. Subsequently, the same pairwise comparisons were done with the individuals acting as a group.

Figure 7 shows the weights obtained for the group and the average for all individuals. Comparison of the results from the group and individual weight definition exercises showed:

- The results for the group and for the individuals when averaged correlated reasonably well. Although there were differences in the absolute weights calculated from the two exercises, there was a level of agreement in absolute priorities for some attributes. The similarity was, however, not convincing enough to suggest either set of weights could be freely interchanged. The group decided that the group-derived weights were the most representative. Notably, the group weights place a much lower priority on the financial performance.

- The level of pairwise consistency[11] varied between individuals, but on the whole lay within reasonable bounds of variation.

---

11. The pairwise technique allows a consistency index to be calculated. This is a useful indicator of how closely the decision-maker has followed the rules of perfect transitivity. For example, if A is twice as important as B and B is equally important as C, then perfect consistency would imply that A is twice as important as C.

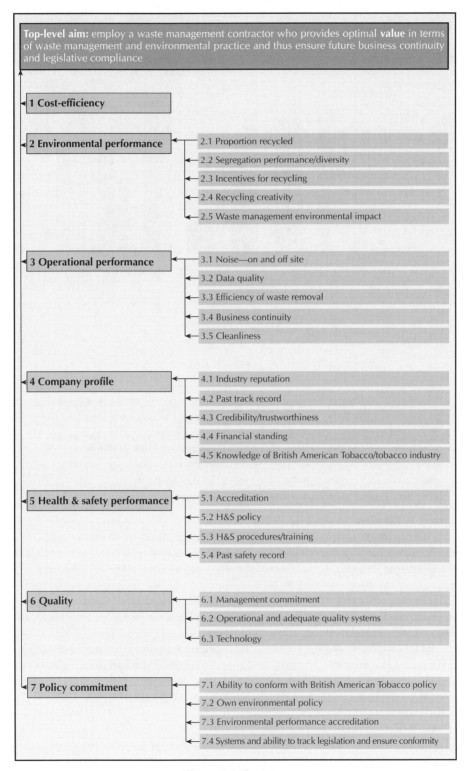

Top-level aim: employ a waste management contractor who provides optimal **value** in terms of waste management and environmental practice and thus ensure future business continuity and legislative compliance

**1 Cost-efficiency**

**2 Environmental performance**
- 2.1 Proportion recycled
- 2.2 Segregation performance/diversity
- 2.3 Incentives for recycling
- 2.4 Recycling creativity
- 2.5 Waste management environmental impact

**3 Operational performance**
- 3.1 Noise—on and off site
- 3.2 Data quality
- 3.3 Efficiency of waste removal
- 3.4 Business continuity
- 3.5 Cleanliness

**4 Company profile**
- 4.1 Industry reputation
- 4.2 Past track record
- 4.3 Credibility/trustworthiness
- 4.4 Financial standing
- 4.5 Knowledge of British American Tobacco/tobacco industry

**5 Health & safety performance**
- 5.1 Accreditation
- 5.2 H&S policy
- 5.3 H&S procedures/training
- 5.4 Past safety record

**6 Quality**
- 6.1 Management commitment
- 6.2 Operational and adequate quality systems
- 6.3 Technology

**7 Policy commitment**
- 7.1 Ability to conform with British American Tobacco policy
- 7.2 Own environmental policy
- 7.3 Environmental performance accreditation
- 7.4 Systems and ability to track legislation and ensure conformity

**Figure 6:** Value Tree

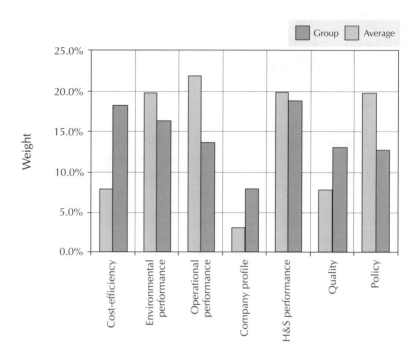

**Figure 7:** Weights for Group and Average for Individuals

- The best consistency ratios were obtained when the individuals were acting as a group, indicating perhaps that the group environment helped to focus the decision-makers into making more consistent pairwise preferences. Indeed, it was noticeable that the group members were prepared to debate the pairwise comparisons and link their reasoning to pertinent examples and strategic considerations driving the investment. In this sense, therefore, the priority weights became a proxy for the company's strategy.

*Phase 3: Performance analysis.* This phase of the decision analysis involved measuring each contractor's performance against the attributes defined in the stakeholder-wide value tree (see Fig. 6). A 'word' model was designed for each performance attribute to facilitate the process. Figure 8 shows the word model used to score 'proportion recycled'.

The scoring was carried out as a group exercise. Members of the group were the same as those at the interview stage except for two additional members, both of whom were members of British American Tobacco's environmental committee.

Scoring was based on analysis of a combination of quantitative data contained in the proposals (for example, the financial appraisal defined the cost per tonne disposed for each contractor's bid) and on the group's experience of the contractors based on personal contact and visits to sites already under management by the bidding contractors.

*Phase 4: Results analysis.*

Top-level scores. The final phase of the decision analysis involves analysing the results from the MADE model. The scores awarded for each contractor's performance

| Word description | Score | Guide |
|---|---|---|
| | **Attribute** | Proportion recycled |
| | **Suggested performance measurement tool** | Contractor performance data |
| | **Metric** | Percentage of waste recycled by contractor |
| Waste industry best recycling performance | 100 | >80% recycled |
| Waste industry above-average recycling performance | 75 | 65% recycled |
| Waste industry average recycling performance | 50 | 50% recycled |
| Waste industry below-average recycling performance | 25 | 35% recycled |
| Waste industry worst recycling performance | 0 | <20% recycled |

**Figure 8:** Example Word Model

were multiplied through by the branch weights to calculate scores at each level in the decision hierarchy. These scores are not only an indication of each contractor's performance against the defined performance attributes, but also an indication of how each contractor compares with the others and with the 'best in the industry'.

Figure 9 illustrates a graphical rosette showing the top-level score calculated for each contractor based on:

1. Each individual's own personal weights
2. The average for all individuals' weights
3. The average for the group-derived weights

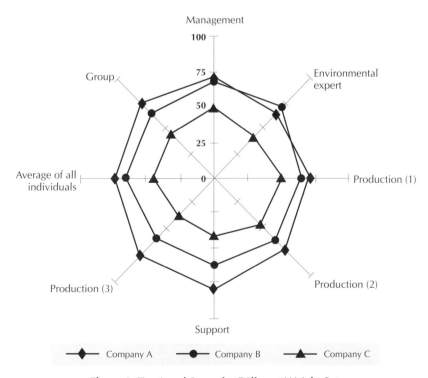

**Figure 9:** Top-Level Score for Different Weight Sets

Reference to Figure 9 clearly shows that Company A is consistently the top performer. There is one exception to this trend (based on weights from the environmental expert). However, the group-based results are the primary source of reference to indicate the company-wide view on attribute weights and hence the top-level performance score for the contractors.

Company C is consistently the worst performer by a substantial margin and therefore the decision could safely focus on the performance scores recorded for Companies A and B.

Sensitivity analysis. Sensitivity analysis can be used to examine the two basic data inputs of MADE, which are the weights and scores. Hence three levels of sensitivity analysis are possible.

1. Contractor top-level score sensitivity with respect to attribute weights

2. Contractor top-level score sensitivity with respect to attribute performance scores

3. Contractor top-level score sensitivity with respect to both attribute weights and scores

In this case, a significant level of quantitative data was used to establish the contractor scores for each performance attribute. As a consequence, a high level of confidence can be associated with the contractor scoring. On the other hand, the attribute weights were derived from judgements made by the group, and confidence in the final output from the MADE model can be gained by analysing the sensitivity of the contractor top-level score to changes in the attribute weights. A sensitivity analysis was therefore carried out to investigate the effect on the top-level score of varying the relative importance (weights) of the performance attributes.

Figure 10 shows the sensitivity of the top-level score to changes in the weight set by the group for the 'cost-efficiency' performance attribute. The group weight is shown in the title of the graph for reference. The sensitivity analysis showed that the 'cost-efficiency' performance attribute was the only one that could reasonably cause a rank reversal of preferred waste contractor.

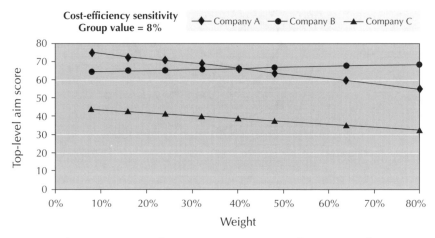

**Figure 10:** Top-Level Score Sensitivity to Cost-Efficiency Weight

In general it was possible to conclude that contractor ranking was not critically sensitive to any one attribute weight. Cost-efficiency is the most critical attribute. This, however, would need to change by a factor of five (from 8% to over 40%, as shown in Figure 10) to cause a rank reversal. The final decision, therefore, proved to be satisfactorily robust.

***Value Added to the Decision from the SVA Analysis.*** The results from the SVA analysis were useful, interesting and of practical help to British American Tobacco, since:

- The relative importance of the cost-efficiency attribute in meeting the company's overall aim for the project was put firmly into context with other project performance attributes. Because of the long-term and environmental performance consequences of the project, the actual weight calculated for the financial performance attribute was not the most significant factor. This could not have been foreseen before the study was carried out.

- The MADE analysis helped the British American Tobacco decision team structure their analysis and understand their objectives and the project performance attributes that were relevant.

- Using the pairwise technique, it was possible to determine the group's relative priorities for the performance attributes in a structured, realistic and auditable fashion.

- The word model produced absolute performance values for each contractor which could be compared with one another and to industry best practice.

- The results from the MADE model provided quantitative justification on which to base the decision, when in fact the preferred contractor, Company A, was not the cheapest.

- The output from the MADE model produced auditable documentation, which gave clear, concise justification for each step in the analysis suitable for future reference.

In this test case, both elements of the SVA Toolkit were used, with a financial analysis of contractor performance feeding directly into the MADE analysis. The major theme recurring in the feedback from the company was the benefit gained from the tool's structure and logical framework and the way these were able to incorporate the company's strategic objectives for the decision into the decision process. Although not certain, it is likely that the SVA analysis was instrumental in justifying to higher management the choice of the higher cost option as the contractor that optimised the total value from the contract.

## ▌ Discussion: An Interpretation of Stakeholder Value Analysis and its Limitations.

It was noted above that SVA relies on identifying stakeholder groups and also individuals who can voice a representative view for each group, and on sorting the plethora of groups to reduce them to a manageable number of stakeholders and stakeholder groups. In the examples given in the preceding section, this step proved to be relatively straightforward. Particularly in the example of selecting a waste contractor, this is probably not surprising, since the individuals concerned all work in the same organisation and therefore inevitably share some concerns and priorities. However, this will not always be the case, and there will inevitably be some applications where it is difficult or impossible to define representative stakeholder groups who are prepared to engage constructively in the SVA process. We therefore conclude with a very preliminary discussion of the circumstances that are likely to limit or prevent the use of SVA.

The circumstances under which SVA will or will not be useful can perhaps be understood by reference to competing conceptions of justice and the cultural theory of risk (e.g. Long 1997; Davy 1996; Thompson, Ellis and Wildavsky 1990; Linnerooth-Bayer and Fitzgerald 1996). Linnerooth-Bayer and Fitzgerald (1996) have argued that theories of justice and cultural theory can be reconciled, and Long (1997) has argued that this analysis can be used to understand and advance the achievement of universally acceptable participation in decision processes. The basic concepts are summarised in Figure 11.[12]

It is important that the four-way classification originally proposed by Douglas and Wildavsky (1982) refers to groups and organisations; the behaviour of any individual person varies according to context and therefore does not always lie within the same category. However, the important point is that different categories of organisation or group generally have different perspectives on the environment (Thompson, Ellis and

| Category | Hierarchist | Individualist | Egalitarian |
|---|---|---|---|
| Concept of justice | Utilitarian | Libertarian | Social |
| Characteristics | Collectivist, but recognising inequality | Individualist, emphasising equality | Collectivist, emphasising equality |
| Attitudes to decision-making | Decisions should be based on some form of cost–benefit analysis. Public participation is useful in explaining to the public the benefits and risks involved in a decision, but a strong central authority is needed based on representative liberal democracy. | Pursuing self-interest promotes efficient use of resources. The best information is needed to inform decisions which individuals are left free to make for themselves, without interference from regulators or pressure groups. | Regulation is needed to ensure equitable distribution of benefits and disbenefits, and to protect the less privileged against the effects of risk and uncertainty. The public must face their responsibilities as consumers. |

**Figure 11:** Cultural Perspectives of Different Stakeholder Groups

*Source: Adapted from Long 1997*

---

12. Figure 11 omits the 'Fatalist' category, whose attitude can be summarised as 'The world is as it is. I am not interested in philosophising about it; others will run the world anyway, regardless of my opinion' (Long 1997). Long has argued that such a group regards the world as inherently unjust, so that no coherent view of justice has formed and involvement in multi-value decisions is for all practical purposes impossible.

Wildavsky 1990). Therefore the classification provides a possible framework for approaching the problem of sorting a plethora of stakeholders into a manageable number of compatible groups, by assigning each to one of the categories in Figure 11, selecting a small number of stakeholder groups from those in each category, and then involving the selected groups in each category together in the SVA process. As noted above, a hierarchy of values can be expected to emerge within each category, enabling the final decision to recognise the most important issues emerging from involvement of stakeholders in the three categories.

Future tests of Stakeholder Value Analysis will attempt to involve broader ranges of stakeholders so that consensus may be harder to achieve, to identify where systematic classification of stakeholder groups is needed and, in the extreme, where SVA ceases to be of practical value.

# ▌Conclusions

Increasingly, decision-makers in industry are being faced with complex investment decisions created and made emotive by a plethora of stakeholder expectations. Investments with the potential to have an impact on a company's environmental performance are a prime example of this and share many of the following features.

- They are complex with multi-dimensional impacts, many of which are difficult to quantify in scientific and monetary terms.
- The decision outcome is stakeholder-sensitive.
- Stakeholders often seem to have contradictory or emotionally driven requirements.
- The final decision often faces regulatory and/or public scrutiny.

Conventional financial models share fundamental problems in gauging the full value implications of such decisions, especially in measuring stakeholder expectations and value trade-offs and, as a result, introduce considerable levels of uncertainty and risk to the decision.

The Stakeholder Value Analysis Toolkit, which is made up from the Paras and MADE models, allows decision-makers to manage uncertainty associated with their decisions by incorporating stakeholder expectations into the analysis. The views of different stakeholders can be canvassed and fed into the SVA models to produce output metrics which can be used by decision-makers to:

- Choose investment options that reflect stakeholder priorities
- Operationalise company strategy through the company's investment decisions
- Define the environmental priorities of the company making the decision, particularly when the regulatory environment is undeveloped or ill defined
- Define and quantify the issues that need to be debated to try to find a mutually acceptable position

# 13 Shared Savings and Environmental Management Accounting

Innovative Chemical Supply Strategies[1]

Thomas J. Bierma, Frank L. Waterstraat
and Joyce Ostrosky

## ▌ Introduction

THE PURCHASE of chemicals appears to be a simple, straightforward process. A company identifies its chemical needs and contracts with a supplier to provide the chemicals at the lowest price. However, this traditional approach to chemical supply creates two major problems: misdirected financial incentives and an arbitrary division of chemical management responsibilities. Together, these two problems can create a large chemical cost iceberg (see Fig. 1).

The alternative is a 'Shared Savings' supply relationship, in which the chemical user no longer purchases chemicals, but purchases chemical performance instead. Chemical suppliers actually make *more money* by helping their customers use *less chemical* and reducing other chemical costs. This is accomplished by sharing the chemical and operating savings between both parties.

The benefits of Shared Savings have been dramatic (Bierma and Waterstraat 1996, 1997; *Green Business Letter* 1997; Williams *et al.* 1995; Zeller and Gillis 1995). In this chapter, we use examples from manufacturing plants belonging to General Motors Corporation, Ford Motor Company, Chrysler Corporation and Navistar International. All of these plants experienced not only reductions in chemical use and waste volume, but

1. The authors are grateful to the many companies and employees who so generously donated their time to our education. This work was supported by grants from the US Environmental Protection Agency and the Illinois Waste Management and Research Center. This article does not necessarily reflect the views or policies of these agencies. The mention of company names does not constitute an endorsement.

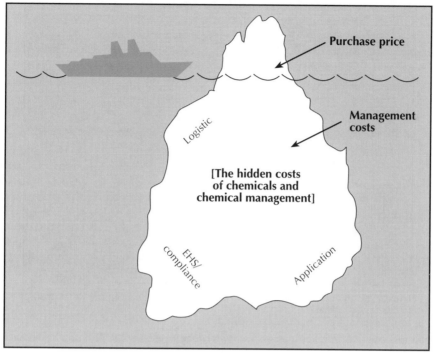

**Figure 1:** The Chemical Cost Iceberg
*Source: Bierma and Waterstraat 1997*

also achieved overall improvements in environmental performance, operations control and product quality. Most importantly, the implementation of Shared Savings was accompanied by improvements in accounting information. In fact, improvements in the chemical supply relationship and in the quality of accounting information are synergistic. Advances in the supply relationship drive and support advances in the accounting system, which promotes further advances in the supply relationship.

## ▌ Chemical Supply

### Misdirected Financial Incentives

In a traditional supply relationship, the supplier is paid per drum (or kg, litre, etc.) of chemical supplied. The greater the volume of chemical supplied, the greater the profit for the supplier (see Fig. 2). This creates a strong incentive for the supplier continuously to find ways to increase sales volume. However, this is in direct opposition to the interests of the chemical user, who is trying to reduce chemical volumes and costs. This 'volume conflict' is a constant source of waste and creates an atmosphere of suspicion and mistrust in the chemical supply relationship.

The price-per-drum sales approach also tends to focus attention on the 'tip of the iceberg', ignoring many of the hidden costs associated with chemical use. In fact, to reduce purchase price, suppliers may 'cut corners' in ways that increase hidden costs. For example, many chemical blends can be made less expensive by reducing the quality of inputs, resulting in equipment problems and product rework. Some companies

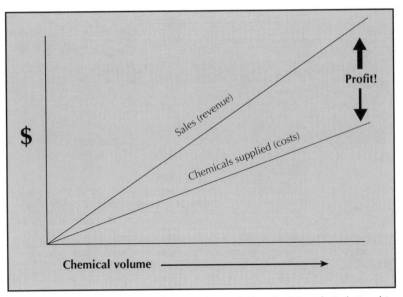

**Figure 2:** Source of Supplier Profit in Traditional Chemical Supply Relationships

may reduce the R&D resources necessary to develop less toxic chemical alternatives if they believe they will not be able to recover that expense in the chemical price.

Paying suppliers on a price-per-drum basis can also contribute to problems with supplier responsibility. Because the supplier receives payment for *the chemicals* instead of for *the performance of the chemical*, this approach tends to disconnect the supplier from the ultimate performance of their chemicals. Though poor chemical performance can eventually result in the loss of business for the supplier, there is no direct link between supplier revenue and chemical performance.

To counter these problems created by the misdirected incentives of traditional supply relationships, chemical users rely on threats. The threat of switching suppliers, whether explicit or implicit, becomes the primary means by which chemical users attempt to minimise price, minimise volume and maximise chemical performance. However, the use of threat creates its own hidden costs. Not only does it inhibit co-operative efforts between chemical supplier and chemical user, the threat is effective only as long as the supplier perceives it to be 'real'. Once the chemical user's demands have been met, there is little reason for the supplier to pursue further improvements.

### Inefficient Division of Responsibilities

Most companies transfer chemical management responsibilities at the loading dock. Once the chemical has been delivered, the chemical user assumes responsibility for its management, use and disposal. This may appear to simplify legal and record-keeping responsibilities, but it does not promote the most efficient division of responsibilities between the supplier and the user.

To manage chemicals, a chemical user must devote considerable time and resources to develop and maintain chemical expertise. For most companies, chemical management expertise is not their core business. Few manufacturers would perceive themselves as being in the 'chemical management' business. Yet resources devoted to chemical expertise are resources that were diverted from core business needs, thus potentially reducing the competitive strength of the chemical user.

For chemical suppliers, however, resources devoted to chemical expertise are not only an investment in their own core business, they can benefit from the tremendous economies of scale. For example, tracking regulations and compliance strategies for their chemicals is much cheaper for chemical suppliers because their expenses are allocated over a much greater volume. Similarly, the cost of developing improved chemicals or technologies can be far lower for chemical suppliers.

## ▌ Shared Savings changes the supply relationship.

Under Shared Savings, the chemical user no longer buys chemicals; the chemicals remain the property of the chemical supplier, at least until they are used in the chemical user's processes. Instead, the supplier is paid for chemical performance. In addition, financial incentives are used to drive continuous chemical cost reductions.

To accomplish this, Shared Savings contracts rely on two key elements: one that promotes chemical performance, and a second that drives continuous cost reduction. Each is discussed below.

### Performance
Chemical performance is promoted through the use of 'performance expectations' and corresponding 'performance fees'.

*Performance Expectations.* Chemical users must begin by defining their chemical performance expectations. These must clearly define what the chemicals are expected to *do*, not just the type and volume of chemicals required. Performance expectations may include product quality expectations, equipment operating characteristics, tool life expectations, corrosion limits, etc. Services, such as inventory management or chemical quality assurance, may be included as well. In some cases, specific cost or chemical reduction targets may be included.

*Performance Fees.* The supplier is then paid a performance fee for meeting the performance expectations. This fee is typically in the form of a fixed fee per month, or fixed fee per unit of production (known as a 'unit price'). The 'Pay-as-Painted' programme at Chrysler Corporation is an excellent example. Their paint supplier, PPG, is paid a predetermined amount for each vehicle that leaves the paintshop with a finish that meets Chrysler's performance expectations for finish quality. If the vehicle does not meet specifications, PPG does not get paid. This connection between supplier revenue and chemical performance focuses suppliers on assuring the performance of their chemicals rather than simply supplying them.

### Continuous Cost Reduction
To drive reductions in the chemical cost iceberg *beyond* the levels set forth in the performance expectations, a mechanism must be used to share these additional savings between both parties. Two commonly used mechanisms are the fixed fee payments and gain-sharing.

**Fixed Fee.** Because supplier revenue is fixed through the performance fee, and not tied to chemical volume, the supplier can increase profit by finding ways to *reduce* chemical volume and management costs (see Fig. 3). In other words, the supplier increases profits by decreasing chemical volumes, just the opposite of traditional supply relationships. Ultimately, some of these savings must be shared with the chemical user so that both parties have incentive to make further cost reductions. One way to do this is to 'rebate' some of these savings to the chemical user. Another way is for the chemical user to reduce the performance fee to reflect the new, lower chemical costs. A third way is to increase the chemical performance or services provided for the existing fees.

**Gain-Sharing.** Another way that savings can be shared is through 'gain-sharing'. Gain-sharing addresses chemical-related costs that are not covered by the performance expectations of the contract. These are costs that are borne by the chemical user, but not the supplier. For example, a supplier may have a technology that would reduce hazardous waste generation. Since the disposal costs are paid by the chemical user, the supplier has no direct financial incentive to implement the new technology. Under a gain-sharing programme, the chemical user agrees to share a portion of these savings generated by the supplier's technology with the supplier, providing incentives for further cost reductions. The manner in which savings are shared may be determined on a case-by-case basis, or established for the term of the contract.

The implementation of gain-sharing can be relatively simple. Where opportunities to produce savings are identified by the supplier, a proposal is submitted to the chemical user outlining the changes and anticipated savings. The two parties must then reach an agreement on the changes to be made, the benefits anticipated and how the costs and benefits will be measured and shared between the parties. As with all aspects of Shared Savings, data collection methods must be commensurate with the

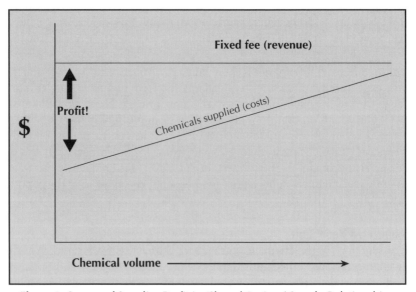

**Figure 3:** Source of Supplier Profit in 'Shared Savings' Supply Relationships

level of trust between the parties. As the level of trust grows, there is less frequent data collection oversight.

### Financial incentives align financial interests.

Together, these two contract elements align the financial interests of chemical user and chemical supplier. As one supplier put it, 'it gets both our arrows pointing the same direction'. The combination of performance expectations and performance fees aligns the interests of both parties toward bringing value to the ultimate consumer. Fixed fees and gain-sharing mechanisms drive continuous cost reduction.

One of the best ways to understand the financial incentives of Shared Savings is to examine its relationship to total chemical costs. As noted above, the traditional supply relationship focuses on chemical purchase price—the tip of the iceberg (Fig. 1). Efforts of the chemical user are concentrated on reducing this portion of the iceberg, but often at the expense of increasing the size of the iceberg below the water.

Shared Savings approaches the chemical iceberg in a very different way. Figure 4 illustrates a chemical cost iceberg that has been divided into two sections. The shaded section represents those costs that have been assumed by the supplier under the performance expectations. We call this the performance expectation component (or PE component) of the iceberg. It includes all of the original chemical purchase costs (since the supplier now owns the chemicals), as well as other responsibilities specified in the performance expectations, such as chemical ordering, inventory, tracking, distribution and selected regulatory compliance activities. The supplier agrees to accept this portion of the chemical cost iceberg in exchange for the performance fees.

The *chemical user* benefits if the performance fee is smaller than the portion of the chemical cost iceberg transferred to the supplier through the performance expectations (in practice, the performance fee is often smaller than the chemical purchase

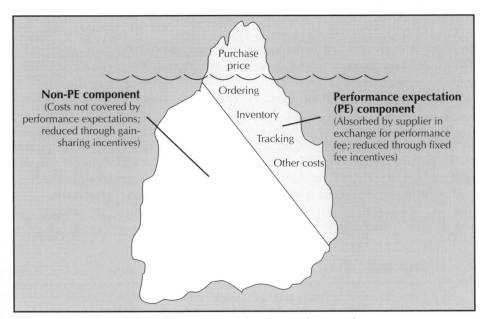

**Figure 4:** Components of the Chemical Cost Iceberg

price alone, representing substantial savings for the chemical user). On the other hand, the *supplier* benefits if the performance fee is larger than the costs they actually incur when absorbing their portion of the chemical cost iceberg. The size of the PE component of the iceberg depends on the efficiency that the supplier can bring to these activities. Because chemical management is the core business of the supplier, costs can often be reduced substantially, resulting in a profit for the supplier.

The use of a fixed fee drives the supplier to work continuously on cost reduction in this PE component of the chemical cost iceberg. The more the supplier reduces these costs, the greater the profits. The added benefits are shared with the chemical user through rebates, reduced fees or increased performance and services.

As illustrated by the unshaded portion of the chemical cost iceberg in Figure 4, a substantial number of hidden chemical costs remain outside the responsibilities assigned to the supplier under performance expectations (we call this the non-PE component). This component can include such costs as waste treatment and disposal, employee health problems related to chemical exposures, potential environmental liability and production downtime or product quality problems stemming from poor chemical performance. Depending on the performance expectations contained in the contract, the non-PE component may contain a large proportion of total chemical costs.

Chemical users may enjoy indirect reductions in these costs as a by-product of the supplier's efforts to reduce the PE component of the iceberg. For example, reductions in chemical volume, which a supplier may have implemented to reduce its own costs, will also reduce waste disposal costs and employee health problems. However, gain-sharing can be used to accelerate the reductions in the non-PE component costs. If the supplier is able to find additional ways to reduce hidden chemical costs, even though it is outside of the PE component area, gain-sharing allows the supplier to participate in the savings. Though most companies have limited gain-sharing in reductions in 'hard costs' (such as materials, utilities, disposal fees, etc.), it can still produce substantial reductions in the size of the chemical cost iceberg.

### Responsibilities are assumed on the basis of core competence.

The natural consequence of linking supplier revenue to chemical performance and financially rewarding the supplier for continuous cost improvements promotes the supplier's interest and involvement in all aspects of chemical management. If the supplier has information or technologies that could improve chemical applications or chemical management activities more efficiently, it is in the supplier's financial interest to bring these to the chemical user for potential implementation. For this reason, suppliers typically place one or more staff on-site to identify such opportunities as well as co-ordinate and in some cases actually perform the full range of chemical management activities.

Having the supplier on-site to monitor and co-ordinate chemical management activities releases the chemical user's personnel from performing chemical management tasks to focus on activities more closely related to the user's core business. For example, one manager at the Ford plant we studied commented:

> This allows us to concentrate on what our specialty is—building quality automobiles and being able to sell those to the public; we certainly want to build what they want. We are managing the manufacturing business, and passing responsibilities on to people who are more experienced in managing certain subsystems.

The on-site supplier representative at the Navistar engine plant put it this way:

> Navistar is in business to make engines, the best engines they can. [We are] in the business of coolants, rust preventatives, cleaners, and that's what we do 100% of the time.

## ▌ Environmental management accounting provides the necessary data.

Mark Opachak, one of the architects of General Motors's Shared Savings programme, summarised it this way: 'In God we trust; all others must bring data.' The link between Shared Savings and quality data is a critical one. Before a Shared Savings programme can begin, detailed financial and operating data are required to define accurate baselines (sometimes known as the 'cost book'). Baselines document the conditions and performance levels that exist at the start of the programme in terms of production volumes, chemical usage and chemical costs. This becomes the basis on which the chemical user develops performance expectations and the suppliers develop their bids. At the four plants we studied, none had adequate baseline data from the accounting systems in use prior to their Shared Savings programmes. Baseline data for the initial chemical 'footprint' was collected through special studies, often as a joint project between chemical user and chemical supplier. These studies were expanded to additional chemicals as the footprint for the Shared Savings programme expanded.

In addition to defining accurate baselines, data are needed to measure the progress in reducing the PE component (Fig. 4, shaded area) of the chemical cost iceberg. Accurately tracking chemical usage per unit of production is critical. Not only does it impact supplier profit, but it also provides data to calculate rebates or reduce supplier fees. Suppliers need to monitor and measure hidden costs such as ordering, inventory and distribution, since these can represent a major portion of their overall project costs.

However, it is also important to identify and measure the hidden chemical costs in the non-PE (Fig. 4, unshaded) component of the chemical cost iceberg. This includes the myriad environmental costs associated with chemical use. Even in Shared Savings programmes that do not use gain-sharing, hidden cost reductions can be used by the supplier to further demonstrate the value of their services to the chemical user. In programmes that use gain-sharing, the identification and measurement of these hidden costs drive supplier revenue and profitability. Of course, some benefits, such as improvements in product quality, are difficult to quantify in monetary terms and are not shared with the supplier. However, such benefits may still be highly valued by both parties.

Under Shared Savings, both the chemical user and chemical supplier have a vested interest in collecting accurate accounting data and promoting the development and implementation of environmental management accounting practices that track total chemical costs. As stated above, improvements in accounting practices drive improvements in Shared Savings, and vice versa.

## ▌ Shared Savings and environmental management accounting drive each other forward.

As we have shown, Shared Savings requires more data and substantially better-quality data than traditional chemical supply relationships. Initially, improvements in materials accounting and cost accounting are essential to establish baselines for the contract. From these baselines, performance expectations are defined and performance fees are ultimately negotiated. Suppliers may be involved in these initial accounting improvements, particularly in tracking the actual application of specific chemicals to specific products and processes. For some companies, baseline data collection may take from several months to a year or more, depending on the quality of existing data and the complexity of the processes. It is not uncommon for the supplier to assist in this data collection process.

Once a Shared Savings programme is established and the most obvious opportunities for cost reductions are implemented, the demand for detailed accounting data grows. Chemical users want to improve and expand supplier performance expectations. Chemical suppliers want to find new sources of revenue and further opportunities to demonstrate their value to the chemical user. Improvements in the accounting process allow the Shared Savings relationship to expand, driving the demand for expanded and better accounting practices.

The co-operative interaction between Shared Savings and accounting practices can be illustrated using the environment-related management accounting pyramid (see Chapter 1, Fig. 3). In a Shared Savings contract, performance expectations are the documented value provided by the chemical supplier for which the supplier receives a performance fee. At the beginning of a Shared Savings programme, value may be stated in terms on the lowest level of the accounting pyramid, such as targets for emission reductions or specific chemical substitutions. In time, as cost accounting data collection practices improve, targets for cost limits or reductions, the next level of the pyramid, can be added. Many companies with advanced Shared Savings programmes include targets for chemical performance such as process requirements and product quality. Throughout the Shared Savings relationship, as the assessment measures of value improve, the activities of the Shared Savings programme improve along with them.

There is no reason why performance expectations cannot ultimately include targets for 'sustainable value added' practices, as long as the accounting system has advanced to the point that it can measure such value accurately. The link between performance fees and performance expectations would drive the supplier to achieve the targets, while gain-sharing opportunities would provide financial incentives for the supplier to exceed the performance targets.

## ▌ Shared Savings Case Studies

Over a two-year period, the authors studied Shared Savings programmes at a number of US manufacturing plants. Below are profiles of four plants with over thirty years of combined Shared Savings experience. The full case histories for these plants have been published elsewhere (Bierma and Waterstraat 1997).

### General Motors Corporation: Truck and Bus Assembly Plant, Janesville, Wisconsin

The Shared Savings programme at General Motors is known as the Chemicals Management Program (CMP). Though GM began implementing CMPs at some of its plants in the mid-1980s, the programme at their truck and bus assembly plant in Janesville, Wisconsin, was begun in 1992. Their supplier, BetzDearborn, is responsible for water treatment chemicals (powerhouse, cooling towers, waste-water treatment, air-houses), paint detackification and booth-maintenance chemicals, lubricants, maintenance paints, commodity chemicals and purge solvents. BetzDearborn relies on tier-2 suppliers for many of these chemicals.

Performance expectations in the contract include a wide range of activities, from the procurement and management of chemicals to employee health and safety training. GM has also specified continuous improvement targets, including an annual cost savings expectation. In return, BetzDearborn receives a fixed payment (performance fee) for each vehicle produced at the plant. In some cases, where the chemicals are used indirectly in the assembly process, an alternative unit of production measure may be used. For example, payment for boilerhouse chemicals is made 'per million pounds of steam' produced, rather than per vehicle produced. For a few chemicals unrelated to production, such as janitorial supplies, payment is made as a fixed monthly fee.

GM-Janesville has experienced substantial benefits from the CMP. Though GM has requested that specific financial data be kept confidential, the overall programme has proved to be so successful that GM is mandating its implementation in all its plants worldwide. The most immediate benefits at Janesville occurred as a result of BetzDearborn assuming responsibility for the PE component of the chemical costs iceberg (the shaded portion in Fig. 4). GM assumes ownership of the chemicals at the time they are used in the process, but never 'buys' them, thus eliminating the 'tip' of the iceberg for GM. In addition, BetzDearborn assumes responsibility for many other chemicals costs, such as ordering, inventory, tracking, distribution, quality control and assisting in regulatory compliance. For GM, these cost savings substantially exceeded the fees that are paid to BetzDearborn.

BetzDearborn, on the other hand, has been able to make a profit by implementing operating efficiencies that substantially reduced the PE component of the iceberg, to the point that the costs are smaller than the performance fees they receive to provide the services. They have been able to reduce the volume of chemicals required per unit of production, as well as the cost of obtaining some of the tier-2 chemicals. In addition, they have been able to introduce efficiencies in their chemical logistics responsibilities.

Though gain-sharing is not yet in practice at Janesville (it is planned for the near future), BetzDearborn has undertaken a number of projects that reduced GM's non-PE component (the unshaded area in Fig. 4). For example, they undertook an extensive study of the paint sludge handling system, providing GM with considerable savings on consulting fees. This resulted in a modification to the waste-water system which reduced sludge clean-out time by 80%, substantially reducing personnel and equipment costs as well as production downtime. BetzDearborn also developed a system for washing the cotton covers used to protect robots from dust. The new system extended the useful life of each cover sixfold, reducing both waste and cost. The implementation of a new gain-sharing provision in the contract is expected to accelerate the development of such cost-saving innovations.

Many activities have been used to improve the quality and accuracy of the data related to chemicals and chemical costs. In the preparation stage of the CMP at Janesville, GM involved its chemical suppliers in studying the various processes to be covered by the chemical contract. Suppliers assisted with the development and definition of baseline chemical use and cost-per-unit data.

It was noted above that one problem with traditional accounting systems is that hidden chemical costs can increase because these costs are posted to different department budgets. Therefore decisions that save money for one department can dramatically increase costs for other departments. To overcome this problem, GM specifically included a performance expectation for the supplier to 'co-ordinate chemical usage with all affected plant departments'.

The nature of the relationship between GM and BetzDearborn has also encouraged the use of accounting data for long-term, rather than short-term, decision-making. Though the contract has a three-year term, GM is quite open about their intent to make this a long-term supply relationship. Mark Opachak, with GM's World Wide Facilities Group, explains.

> Look at the Chemical Management Program as a marriage. Our contract has a 'divorce clause', but in a healthy marriage you don't get married with plans for a divorce, and you don't stay married out of fear of divorce. Fear is not a good motivating factor. There has to be a positive, fair relationship with a desire to be together. The divorce clause is one millionth of the contract's importance. As far as we are concerned, a CMP contract is forever! Divorce is not an option.

Finally, BetzDearborn has worked extensively with GM-Janesville to identify hidden chemical costs, including traditional overhead items that are linked to chemical use decisions. In their annual reports to GM, BetzDearborn has broken out GM's savings for those chemical costs which they have assumed in the contract (PE component costs), including items such as the elimination of purchase order and inventory costs, and the value of employee training provided by BetzDearborn. However, they have also included many non-PE component cost savings including:

- Secondary benefits from the reduction or elimination of chemicals (such as reduced EHS compliance costs and expanded floor space previously used for inventory)
- Process improvements that reduced the cost of making de-ionised water for the plant
- Process improvements that reduced the time needed to clean paint booths
- Identification and repair of malfunctioning production equipment

The purpose of BetzDearborn's annual report to GM is to demonstrate the value that they have brought to the plant. It is in their best interest to continuously improve the quality of the accounting data available to identify, document and communicate hidden chemical costs to GM.

### Chrysler Corporation: Neon Assembly Plant, Belvidere, Illinois

The Shared Savings programme at Chrysler is known as the 'Pay-as-Painted' programme, implemented at the Belvidere, Illinois, plant in 1989. Today, Chrysler's paint supplier, PPG Industries, has assumed responsibility for all chemicals in systems related to cleaning, treating and coating the autobody. The programme is implemented through a Pay-as-Painted team, made up of PPG personnel, as well as Chrysler staff from many departments throughout the plant. Though PPG uses many of their own chemicals, they rely on a number of tier-2 chemical suppliers as well.

Chrysler's performance expectations are quite unique, since their targets are defined closer to the ultimate value of the consumer product, the Chrysler Neon, than in any other Shared Savings programme we studied. These include specifications for the quality of coatings and final finish for the autobody of the Neon. If the autobody does not pass Chrysler's quality inspections, PPG does not get paid (unless the problem can be linked to a Chrysler error). Such performance expectations make sense, given that their programme focuses on direct chemicals—those chemicals that become part of the final product. However, the contract also includes more traditional expectations, such as chemical ordering, inventory, tracking and quality assurance. As with other Shared Savings programmes, Chrysler does not own the chemicals until they are used in the production process. PPG is paid a fixed fee per quality vehicle produced.

The Belvidere assembly plant saved over US$1 million in the first year of their programme with PPG. There have been considerable additional savings in the years since then, many of which have involved improved chemical use efficiency. For example, the Pay-as-Painted team researched methods to reduce waste from the electro-coat process by the more efficient placement of the electro-coating electrodes. The improvements reduced the amount of electro-coat chemical needed for the process (the tip of the iceberg in Fig. 4), which saved PPG money. At the end of the model year, such savings are shared with Chrysler through a rebate programme; and, if warranted, performance fees may be reduced to reflect the new, lower costs of production.

However, many of the benefits enjoyed by Chrysler have been in the form of improved product quality and environmental performance. For example, the Pay-as-Painted team implemented a powder coat anti-chip application which increased finish durability while dramatically reducing emissions of volatile organic compounds (VOCs). They also implemented a water-borne primer system which reduced VOCs. Recently, they developed an improved technology for paint repairs. Previously, the paint touch-up process required the use of high-VOC materials. However, the new process makes use of the same low-VOC materials used in the original painting process, and produces a better finish.

Some changes, such as the powder coat anti-chip, were more expensive than the systems they replaced. In other words, the tip of the iceberg grew, and since PPG assumes that cost (see Fig. 4), the performance fee paid by Chrysler to PPG increased to cover it. However, in addition to improving product quality, the powder coat anti-chip system reduced the non-PE component of Chrysler's chemical cost iceberg. In particular, it reduced or eliminated many environment, health and safety costs associated with handling, storing, using and disposing of the high-VOC coatings used in the previous anti-chip process. Though Chrysler maintains the confidentiality of the financial details, it is clear that they consider the increased performance fee paid to PPG to be offset by the improved product quality and reduced hidden chemical costs.

Improved materials and cost data are central to the success of the Pay-as-Painted programme, and the supplier played an integral role in developing both these data systems. The Shared Savings programme at Belvidere began with production areas such as phosphating, where the development of the material use and cost baselines was relatively straightforward. They also began by studying the painting process. Ultimately, they required three years of data to develop paintshop cost baselines for each of the different colour options. They are now in the process of studying solvent usage and costs throughout the plant. The plan is to develop a programme for solvents that parallels

Pay-as-Painted. PPG believes the advanced data system expertise they have developed in the Pay-as-Painted programme gives them a competitive edge over other chemical suppliers.

To Chrysler, good data are essential to continuous process improvements and cost reductions. A comment from Bob Godare, Environmental Co-ordinator for the Belvidere plant, summarises this well:

> We know what we have been spending. We know the defect rate, the scrap rate, the waste rates. You just have to figure out exactly what it's costing you to paint that piece of material. That's critical if you are going to use a program like Pay-as-Painted. You have to know exactly what it's costing you. It doesn't take as much work as you think to figure that out. If you've got the information, you just have to use it.

In fact, the Pay-as-Painted programme fits nicely into Chrysler's overall strategy to implement activity-based costing (ABC) at its plants. As ABC is phased in, the Pay-as-Painted programme will be positioned to expand with it.

### Ford Motor Company: Taurus Assembly Plant, Chicago, Illinois

The Chicago assembly plant began its first Shared Savings chemical supply programme in 1988 with the detackification of overspray from the paint booths. Today, this has evolved into two major Shared Savings programmes, known as 'Total Fluids Management' and 'Total Solvent Management'. These two programmes cover most chemicals in the plant, with the exception of paints, sealers and lubricants. Their chemical supplier, PPG/Chemfil, relies on tier-2 suppliers for many of these chemicals.

Ford's performance expectations in the contract include many standard Shared Savings features, such as the ordering, inventory, tracking, distribution and quality control of chemicals, which are owned by PPG/Chemfil until used by Ford. As with GM, Ford also specifies an annual savings rate in the performance expectations. However, because the plant is located in a large metropolitan area, VOC reduction is a high-priority performance expectation. PPG/Chemfil's performance fee is a combination of fee per vehicle produced for most of the chemical systems covered by the contract and a fixed monthly fee for a few chemicals that are unrelated to production.

The Chicago assembly plant benefited immediately from their Shared Savings programmes by negotiating a performance fee that was smaller than the PE component of Ford's chemical cost iceberg. However, the various programmes have continued to provide ongoing financial and operating benefits. For example, PPG/Chemfil and a tier-2 water treatment chemical supplier were instrumental in modifying waste-water treatment operations to reduce sludge generation by 27%. Ford's decision to introduce certain aluminium body panels meant that sludge from waste-water treatment would become legally 'hazardous', greatly increasing waste disposal costs (a non-PE component cost). The reduction in sludge generation saves Ford US$50,000 annually.

Probably one of the most important benefits has been the reduction in VOCs. Ford faced severe regulatory restrictions in the Chicago ozone 'non-attainment' area. Under the Total Solvent Management programme, PPG/Chemfil worked extensively with Ford personnel to study the usage of VOC-containing solvents throughout the plant. Together they were able to reduce VOC emissions by 57% within the first eighteen months of the programme. Most of these reductions were produced by finding more efficient ways to use the solvents. This significantly reduced hidden chemical costs for Ford by reducing compliance costs and avoiding potential legal costs. The reduced emissions also reduce employee exposures to VOCs. These improvements benefited PPG/Chemfil as well. The increased efficiency of solvent use meant that PPG/Chemfil had to supply less solvent, reducing their own costs.

As with the other Shared Savings programmes we studied, improved data have played an important role. At Ford's Chicago assembly plant, PPG/Chemfil's chemical tracking system was extremely valuable in identifying and tracking Ford's hidden chemical costs. It has significantly reduced the cost of Ford's annual reporting of environmental wastes ('Form R'). Dan Uhle, Environmental Control Manager at the plant, explains it this way:

> Chemical tracking has improved environmental reporting substantially. That used to be the hardest part in doing my Form Rs—coming up with good chemical usage data. Now they tell me how much they've used. I no longer have to worry about getting the data from purchasing or our own inventory records. Take solvents for example. I had to look at what we bought and get records from the suppliers of all of the different materials that had VOCs in them. I had to make the assumption that the inventory at the beginning of the year was the same as at the end. I don't know how much was scrapped and actually went out as waste paint solvent, where we had some recovery, or how much was used. So I had to make assumptions about all of that to the best of my ability using engineering judgment. Now PPG/Chemfil keeps daily records of what they use and the VOCs emitted, and they not only do it by product but they do it by process. This program provides much more help in maintaining control than I've ever had (Bierma and Waterstraat 1997).

### Navistar International: Engine Plant, Melrose Park, Illinois

Navistar's engine plant in Melrose Park has had a Shared Savings programme longer than any of the other plants we studied. Known as Chemical Management Services (CMS), the programme was started by Navistar and their supplier, Castrol Industrial, in 1987.

The CMS programme at Navistar is one of the simplest, yet most productive, Shared Savings relationships we have seen. The programme is limited to machining coolants, cleaners and associated additives. However, these chemicals are critical to all of the machining operations at the plant. Navistar's performance expectations in the contract include the standard Shared Savings features, such as the ordering, inventory, tracking, distribution and quality control of chemicals, though these were phased into the contract over time. Rust prevention targets were also established. Castrol continues to own the chemicals until used by Navistar. In return, Castrol is paid a fixed monthly performance fee, which can be adjusted for unusually large fluctuations in production levels.

The benefits for Navistar have been dramatic, though they have changed as the relationship has evolved. The first area on which Castrol focused was process applications. Gaining greater control of process chemistry and the compatibility between coolants, cleaners, equipment and metals reduced production downtime and improved product quality. Rework on engine blocks and heads was reduced by 93%. These benefits represent a decline in non-PE component costs. As additional chemical logistic responsibilities were phased in, Navistar benefited from the many PE component costs assumed by Castrol, such as chemical inventory management and quality control. Not only were costs reduced, this freed the EHS staff and others to concentrate on issues more central to Navistar's core business.

Following this, Castrol personnel focused on reducing chemical usage and waste. Ultimately, coolant usage was reduced more than 50%. This decreased the tip of the chemical cost iceberg, decreasing Castrol's coolant costs. These benefits were shared with Navistar through annual rebates and also through the reduction of hidden costs associated with coolant maintenance and management. Coolant waste was reduced by more than 90%, further decreasing Navistar's non-PE component costs.

The CMS programme has driven continuous improvements in the quality of data related to chemical use and costs. Bob Hendershott, Castrol on-site manager, explains one such benefit:

> One washer had a problem with the flow control for the automatic make-up. Every morning when they would turn it on it would overflow. Nobody saw anything: they'd turn it on and it would seem fine. Now we can track usage by machine…so it was a case where we could see that one month's usage on this particular washer was very high. So I started questioning people, going out there and looking into it. Next month it continued to be high and we were able to get the manufacturer in and found the problem. The usage went down below what it was before the problem.

Rudy Bernath, Navistar's plant chemist, added:

> It would've probably gone on for years [without Castrol's chemical tracking system]. Even if we found it when looking at the year-end numbers, we might not have pinpointed that machine—the whole department's usage would've been high.

An interesting result of the improved data quality, and the overall success of the CMS programme, has been a recent change in the Navistar–Castrol contract. Navistar has returned to buying coolants and cleaners on a cost-per-drum basis. Castrol is now paid a management fee to perform the activities they previously performed under the CMS contract. However, a gain-sharing clause has been added to the contract. What initially seems like a step backwards in the relationship may actually be one logical progression for Shared Savings programmes. Viewed from the perspective of the chemical cost iceberg, Castrol dramatically reduced the PE component of Navistar's chemical cost iceberg. As a result, non-PE component costs became an increasingly large component of total chemical cost iceberg. Both parties decided to shift the focus of the contract away from the small PE component costs to the much larger non-PE component portion of the iceberg.

## ▌Conclusions

Many of the chemical-related costs experienced by companies today are due, in part, to the use of both traditional accounting and traditional chemical supply practices. Shared Savings chemical supply programmes go hand-in-hand with environmental management accounting practices to continuously reduce chemical-related costs and improve environmental performance. Examination of these four manufacturing plants with over thirty years of Shared Savings experience illustrates that the two key components of Shared Savings, which drive basic chemical performance and continuous cost reduction, require better data than traditional accounting systems can provide. The improvements in accounting data at these four plants promoted the successful implementation of Shared Savings programmes, which generated the need for even better accounting data. This synergistic relationship between chemical supply and accounting practices helps explain why many companies experience nearly intractable 'chemical chaos' while others are making dramatic progress in improving environmental performance while reducing chemical costs.

# 14 Environmental Accounting in an Investment Analysis Context

## Total Cost Assessment at a Small Lithographic Printer[1]

Edward D. Reiskin, Deborah E. Savage
and David A. Miller

## ▮ How Can Environmental Accounting Support Business Decision-Making?

TO THE EXTENT that environmental costs exist in almost every phase of a business's operations, environmental accounting practices can support improved decision-making in many different applications throughout an organisation. The broad range of business decisions that can benefit from the adoption of environmental accounting principles is shown in Figure 1. A forthcoming EPA report explores the applications of environmental accounting relating to capital investments, process-costing and strategic planning via brief matrix descriptions of 37 case studies found in the literature (Graff *et al.* forthcoming).

### EA Informs Capital Investment Decisions

Companies develop and grow their businesses by investing in their human and physical capital. Their long-term financial viability hinges on the strength of these investments. Generally, a company's investors demand at minimum a return comparable to that which they can obtain through other investments. This demand places pressure

1. The authors gratefully acknowledge the United States Environmental Protection Agency (EPA)'s Environmental Accounting Project and the Illinois Waste Management and Research Center (WMRC) for their funding of the research that formed the basis for this chapter. Thanks also to Allen L. White of Tellus Institute for comments on an earlier draft of this article and to Karen Shapiro of Tellus Institute for contributions to the case study. This material borrows from the forthcoming document prepared by Tellus Institute for the EPA, *Snapshots of Environmental Cost Accounting* (Graff *et al.* forthcoming) and the document prepared for the WMRC, *Strengthening Corporate Commitment to Pollution Prevention in Illinois: Concepts and Case Studies of Total Cost Assessment* (Tellus Institute 1997).

| | |
|---|---|
| Product design | Capital investments |
| Process design | Cost control |
| Facility siting | Waste management |
| Purchasing | Cost allocation |
| Process-costing | Product retention/mix |
| Risk/liability management | Product pricing |
| Strategic planning | Performance evaluations |
| Supplier selection | Plant expansion |
| Environmental programme justification | |

**Figure 1:** Business Decisions Supported by Environmental Accounting
*Source: Adapted from US EPA 1995a: 6*

on companies to invest their limited capital funds wisely. Environmental costs are often a significant component of capital and operating costs. Often, there is a considerable financial return available to companies that can reduce these costs. When environmental costs are properly accounted for, investment analyses of environmental performance improvements provide managers with information to determine whether and to what extent the benefits of such investments will exceed the costs. But, to achieve these results, managers must first be able to define and measure these environmental costs in a systematic and consistent fashion.

One specific application of environmental accounting for capital investment analysis is total cost assessment (TCA), a method by which investments, particularly environmental investments, can be evaluated in a way that more accurately reflects their profitability potential. The four basic elements of TCA that make it more informative than conventional analysis are:

1. A more comprehensive cost inventory that includes less direct, less tangible costs[2]

2. Allocation of costs that are typically assigned to overhead accounts, and either allocated on the basis of an inappropriate cost-driver or not allocated at all

3. Evaluation of projects using longer time-horizons in order to better capture the full benefit of the investment, a significant portion of which may be realised after the first two to three years

4. Profitability indicators that account for the time-value of money, making the results more realistic and reflective of an investment's true cost or benefit

Evaluating environmental projects using TCA helps put them on equal footing with other projects competing for capital funds. Projects that appear to be financially weak using conventional analyses may look considerably stronger and more competitive once their true return has been identified. Consider an expensive investment in a process change to accommodate a switch to an aqueous solvent that may appear to be a poor investment with a long payback if only direct labour and material costs are considered over a three-year time-period. Now consider the alternative by allocating the full

---

2. See Chapter 2 and US EPA 1995a for a discussion of environmental costs.

environmental costs of the existing process—such as solvent disposal costs, regulatory permits, worker health and liability for accidental spills or leaks—to the subject process. When included in the analysis, the less visible cost-savings associated with the switch, considered over a longer, 7–8-year period, may well yield an impressive rate of return and a shorter-than-expected discounted payback. Of course, TCA *a priori* does not ensure profitability. It does, however, ensure greater transparency, clarity and rigour in making capital investment decisions.

## Capital Investment Decision Case Studies

The largest group of environmental accounting case studies identified in the EPA *Snapshots* report (US EPA forthcoming) is the group of 22 capital investment decisions (Fig. 3). These decisions were informed, at varying levels, using environmental accounting principles in the vein of TCA. The case study firms represent printing, metals processing, electronics manufacturing, chemicals manufacturing, pulp and paper, tools/parts manufacturing, wood products manufacturing, jewellery manufacturing/distribution, and electric power generation. Firm sizes range from 15 to approximately 1,000 employees, from annual revenues of $1 million to the hundreds of millions.

Nine of these cases studied investments in technologies that would enable recovery of a non-product output—e.g. solvents, powder coatings, rinse-water, oil, by-product—to minimise or eliminate its discharge as a hazardous or liquid waste. Other projects studied included parts degreasers, waste-water treatment enhancements, material substitutions, and other technological investments that would reduce material input requirements. Motivations for the projects varied, and, while all had an environmental component, some were driven more by other business considerations.

Our case studies evaluated the profitability of both past and proposed investments, or compared the economics of several pollution prevention (P2) investment proposals. The financial results reported in Figure 3 represent various degrees of total cost assessments; all incorporate a broader inventory of costs, but they vary with regard to the extent that the four components of TCA mentioned above are in place. Almost all of the analyses calculated profitability in terms of net present value (NPV) for the projects under consideration; these values ranged from –$1.4 million to $11 million, with most in the range of $10,000 to $100,000. Some of the highest include: a 5-year NPV of $310,255 for a screen printer; an 8-year NPV of $352,814 for an electronic equipment manufacturer; and a 15-year NPV of $11,633,835 for a diversified chemical company. These numbers give a general sense of the magnitude of values calculated for these projects but they are not strictly comparable because each study was conducted with a different discount rate. The rate chosen varies from firm to firm and is often quite high, reflecting businesses' traditional short-term focus and high demands for returns from all investments.

Two of the 22 analyses—including the paper-coating mill study shown also in Figure 3—yielded negative NPVs, but both of these projects contained significant qualitative benefits. One of the projects was approved on the basis of these qualitative benefits. Many of the analyses for which a discounted payback was calculated (and for some it is not clear if it was a discounted or simple payback) would pay for themselves in under three years; all but two had paybacks under five years. In a few cases, however, the encouraging financial analysis was insufficient to over-ride doubts about unproven technology, so project implementation was put on hold.

A number of the studies provided a comparison of TCA to more conventional investment analyses. These comparisons demonstrate the value of TCA as a more comprehensive analysis method. Figure 2 summarises and highlights the differences in these cases. TCA, of course, does not guarantee improved profitability of an investment, but does provide a clearer picture of the profitability by eliminating some of the biases inherent in conventional analyses. One of the entries in Figure 2 provides an example in which a TCA showed an investment to be less profitable than calculated through conventional means.

While Figure 2 demonstrates the impact that TCA can have on a profitability analysis relative to a less comprehensive, conventional analysis, Figure 3 demonstrates more generally that TCA can be used in a diverse array of industries to provide analyses of a wide range of projects. This collection of case study results also demonstrates that, in many cases, those projects that improve a firm's environmental performance can also contribute positively and significantly to a firm's financial performance. This empirical evidence supports the premise that there can be high costs associated with poor environmental performance and therefore it can be cost-effective to make upstream investments that enable improvements and thus lower those costs. In these instances, TCA facilitated these investments by rendering their profitability more apparent.

We turn now to a closer look at the first case in Figures 2 and 3—the small commercial lithographic printer—as a good example of the value of environmental accounting to business decision-making. The following pages contain the case study itself and document the TCA analysis. This particular investment did not have large environmental ramifications, but TCA was nevertheless useful in providing a more

| Description of firm | Investment | Conventional results | TCA results |
|---|---|---|---|
| Small commercial lithographic printer | Computerised pre-press system | 5-year NPV = $58,358<br>5-year IRR = 51%<br>Discounted payback = 2.1 years | 5-year NPV = $187,700<br>5-year IRR = 132%<br>Discounted payback = 0.8 years |
| Large commercial printing facility | Waste-water treatment system enhancement | 10-year NPV = $51,887<br>10-year IRR = 14.7%<br>Discounted payback = 6.9 years | 10-year NPV = $81,152<br>10-year IRR = 17.8%<br>Discounted payback = 5.7 years |
| Private company with two metal fabrication job shops | Paint/water separator | 15-year NPV = $9,332<br>15-year IRR = 20%<br>Discounted payback = 4.3 years | 15-year NPV = $12,436<br>15-year IRR = 23%<br>Discounted payback = 3.8 years |
| Small circuit-board manufacturer | Plastic-coated plating racks | 5-year savings = $555 | 5-year NPV = $33,589 |
| Privately held paper-coating mill | Heavy-metal-free, aqueous base-coat | 15-year NPV = –$203,600<br>15-year IRR = 11%<br>Discounted payback = 7.6 years | 15-year NPV = –$395,600<br>15-year IRR = 6%<br>Discounted payback = 11.7 years |
| Speciality paper mill | White-water collection system modification | 15-year NPV = $360,301<br>15-year IRR = 21%<br>Discounted payback = 4.2 years | 15-year NPV = $2,800,000<br>15-year IRR = 48%<br>Discounted payback = 1.6 years |

**Figure 2:** Comparison of Conventional and TCA Analysis Results

| Description of firm | Investment project | TCA financial results | Other benefits |
|---|---|---|---|
| Small commercial lithographic printer; 15 employees and annual revenues of $1 million | Computerised pre-press system | 5-year NPV is $187,700 5-year IRR is 132% Payback is 0.82 years | Improved product quality and reduced turnaround time |
| Large commercial printing facility; 650 employees and annual revenues of $95 million | Continuous membrane-filter waste-water treatment system | 10-year NPV is $81,152 10-year IRR is 17.8% Payback is 5.66 years | Less hazardous waste generation |
| Private company with two fabrication job shops; 200–300 employees | Paint/water separator | 15-year NPV is $12,436 15-year IRR is 23% Payback is 3.8 years | Reduce waste metal generation |
| Small circuit-board manufacturer; 30 employees | (a) Plastic-coated plating racks (b) Waste-water treatment enhancement | (a) 5-year NPV is $33,589; 5-year IRR is 66% (b) 5-year NPV is $62,824; 5-year IRR is 1,886% | (a) Increased worker morale and product quality; reduced future liability (b) Improved worker safety; reduced volume/toxicity of treatment materials and waste |
| Privately held paper-coating mill | Switch from solvent/ heavy-metal base-coat to an aqueous/ heavy-metal-free formulation | 15-year NPV is –$395,600 15-year IRR is 6% Payback is 11.7 years | Reduced flammability and explosivity hazards; employee exposure to solvent, volatile organic compound (VOC) emissions, hazardous waste and solvent/ heavy metal usage |
| Speciality paper mill | Modifications to a white-water collection system | 15-year NPV is $2.8 million 15-year IRR is 48% Payback is 1.6 years | Reduced freshwater usage, fibre and filler loss, waste-water generation, heating of freshwater, and use of raw materials |
| Small commercial flexographic printer; 36 employees and annual revenues of $15 million | In-line solvent recovery still | 15-year NPV is $99,879 15-year IRR is 57.34% Payback is 2.06 years | Reduced solvent consumption and still bottom generation |
| Small commercial screen printer; 40 employees and annual revenues of $3 million | Dry-film imaging system | 5-year NPV is $310,255 5-year IRR is 282% Payback is 0.4 years | Reduced generation of darkroom wash-water and used fixer |
| Manufacturer of precision sheet-metal products; about 100 employees work in two shifts | High-volume, low-pressure spray guns | 8-year NPV is $140,900 8-year IRR is 906% Payback is 0.12 years | Reduced worker exposure to VOC emissions; new customers attracted by P2 publicity |
| Manufacturer of all-welded lockers; 80 employees and annual sales of $6 million | Polyester powder coatings; insulation of paint drying and curing ovens | 10-year NPV is $264,865 10-year IRR is 26.8% Payback is 4.17 years | Increased quality and productivity; reduced liability and worker exposure to emissions |
| Metal finisher; employees and annual sales of $5 million | (a) Powder recovery unit (b) Rinse-water recycling option | (a) 10-year NPV is $32,368; Payback is 1.2 years (b) 10-year NPV is $168,697; Payback is 2.3 years | Less materials handling, waste powder and risk of accidents; better employee morale, air quality, corporate image and market share |

| Manufacturer of aluminium reflectors; 500 employees and sales of $110 million | Vapour degreaser | 10-year NPV is $101,292 | n/a |
|---|---|---|---|
| Manufacturer of precision metal parts; 16 employees | Three alternatives to a CFC degreaser | 8-year NPVs were: $77,400; $73,531; $76,926 | Minimised well-water use |
| Manufacturer of printed circuit-boards; 80 employees and annual sales of over $6 million | (a) Single-speed motors ventilation system (b) Dual-speed motors ventilation system | (a) 15-year NPV is –$171,782 (b) 15-year NPV is –$143,424 | Eliminated compliance fines; reduced worker exposure; allowed increased production |
| Medium-sized manufacturer of military and civilian electronic equipment | (a) Bench-top degreasers using hydrochloro-fluorocarbons (HCFCs) (b) An ultrasonic cleaning system using terpenes | 8-year NPVs were: (a) $93,446 (b) $352,814 | Reduced reporting, improved cleaning efficiency |
| Large multinational chemical manufacturing company | Closed-loop batch still solvent recovery system | 12-year NPVs ranged from –$1.4 million to $3.4 million for scenarios representing different waste stream mixes, capital expenditure, etc. | Strengthen firm's strategic objectives in terms of production and product flexibility, long-term self-reliance in solvent production, and meeting toxics use reduction goals |
| Diversified chemical manufacturer with 450 facilities in 40 countries; annual sales of $12 billion | Process to convert by-product into usable input material | 15-year NPV is $11,633,835 15-year IRR is 41% Payback is 2.2 years | Reduced waste sent to landfill; improved system operation and energy use; and enhanced company's ability to respond to market development |
| Family-owned tool manufacturer; ~250 employees | Water and oil recycling system | 10-year NPV is $14,601 | Reduced water use and potential liability of corporate officers |
| Manufacturer of precision industrial valves; annual sales in US of $150 million; 1,000 employees | Closed-loop system on zinc phosphating line to eliminate plant's discharge to sewer | 10-year NPV is $54,913 | Reduced water use and potential liability from future, more stringent effluent limits |
| Medium-sized manufacturer of wood doors and window trim; 50 employees | Conversion from petroleum-based coating process to a method using UV-cured lacquer | 10-year NPV is $1,409,270 10-year IRR is 69% Payback is 1.3 years | Improved quality; reduced air emissions; and reduced workplace toxicity |
| Manufacturer and distributor of jewellery, leather goods and accessory items; ~500 employees | Ethyl acetate still to reduce use of the toxic chemical | 5-year NPV is $28,279 | Reduced use of ethyl acetate |
| Fossil-fuel electric power generator | Co-firing a fossil-fuel-generating plant to produce bioenergy | 5-year NPV is $792,795 | Compliance with pending air emission regulations; mitigation of externalities associated with burning fossil fuels |

**Figure 3:** Summary of Capital Investment Decision Cases

comprehensive analytical framework. Like most environmental investment decisions, this one was driven by multiple factors and yielded multiple benefits.

## ▌ Case Study: Computer Pre-Press System Purchase by a Small Lithographic Printer

### Company and Facility Background

This company is a small lithographic printer, printing one- to four-colour posters, cards and booklets. Many of the company's customers are community groups, political groups or national non-profit organisations, but the company also serves commercial clients.

The company, started in the late 1960s at an Illinois university, later moved to a Chicago location and became a non-profit organisation. In the mid-1980s, the company became incorporated as a for-profit co-operative in order to gain access to financing through avenues that are unavailable to non-profit organisations. There are eight partners who make business decisions collectively, plus seven other employees. The company moved to its current location in late 1994. The building, which the company owns, is about forty years old.

### Project Background

This printer provides traditional lithographic printing, including pre-press, printing and some post-press processing. In the past, the company's computing abilities were limited, so disk-based jobs requiring computer pre-press operations were sent to a service bureau for production of film for plate-making (see Fig. 4 for a schematic of this pre-press operation). These jobs represented about two-thirds of the company's total jobs, a percentage that was rapidly increasing. The costs of using the service bureau were increasing even faster, from under $1,500 in 1994 to a projected $48,000 in 1996.

The remaining third of the business came in on conventional media. This original art was photographed and the film was sent to an in-house darkroom for silver-film open-tray developing. Film-developing produced spent fixer and spent film developer. Spent fixer, about six gallons per year, was poured through a small silver recovery unit. The Production Manager suspected that the recovery unit was too old to recover very much silver from the fixer. The spent developer, about 24 gallons per year, was washed down the drain. As this darkroom process was an open-tray developing process, it produced fugitive vapour emissions of unknown quantity and composition.

The Production Manager was concerned about the fate of the spent film developer. Although it was aqueous, and the vendor assured him that it was safe to pour down the drain, he was not sure about the regulations or the possible environmental effects.

From the darkroom, the prints were laid out on a stripping table where manual adjustment and formatting was done. Scrap material from this process was thrown in the garbage. A rag service collected dirty rags and provided clean ones. The rag service also collected the company's dirty citrus-based solvents. The Production Manager did not know what the rag service did with the rags or the solvent waste.

### Project Description

The company recently purchased a computer pre-press system. The system is able to directly process jobs that customers submit on computer disk; jobs submitted as camera-ready art can be scanned and then processed. The film processor that is part of

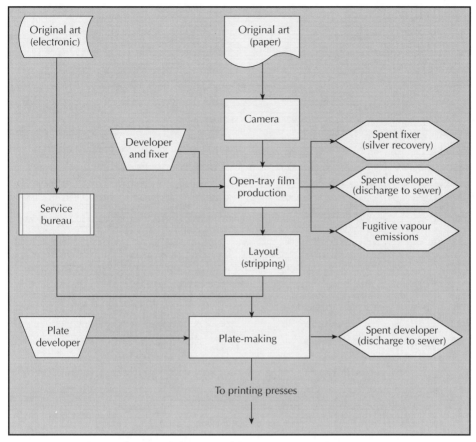

**Figure 4:** Schematic of Former Pre-Press System

the system produces film that can be used in plate-making, thus bypassing the dark-room entirely (see Fig. 5).

The company sought to purchase an integrated turnkey system from a single vendor to avoid product incompatibilities and to facilitate maintenance and technical support. After soliciting equipment bids from a number of vendors and considering the processing speed and features required, the company purchased a nearly complete system for approximately $46,000. This price included a PowerMacintosh for digital artwork and publishing, a Pentium PC for image processing, various software packages, accessories (scanner, modem, zip drive, etc.), installation and configuration. Separately, the company purchased a used film processor from the company's primary film vendor for $3,000. Finally, some miscellaneous electrical and plumbing work was done to accommodate the new system.

The computer pre-press system allows the company to bring disk-based jobs—which were previously all sent to a local service bureau for processing—back in-house. Savings from avoided service bureau and courier expenses could potentially reach $3,000–4,000 per month. In addition to these direct savings from avoided costs, there are other benefits that are expected to contribute savings indirectly as well.

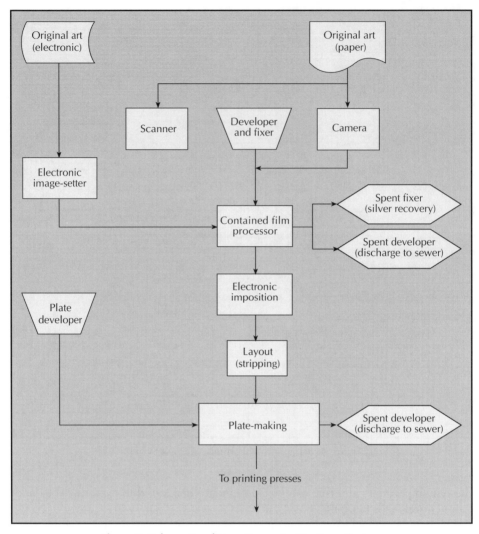

**Figure 5:** Schematic of New Computer Pre-Press System

Conducting pre-press operations in-house reduces job turnaround time by a minimum of 24 hours on disk-based jobs. As printing operations generally become more electronic, customer expectations and demands increase, and lead times that used to be acceptable are now considered too long. Thus any reduction in the time it takes to turn around a job for a customer creates a competitive advantage.

In-house processing also provides more control over pre-press decisions and facilitates easier rework. Maintaining control over this critical aspect of the printing process gives the facility the means to greatly enhance customer satisfaction via production of quality work. Furthermore, the new system eliminates the use of the darkroom on most jobs, as the new film processor allows camera-ready jobs to be processed without use of open-tray developing.

The primary motivations for this project were to streamline pre-press operations, reduce service bureau costs, and provide new capabilities and flexibility that would increase product quality and services offered and attract new business. Avoiding darkroom use and minimising chemical use in the pre-press process for both environmental and worker safety reasons were significant secondary motivations.

## The Company's Financial Analysis of the Project

The company does not use financial indicators beyond simple payback, and does not collect the data necessary to conduct rigorous financial analyses. When evaluating capital expenditures, the partners estimate the direct quantifiable costs and savings, consider less tangible benefits, and make a conservative but somewhat intuitive decision. The evaluation of the computerised pre-press system was done in this fashion. The potential effects of taxes, depreciation, inflation and discounting were not considered.

For this project, the company made a quick analysis of its expenditures to the service bureau. It was estimated that upwards of two-thirds of current service bureau costs could be eliminated with the installation of the new pre-press system. The Production Manager assumed that, at first, that fraction might be lower as the facility moved along the learning curve of the new equipment. He further estimated that roughly one-third of the work that currently comes in electronically would still be beyond the capabilities of the new system for either technical or logistical reasons. One such reason is that the processor that is in the facility's price range is too small for large posters. Also, the more complex four-colour work would be, in the near term, beyond the technical capabilities of the staff. Similarly, some of the work may remain in the darkroom due to limitations of electronic processing and specific customer needs. The final constraint of the new system would be one of resources: during very busy periods, one operator working on one processor will not be enough. This constraint will relax in time as operators move along the learning curve and require less time per job.

Because of the anticipated learning curve with the high-tech equipment in the otherwise traditional printing environment, the effects of the project on labour costs were not considered. The Production Manager felt they would be too difficult to quantify and would change over time. Other costs, such as raw materials, utilities and waste disposal were also not considered.

In summary, the costs considered by the company were:

- Initial purchased equipment costs, as specified in vendor quotes ($49,385)
- Annual savings from a reduction in service bureau charges ($22,500)
- Annual savings from reduced courier charges ($1,500)

## Total Cost Assessment of the Project

A total cost assessment was performed in order to provide a more accurate economic picture of the project. An essential element of a TCA in analysing the true costs of a project is an enhanced cost inventory. Direct project investment costs and operating cost-savings are typically included in a standard project analysis, but many other relevant costs may be omitted because they are less apparent. Indirect labour, compliance and waste disposal are examples of actual costs that can materially affect the potential profitability of a project. In this case, the TCA includes many other costs and savings relevant to the purchase and operation of the computerised pre-press (see Fig. 6).

| | Company | TCA |
|---|:---:|:---:|
| **Initial investment costs** | | |
| Purchased equipment | ✓ | ✓ |
| Electrical work | | ✓ |
| Plumbing | | ✓ |
| **Annual operating costs** | | |
| Materials:   film | | ✓ |
| chemicals | | ✓ |
| Service bureau work | ✓ | ✓ |
| Typesetting | | ✓ |
| Courier | ✓ | ✓ |
| Revenues | | ✓ |

**Figure 6:** Comparison of Cost Items Included

One category of costs not included in this preliminary analysis is labour. The facility did not intend to change overall staffing levels as a result of this investment. If production volume continued to grow, future hiring would be likely, but in the short term, no changes in labour were anticipated. A follow-up analysis, discussed later, did take into account costs of investment labour and shifts in operating labour resulting from the investment as well as changes in the allocation of the Production Manager's time.

The TCA included the following costs and cost-savings that the company's analysis did not.

### Initial Investment Costs (summarised in Fig. 7)

1. Since the system was not purchased as an entirely turnkey package, there were contractor costs associated with the installation. Wiring and piping to accommodate the new system cost $1,400 and plumbing to accommodate the film processor cost an additional $275.

| Cost item | Description | Cost |
|---|---|---|
| **Purchased equipment** | From vendor invoice for PowerMacintosh, image-setter, software and associated hardware | $46,310 |
| | From vendor invoice for film processor | $3,075 |
| **Electrical work** | From electrician invoice for materials and labour to wire the new system | $1,400 |
| **Plumbing** | From plumber invoice to relocate drain and water supply and install shut-off valves to accommodate the film processor | $275 |
| **Total** | | **$51,060** |

**Figure 7:** Initial Investment Costs

*Annual Operating Costs and Savings (summarised in Fig. 8)*

1. **Increased film costs.** Film for the new processor is considerably more expensive on a unit basis than the darkroom film, but it is still less than film processed by the service bureau. Annual film costs, currently $2,100 per year, are expected to increase to $14,700. This figure does not account for the reduction of service bureau film, since this cost is bundled into the total service bureau cost.

| Cost category item | Previous—service bureau | Computer pre-press | Annual savings (costs) |
|---|---|---|---|
| **Materials** | | | |
| Chemicals | $2,500/year Current cost | $1,000/year Reduction of developer and fixer use, expected to decrease 60% based on preliminary use data | $1,500 |
| Film | $2,100/year Current cost | $14,700/year Increase in film costs based on assumption that one-third of business comes on conventional media, so one-third of film usage will be at current costs, two-thirds will be ten times that figure | ($12,600) |
| **Service bureau** | | | |
| Service bureau | $45,000/year Current cost | $22,500/year Reduction by half of current costs of service bureau—this cost estimated to fall by 10%/year as in-house proficiency increases | $22,500 |
| Courier | $3,000/year Current cost | $1,500/year Reduction by half of current costs of courier corresponding to service bureau use—this cost also estimated to fall by 10%/year | $1,500 |
| Typesetting | $2,820/year Current cost | $1,110/year Reduction by 60% of vendor typesetting costs | $1,710 |
| **Revenues** | $1,100,000/year current projected revenues | $1,210,000/year Increase in total revenues by roughly 10% resulting from increased capabilities, faster turnaround and better process control | $110,000 |
| **Net revenues** | **$1,044,580** | **$1,169,190** | **$124,610** |

Note: Current costs of materials and service bureau are based on invoices from the first seven months of 1996, before the new system was installed. All monetary amounts are in 'Year 0' dollars, before any adjustments for expected price changes in Year 1 and subsequent years.

**Figure 8:** Annual Operating Costs/Savings

2. **Reduced use and disposal of darkroom fixer and developer.** Moving film developing from the darkroom to the enclosed processing unit is expected to reduce chemical cost by 60% from its current cost of $2,500 per year. This reduction is achieved because less volume is required for a given quantity of film, and chemical life is increased by a water filtration system inside the processor. With open-tray developing, chemicals exposed to the atmosphere lose their integrity after four or five hours and must be disposed of. (Since the chemicals are currently disposed of without direct cost, no savings from avoided disposal cost was included.)

3. **Lower costs of service bureau charges and of ferrying disks between the facility and the service bureau.** Service bureau and shipping costs, previously projected at $45,000 and $3,000 respectively for 1996, are expected to drop by a half to two-thirds. The initial drop is assumed to be by one-half with costs continuing to decline by 10% per year as the facility moves along the learning curve.

4. **Lower typesetting costs from vendor.** Much of the work previously sent out for typesetting (e.g. business cards, letterhead) can now be done with the digital imaging software. The expected 60% drop in this service will save $1,710 annually.

5. **Improved customer response by eliminating the service bureau's 24-hour turnaround time.** This improvement is one of the more difficult to quantify. Even without this project, the company's revenues have been growing rapidly. The Production Manager estimates that digital capabilities will enable an additional 10%–15% sales increase over 1996 projected revenues of $1,130,000. This incremental revenue stream is assumed to decline at 10% per year to conservatively reflect diminishing returns from the system and aggressive competition. Because the facility is not currently at capacity, it is expected that this increase in business can be handled with existing resources. Before long, however, additional staff will be required as the business continues to grow.

In addition to these costs and benefits uncovered by the TCA approach, the impact of taxes, the equipment depreciation tax break and the time-value of money were included to better characterise the project's profitability. As this firm is in the United States, the particulars of the tax and depreciation used in the analysis relate to rules specified by the US Internal Revenue Service. While the foregoing specifics are not directly applicable to tax schemes employed in other countries, they are included to demonstrate the types of consideration given to these items and to support the numbers used in the actual analysis.

Physical investment items were depreciated using a double declining balance method to take advantage of the tax effects of accelerated depreciation. All of these items, except the film processor, were assumed to be in the five-year class of goods that includes computer equipment. The film processor was placed in the seven-year class that includes assets used for lithographic printing, and was assumed to have an equipment lifetime of five years with a salvage value of roughly half its cost. This relatively high salvage value estimate was used because the processor was purchased as used equipment, therefore its initial devaluation had already occurred. The two computers were assumed to have equipment lifetimes of four years and salvage values of roughly 20% of their purchase price.

Inflation for the analysis was set at 3%, and income taxes were 37.94% for federal (this number is an effective rate based on 15% tax on the first $50,000 and 39% on the balance) and 7.3% for the state of Illinois. The projected cash flows were discounted at 12% for the analysis. This rate was chosen as a conservative estimate of the facility's cost of capital.[3] The relatively short five-year project lifetime reflects the rapidly changing technology of lithographic pre-press operations. Advances in direct-to-plate processes and imaging software may shorten the lifetime of any current capital purchase.

## Summary of Results

The more comprehensive cost analysis highlighted a number of costs and some savings that resulted from the purchase of the computerised pre-press system. Figure 9 compares results of the company's analysis with those of the TCA. In order to make a direct comparison of net present values and discounted paybacks, both analyses were run using a 12% discount rate. Although the more thorough treatment of the initial investment showed its true cost to be more than $2,500 above the estimate used in the company analysis, and the labour and film costs associated with the new system represented an increase, the overall project is more profitable due to the savings from decreased chemical cost and increased product revenues.

The new annual operating costs identified by the TCA, comprising the increase in film costs, total $12,600. Meanwhile, the cost-savings, from reduced chemicals purchases, reduced service bureau and courier costs, and reduced typesetting costs, total $27,210. Therefore, the net effect of the operating items in the TCA is an annual saving of close to $15,000, exclusive of the expected increase in revenues, which brings the total in Year 1 to almost $120,000. Inflation notwithstanding, this incremental saving grows over the life of the project as the operators become more adept at using the computer pre-press.

## Opportunity Costs and Allocation

While overall staffing levels were not expected to change as a result of the computer pre-press system, the new system would necessitate changes in operating labour. The facility has two operators for the pre-press process activities of stripping (layout work) and film-processing in the darkroom. The new system will shift time away from the darkroom to operation of the computer-based system. Due to the high volume of pre-press work, the Production Manager spends some of his time supplementing the work of the two operators.

|  | Company | TCA |
|---|---|---|
| Initial investment costs | $49,385 | $51,060 |
| Annual operating savings: Year 1 | $26,400 | $119,852 |
| Net Present Value: Years 0–5 | $58,358 | $200,074 |
| Internal Rate of Return: Years 0–5 | 51% | 140% |
| Discounted payback | 2.14 | 0.77 |

**Figure 9:** Comparative Summary of Cost Data and Profitability

3. This analysis could be run with different discount rates to assess the sensitivity of the profitability to the rate chosen.

The broader inventory of costs and cost-savings used in the analysis can be enhanced further to encompass another element of TCA: more precise allocation of costs to specific projects or processes. Facilities incur many costs—including many environmental costs—that are placed in overhead accounts and then broadly allocated across the facility. Misallocation distorts the costing of individual processes, making some seem more profitable than they really are and others less. By correcting this imprecise allocation, TCA more realistically reflects the costs of a process. For this project, the Production Manager's managerial time was allocated to the pre-press in proportion to the relative share of time demanded by that process.

Accounting for the changes in investment and operating labour, the analysis was re-run to include the following labour costs/savings.

### Initial Investment Costs

1. The Production Manager and others at the facility spent time carefully considering the purchase of this system. After the group had decided to seek vendor quotes, time was spent working with the vendors, selecting the system and co-ordinating the installation. All of this time was estimated to have taken, in total, the equivalent of one person's time for one week, costing roughly $650.

2. The new system, unlike any equipment currently in-house at the facility, required some training time. Training required roughly 14 hours each for the Production Manager and the operator, costing $190. (This figure includes only direct wages and does not try to account for lower productivity due to lost time.)

### Annual Operating Costs/Savings (summarised in Fig. 10)

1. **Reduced labour for pre-press darkroom and stripping operations.** The new system virtually eliminates the need for darkroom work and reduces the overall workload on stripping operations. As shown in Figure 10, the darkroom previously required approximately 16 hours per week of labour costing $10,103. These darkroom costs essentially were reduced to $2,799 for the new digital pre-press system.

| | Wage | Service bureau pre-press | | | Digital pre-press | | | |
| | | Stripping | Darkroom | Supervision | Stripping | Darkroom | Computer | Supervision |
|---|---|---|---|---|---|---|---|---|
| **Production Manager** | $16.28 | 50% | | 17% | 75% | | | 8% |
| **Operator 1** | $12.20 | 75% | 25% | | 90% | 10% | | |
| **Operator 2** | $9.75 | 85% | 15% | | 5% | | 95% | |
| | | $54,369 | $10,103 | $5,260 | $49,890 | $2,799 | $19,672 | $2,630 |

Note: The balance of the Production Manager's time is spent managing the press and post-press operations. Pre-press labour represents approximately one-third of the facility's total labour. Using this labour figure as an allocation base, one-third of the Production Manager's supervision time is allocated to pre-press. In total, labour costs for pre-press operations at the facility have increased with the new system because, although the operators already spent all of their time in pre-press, the Production Manager now devotes more of his time (83%, up from 67%) to pre-press operations.

**Figure 10:** Changes in Allocation of Labour for Pre-Press Operations

2. Similarly, pre-press stripping operations previously required two workers for three-quarters of their time plus half of the Production Manager's time. Total stripping costs previously were $54,369 per year and have dropped to an estimated $49,890 due to the increase in digitally processed art.

3. **New computer pre-press labour.** The electronic imposition work, previously done by the service bureau, requires almost a full-time employee in-house, costing $19,672 per year. As shown in Figure 10, an employee who previously spent 85% of his time on pre-press stripping operations and 15% of his time on darkroom operations was shifted almost full-time to this electronic imposition work.

4. **Reduced supervision cost.** With the previous system, the Production Manager spent half of his time performing pre-press stripping operations and half performing managerial duties at the facility. As shown in Figure 10, one-third (17%) of the Production Manager's managerial time has been allocated to the pre-press process since it accounts for roughly one-third of the facility's labour cost.

5. With the new digital pre-press system, the Production Manager needs to spend 75% of his time performing pre-press stripping operations. The remaining 25% of his time is available for managerial duties at the facility, and one-third of this (8%) again is allocated to pre-press operations. The end result is a decrease in supervision time for the facility's pre-press operations (from 17% to 8% of the Manager's time). This corresponds to a cost reduction from $5,260 to $2,630.

The analysis endeavoured to account for and allocate all relevant costs relating to both the former process and the new one in order to perform a more precise TCA analysis. Allocations of labour within the larger framework of a fixed labour pool (there was no change in overall headcount as a result of this process change) were included to capture the opportunity costs of labour. In other words, reassigning people to make better use of their time can yield significant benefits, despite the fact that bottom line costs are not changing. The TCA analysis was re-run to include annual operating labour costs. Because operating labour costs associated with the new system are higher, the profitability of the investment is slightly diminished with their inclusion, as shown in Figure 11. The labour allocation, in this case, made the project appear less profitable because it accounted for labour that was drawn from other facility activities. A more complete labour allocation would have accounted for the fact that the Production Manager's time in non-pre-press operations changed as well.

### Sensitivity Analysis

The main assumption in the analysis concerned the increase in revenues that would result from the investment. Since the system has been on-line for just a short time, no data were yet available to support the assumption that revenues would increase by 10%. This assessment is further complicated by the fact that revenues are currently increasing for reasons not related to the new system, and that cost increases necessary to generate the increased revenues are unknown. Separating out the commingling effects would be difficult if not impossible. The sensitivity of the analysis was tested against varying assumptions on the magnitude of the revenue increase. As shown in Figure 12, if rev-

| | Company | TCA excluding labour costs | TCA including labour costs |
|---|---|---|---|
| Initial investment costs | $49,385 | $51,060 | $51,900 |
| Annual operating savings: Year 1 | $26,400 | $119,852 | $114,435 |
| Net Present Value: Years 0–5 | $58,358 | $200,074 | $187,701 |
| Internal Rate of Return: Years 0–5 | 51% | 140% | 132% |
| Discounted payback | 2.14 | 0.77 | 0.82 |

**Figure 11:** Comparative Summary of Cost Data and Profitability

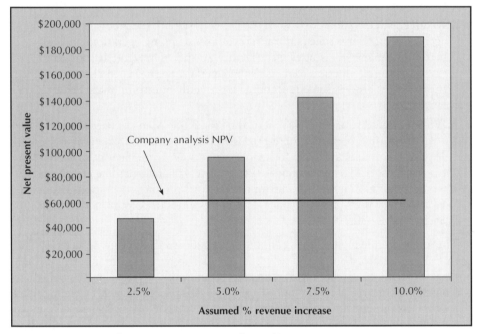

**Figure 12:** Sensitivity Analysis of Revenue Increases

enues increase by only 5%, the discounted payback calculated via the TCA climbs to 1.43 years and the five-year NPV falls to $93,963. Thus, the analysis is quite sensitive to changes in this assumption.

## Conclusion

This case study demonstrates how environmental accounting can directly and positively impact business decision-making. In both the company analysis and the TCA, the project is profitable for the facility. The TCA, however, strengthens the intuitive assessment of the project's value to the firm and helps the company understand where it can expect cost increases and where the savings are likely to originate. Had the facility been less certain in its decision to move forward with the investment due to a perceived payback of more than two years, the more revealing analysis of the

TCA might have induced management to pursue the investment when it would not have without the expectation of higher profitability. In this respect, environment al accounting would have helped the firm make a better investment decision.

This particular investment decision did not have a large environmental impact and the use of TCA did not serve the purpose of merely including environmental costs that had been omitted. The definition of what constitutes an environmental cost is unclear and is not terribly important *per se*. As TCA has evolved over the past decade, it has grown from being a vehicle for ensuring the inclusion of environmental costs to being, more generally, a better financial analysis of projects that have an environmental impact. TCA thus adds value by providing a better framework for investment analysis in which environmental accounting contributes to better decision-making and, with it, improved environmental performance.

The principles of environmental accounting can similarly be applied to a wide range of business decisions. Our case demonstrates just one of many ways in which environment accounting can play a key role in assessing business operations while simultaneously improving both environmental performance and bottom line results.

# Part Three

## Case Studies

# 15 Making Environmental Management Count

## Baxter International's Environmental Financial Statement[1]

### Martin Bennett and Peter James

## ■ Introduction

THE AMERICAN CORPORATION Baxter International is a leading producer, developer and distributor of medical products and technologies. It makes thousands of different products at over 100 facilities around the world and has revenues of over $5 billion per annum.

The present Baxter International Corporation is the outcome of a corporate reconstruction in 1996, which spun off the US distribution, medical kits, gloves and surgical instruments businesses into a separate company, Allegiance Corporation. Prior to this, the environmental function was separate from the health and safety function. Concurrently with this reconstruction, they were merged in the new Baxter International under the leadership of Bill Blackburn, who had been Vice-President for Corporate Environmental Affairs and Chief Environmental Counsel in the old corporation, and now became Vice-President and Chief Counsel, Corporate Environmental, Health and Safety.

Baxter's most significant environmental impacts are emissions to air and water, use of energy and other natural resources, packaging and waste generation. Historically, it has also been a large user of ozone-depleting substances, but these have now been phased out from all but a few essential medical uses.

Baxter's long-standing environmental programme was redesigned on total quality management principles in 1990 under the leadership of Bill Blackburn. This went beyond

---

1. An earlier version of this case, in more depth, based on the 1994 Environmental Financial Statement, can be found in Tuppen 1996.

compliance to include: the development of standards for 'state-of-the-art' environmental management systems that can be used to assess and compare the strength of the infrastructure of the environmental programmes at individual facilities and divisions; an extensive environmental benchmarking programme; and ambitious targets for reduction of packaging, reduction of hazardous and non-hazardous waste and removal of underground storage tanks.

Part of the new programme was the creation over time of a financial statement (generally referred to within Baxter as the 'environmental balance sheet') to itemise and report total environmental costs and savings. The purpose of the Environmental Financial Statement (EFS) was to collect together in a single report, annually, the total of the financial costs and benefits that could be attributed not only to the environmental programme itself but to any environmentally beneficial activities across the corporation. Its aim was to demonstrate that, contrary to the preconception of many, environment need not be only a burden on business performance but could make a positive contribution. This would mean that the EFS could provide a focus within the corporation that would attract attention to the environmental programme, stimulate discussion in internal meetings and encourage motivation.

The emphasis is deliberately on the value generated for the business, rather than attempting to measure the company's environmental performance as such. The principle behind the EFS is that environmental costs are to be treated in the same way as any other costs incurred by the business: to be minimised so far as possible, and where possible to generate maximum financial benefits.

A few elements of the costs and savings that are now found in the corporate EFS were originally identified and circulated internally within the corporation in 1990. Since then, the format and content of the EFS have not remained static but have evolved over time. In particular, the number of items of cost and savings that are included in the EFS have been increased substantially over this period. Since 1992, it has also been published in the annual Environmental Performance Report, though its main function continues to be internal. Following the reconstruction in 1996, the new Baxter International has continued to prepare and report annually an EFS, after adjusting the amounts reported as comparative figures for previous years in order to exclude what related to the businesses taken over by Allegiance. Baxter is currently planning how to extend this to include also the health and safety responsibilities which were combined with the environmental management function after the reconstruction (see Section 6, 'Future Plans', below).

The EFS reflects a more focused purpose than is often apparent from other similar experiments, and is probably the most advanced example in the public domain at this time of a corporate environmental accounting statement. The following pages therefore examine its components in detail. Section 1 deals with Baxter's definitions of environmental costs, and Sections 2, 3 and 4 with benefits. Section 5 examines the processes of data collection, Section 6 outlines plans to develop the statement further, and Section 7 provides evaluation and conclusions.

## 1. Costs

Baxter's 1996 EFS, published in the 'Business Integration' section of their Environmental, Health and Safety Performance Report (see Fig. 1), reported total environment-related costs of $19.6 million, in two categories:

- The costs of the 'basic' programme (totalling $15.1 million). These include compliance-related costs such as environmental audits as well as costs incurred proactively in order to generate future improvements in performance.

- 'Remediation, waste and other response' costs (totalling $4.5 million). These are reactive rather than proactive: the aim is to reduce these to zero (though in recent years they have temporarily increased, due to a more proactive remediation/clean-up programme and the need to address promptly some problems at newly acquired sites).

This section examines how environment-related costs are defined and measured, based on the main items included. For each cost item there are two basic data elements to be quantified:

- The quantity of whatever **resource** is being used, whether this be materials, equipment or people's time

- The price at which the company is purchasing this resource or, to be more strictly correct, the **rate** at which the physical consumption is translated into money terms.

### Pollution Controls: Operations and Maintenance ($3.6 million)

Each facility in the US is required to file with the Bureau of Census annually a report of the costs of pollution abatement. There is no problem in identifying what is or is not a pollution control, which is self-evident—usually major items of end-of-pipe equipment, acquired exclusively for environmental reasons, such as waste-water plants and ethylene oxide scrubbers. These environmental reasons are not necessarily limited to what is required for strict legal compliance: they can include any expenditure needed to achieve Baxter's internal 'state-of-the-art' goals, or its policy of generally following the same environmental standards in all countries in which it operates (since compliance-based standards in the US often represent 'compliance-plus' standards abroad). However, the equipment would not include anything purchased for conventional commercial reasons of profitability, e.g. through meeting Baxter's usual investment targets of discounted cash flow present values or payback periods.

The quantity of **resource** is therefore non-problematic. These items are clearly single-use, and do not have any commercial value other than pollution control. The **rate** is similarly non-problematic, since the total costs of each of staff pay, materials, energy, etc. are taken directly from the general ledgers. The rates are therefore the actual ones that Baxter is currently paying for each of the items of resource.

Of the total of $3.6 million, $3.2 million relates to US facilities, which was based on data reported annually to the Bureau of Census, as adjusted based on internally collected data. The balance relates to facilities outside the US. In the early years of the EFS, this had to be estimated, but these estimates are being replaced in time by more accurate measurements by facilities as data collection systems are developed. Until this is complete, the Census Bureau data continues to be used as a reference.

### Pollution Controls: Depreciation ($1.4 million)

The amounts included here are identical with the amounts actually charged against corporate profits for environmental equipment (defined as in the last category), though

| ENVIRONMENTAL COSTS | | | |
|---|---|---|---|
| **Costs of basic programme** | **1996[1]** | **1995[1]** | **1994[1]** |
| Corporate environmental affairs and shared multi-divisional costs | 1.4 | 1.4 | 1.3 |
| Auditors' and attorneys' fees | 0.5 | 0.3 | 0.4 |
| Corporate environmental engineering/Facilities engineering | 0.6 | 0.7 | 0.8 |
| Division/regional/facility environmental professionals and programmes | 6.6 | 6.8 | 7.6 |
| Packaging professionals and programmes for packaging reductions | 1.0 | 2.3 | 1.6 |
| Pollution controls: operations and maintenance | 3.6 | 3.6 | 3.3 |
| Pollution controls: depreciation | 1.4 | 1.7 | 2.1 |
| **Total costs of basic programme** | **15.1** | **16.8** | **17.1** |
| **Remediation, waste, and other response costs** (proactive environmental action will minimise these costs) | | | |
| Attorneys' fees for clean-up claims, notices of violation (NOVs) | 0.1 | 0.2 | 0.2 |
| Settlements of government claims | 0.1 | 0.0 | 0.0 |
| Waste disposal | 3.1 | 2.5 | 2.4 |
| Environmental taxes for packaging | 0.8 | 0.3 | 0.0 |
| Remediation/clean-up: on-site | 0.3 | 0.3 | 1.0 |
| Remediation/clean-up: off-site | 0.1 | 0.6 | 0.0 |
| **Total remediation, waste and other response costs** | **4.5** | **3.9** | **3.6** |
| **Total environmental costs** | **19.6** | **20.7** | **20.7** |
| ENVIRONMENTAL SAVINGS (Income, savings and cost avoidance from 1996 initiatives) | | | |
| Ozone-depleting substances cost reductions | 0.6 | 0.5 | 1.9 |
| Hazardous waste: disposal cost reductions | (0.2) | 0.1 | 0.4 |
| Hazardous waste: material cost reductions | (0.2) | 0.1 | 0.2 |
| Non-hazardous waste: disposal cost reductions | (0.9) | 1.3 | 4.6 |
| Non-hazardous waste: material cost reductions | 1.3 | (0.5) | 4.6 |
| Recycling income | 5.6 | 5.2 | 3.0 |
| Energy conservation: cost savings[2] | 2.5 | 1.0 | 0.5 |
| Packaging cost reductions | 2.4 | 5.6 | 5.7 |
| **Total 1996 environmental savings[3]** | **11.1** | **13.3** | **20.9** |
| **As a percentage of the costs of basic programme** | **74%** | **79%** | **122%** |
| SUMMARY OF SAVINGS | | | |
| Total 1996 environmental savings[3] | 11.1 | 13.3 | 20.9 |
| Cost avoidance in 1996 from efforts initiated in prior years back to 1989[4] | 93.5 | 81.3 | 73.7 |
| **Total income, savings and cost avoidance in report year** | **104.6** | **94.6** | **94.6** |

Notes
1. These amounts have been adjusted to reflect the spin-off of allegiance operations.
2. Cost savings for 1996 calculated from data. Amount for 1995 is conservatively estimated. Amount for 1994 is only for us energy-conservation projects involving lighting.
3. Cost avoidance from initiatives completed in prior years is listed as a separate line item in this report.
4. No carry-forward savings are included for 1995 and 1996 for reductions of those ozone-depleting substances (ODSs) that are no longer available to Baxter. This year's report used a more accurate method of calculating this line item, taking into account current-year pricing and compounded growth in business activity. This produced a significant increase in these sums over that which would have resulted from the previous method.

| DETAIL ON INCOME, SAVINGS AND COST AVOIDANCE FROM 1996 INITIATIVES ($ IN MILLIONS) | | | |
|---|---|---|---|
| | Savings and income | Cost avoidance | Total financial benefit |
| Ozone-depleting substances cost reductions | 0.3 | 0.3 | 0.6 |
| Hazardous waste: disposal cost reductions | (0.3) | 0.1 | (0.2) |
| Hazardous waste: material cost reductions | (0.3) | 0.1 | (0.2) |
| Non-hazardous waste: disposal cost reductions | (1.9) | 1.0 | (0.9) |
| Non-hazardous waste: material cost reductions | (1.3) | 2.6 | 1.3 |
| Recycling income | 5.6 | * | 5.6 |
| Energy conservation—cost savings | 2.5 | * | 2.5 |
| Packaging cost reductions | 2.4 | * | 2.4 |
| * Not applicable | | | |
| **Total savings** | **7.0** | **4.1** | **11.1** |

The following undetermined costs are not included in the financial summary:

- Environmentally driven materials research and other research and development. This is typically offset by increased sales and other non-environmental benefits not presented in the environmental financial statement.
- Capital costs of modifying processes other than adding pollution controls. This is typically offset by increased production rates and efficiencies and other non-environmental benefits not presented in the environmental financial statement.
- Cost of substitutes for ozone-depleting substances and other hazardous materials. This is estimated to be relatively minor.
- Lost sales from environmental issues.

A proactive programme, such as Baxter's, produces savings that are not easily measurable. Examples of such undetermined savings are:

- Reduction in liability exposure resulting from tank removals, waste site evaluations and other risk management programmes.
- Reduced record-keeping and administrative costs
- Increased goodwill, sales and employee morale
- Avoidance of costs for environmental problems that did not occur because of Baxter's proactive efforts

**Figure 1:** Baxter's Environmental Financial Statement: Estimated Environmental Costs and Savings Worldwide ($ in millions)
*Source: Baxter 1997*

this means that (as in the financial accounts) the total amount therefore does not represent a uniform depreciation policy across the corporation.

The main distinction is between the equipment used in facilities in the US and that used abroad. Depreciation on the former is calculated on the 'six-year accelerated cost recovery' method, which is advantageous for tax reasons, and which Baxter uses in its financial accounts. On overseas environmental equipment—which represents only 10% of the total depreciation charge—a straight-line depreciation policy is applied, with an expected asset life of ten years.

In reporting capital environmental costs, some companies prefer to report the actual amount of the expenditure in the current year. Baxter has deliberately chosen to report instead the depreciation expense in order to relate the EFS as closely as possible to the quantities reported in the Income Statement, on which profit is based.

### Division/Regional/Facility Environmental Professionals and Programmes ($6.6 million)

Baxter has over 200 full- or part-time environmental staff, which equates to 85 'full-time equivalent' staff. The definition of 'environmental responsibilities' on which the apportionment of part-time staff is based is the tasks identified in the corporation's standard job description for an environmental manager. However, no attempt is made to estimate and include time spent by production and other staff on environment-related tasks, though some of this will be included under other headings (for example, the staff costs of waste-water plant operators are included in 'Pollution Controls: Operations and Maintenance').

Because of this, there is no problem in identifying the individuals whose time might be included, in whole or part. In calculating the quantity of the **resource**, the proportions used in the apportionment may be less reliable. It can be difficult for people to report accurately on even the past use of their own time unless a structured activity analysis is carried out, as several exercises carried out as part of either activity-based costing programmes or time-management programmes have shown.

Some effort is put into overcoming these problems and creating accurate data, since this is used for other purposes also. Time availability is a useful control measure in its own right, since Baxter believes that environmental under-performance in specific units of the business can often be explained by it. It is also useful in internal benchmarking.

The amount of $6.6 million that is reported in the EFS is the aggregate of the staff costs reported annually to the corporate centre by facilities and divisions in the 'State of the Environmental, Health and Safety Programme' (SOEP) report, and is therefore the actual amount without any approximation or estimate (see Section 5 below for more on the SOEP reporting process). There is therefore now no problem with determining the appropriate **rate** at which staff costs should be included. The figure has been included on this basis only in 1996. Before this, the facilities were not reporting this as a matter of course, and in order not to overload them too quickly with requirements for additional data the amount reported was estimated. This was done by multiplying the number of 'full-time equivalent' people by $60,000 per person, which was the estimated average annual full employment cost across the corporation as a whole.

### Packaging Professionals and Programmes for Packaging Reductions ($1.0 million)

Since reductions in packaging costs are included in the statement as a benefit, the corresponding costs of the time incurred by packaging designers and others in achieving these savings are included as a cost. (Only time costs are included; more indirectly associated costs such as the capital costs of modifying processes are not included.) Figures were collected directly from the packaging designers with whom the environmental affairs unit deals most frequently, who were asked to estimate the costs of their time incurred in the past three years on these projects.

This collection process is an ad hoc one at present, since these figures have been collected only since 1994. The measurement of benefits arising from packaging reductions is being re-evaluated, and it is planned to incorporate this, as well as the costs for packaging professionals, into the SOEP reporting system next year.

### Corporate Environmental Affairs and Shared Multi-Divisional Costs ($1.4 million)

This is the cost of running the environmental portion of the corporate Environmental, Health and Safety unit, and the costs for regional training, waste site evaluations and other environmental initiatives undertaken jointly by the divisions. The costs included are the salaries and 'salary overhead costs', such as social security payments, of the staff of the unit; administrative, travel and other miscellaneous departmental costs; and outside management consultants and publications. The amount stated is extracted directly from the general ledger.

The other cost items reported are calculated with similar combinations of sources and methods: directly from ledgers, calculated from physical quantities reported to the corporate centre, and where necessary with some estimation.

## 2. Benefits

The benefits flowing from each of several items are a combination of (in Baxter's terminology) 'income', 'savings' and 'cost avoidance' (these terms are more fully described in the next section). They totalled $11.1 million in 1996, which represents 74% of the costs of the 'basic' programme.

Like costs, a few large items predominate, and these are described here to demonstrate the general principle.

### Recycling Income ($5.6 million)
The amount reported is the sum of the recycling income earned at facility level for the sales of wastes and scrap to outside buyers. Each facility reports its recycling income to the corporate centre through the annual SOEP reporting form.

This item has increased significantly in recent years due both to improved recycling procedures, which increased the quantities collected, and to improved unit prices obtained for those quantities.

### Energy Conservation ($2.5 million)
Up to 1995, amounts reported in respect of energy savings came only from energy used in lighting, mainly under the US Green Lights energy conservation programme. The corporate environmental unit has now taken responsibility for tracking the effect of projects to reduce energy consumed by facilities in production, increasing several-fold the potential savings available. The savings arising from this are expected to grow as more attention is provided to the programme in the future through setting goals, running workshops and writing articles, etc.

This has required some work to develop data collection systems for energy consumption and costs that are reported by facilities. The corporate EH&S unit has collected total energy data since 1993, but it was only in 1996 that they felt confident that this was sufficiently reliable for their purposes. They are continuing to develop the system in areas where they feel that data collection can be improved, such as energy used in transportation.

### Packaging Cost Reductions ($2.4 million)
With a high number of customer locations to service and thousands of products, many of which require careful packaging for protection during transit, packaging is both a significant cost to the corporation and a significant environmental impact. The environmental programme has stimulated a search for creative ways to redesign packaging, which has been reinforced by government pressures in some of the countries in which Baxter operates, such as Germany.

The quantity of **resource** represents that part of total packaging costs saved, comparing the total cost incurred for packaging in 1996 with that in 1995. The costs to which this proportion is applied are primarily savings in the costs paid to suppliers for materials, though some downstream savings are included, if significant and identifiable.

The **rate** is non-problematic: this is the current cost, per pound, that the company is paying for its packaging materials. This is available from the accounting and procurement systems, and is well known by those working in either packaging or environment.

The amount stated of $2.4 million represents an annualised figure rather than the actual savings that had been realised by 31/12/96. For a change introduced in November 1996, for example, the expected savings for the following twelve months would be estimated and included in the total, rather than only the small amount that could be realised before the end of the year. This is because, unlike the other items reported, the savings measured are those that are achieved through specific projects, rather than those determined from a year-to-year comparison of total quantities wasted or consumed. In the first place this is pragmatic—the information is already available in this form, since this is the corporation's method of evaluating the success of projects. It also permits internal benchmarking between sites and divisions, which has led to faster sharing of new ideas.

Most important, it is consistent with the aim of motivating people to achieve benefits through the project. It might be that savings are apparent from comparing current spending against the previous year's, but that these are in fact due to other causes such as changes in the volume of production, or falls in the market prices for packaging materials. This could reduce the motivation to look for genuine savings since, in any case, 'the numbers will look after themselves'. As Bill Blackburn points out, this is precisely what is *not* wanted: 'this is a situation where you don't *want* the numbers to look after themselves! You want to motivate people to do projects.'

This practice, of reporting annualised figures, applies only to savings in packaging costs. Other items are calculated by comparison between the current and previous years, so that the benefit reported is always the actual amount arising in the business year.

Ideally they would like to calculate energy savings (see previous section) on the same basis as packaging, by projects. However, the infrastructure for capturing energy data by project is not as fully established as it is for packaging, and the task became much more complicated when the scope of measurement was extended from only lighting to all the energy consumed in production. Since the systems for measuring energy consumption and costs in total, on the other hand, are fairly reliable, this continues to be the method used here.

### Waste (Hazardous and Non-Hazardous): Material Cost Reductions ($1.1 million)

One of the main focuses of the environmental programme has been the management of wastes, both hazardous and non-hazardous. As well as savings in the actual costs of disposal, waste also represents a loss of the original resource, so waste reductions also generate reductions in materials costs which would otherwise be lost to scrap.

The amount reported is the reduction in the amounts paid to suppliers of the materials that were previously being lost through waste which is now being avoided. Through the annual data collection process, divisions and facilities are asked to report wastes, by different waste streams, in both physical quantities and monetary amounts. A corporate quality control procedure checks these returns for reasonableness, to identify any reported results that appear implausible. If any are found, the division/facility is contacted to ask them to re-check the amount and confirm that they are interpreting the definition as intended. In exceptional cases where necessary, the amount is re-estimated by the corporate EH&S unit before being included in the total.

In addition to waste-material cost reductions, a separate amount of $1.1 million (negative in 1996) is reported for savings and cost avoidance in waste disposal costs. These represent only the outlaid costs of payments to suppliers and do not include any internal savings in in-house waste storage and management. Although the amounts reported in the latest year were negative, this masks an underlying improvement since the rate of waste disposal per unit of production in fact decreased during the year. However, special factors combined to increase the total amount, such as the start-up of new facilities in the US and South America, which inevitably means that extra wastes are generated before increased production volumes come fully on-stream; and wastes created outside the company's control when a plane crashed into one of their Californian facilities.

### Ozone-Depleting Substances (ODSs) Cost Reductions ($0.6 million)

The amount of $0.6 million represents the reduction in the quantities of ODSs (CFC-12, CFC-113 and TCA) actually consumed by the corporation in 1996 compared with 1995, multiplied by the cost per unit. The main element is a reduction compared with 1995 of 56,000 lbs (18%) in the quantity of CFC-113 used, at $8.01 per lb (compared with less than $1 per lb in 1988). However, the cost of whichever materials have been substituted for the ODSs previously used, or the costs of changing processes and/or redesigning products, are not netted off from the calculated saving but are left as 'undetermined costs'. In practice, the cost of substitutes has been minimal when compared with current CFC prices.

Given the rapid rise in the price of ODSs in recent years, reductions in their usage have generated substantial savings for several years. However, as the total quantity used reduces towards zero (barring only essential medical applications where complete elimination is not possible, as the Montreal Protocol recognises), in line with regulation in the US and Baxter's state-of-the-art policy elsewhere, the amounts of benefits reported annually have declined in recent years, and the scope for future savings is clearly limited.

### 3. Income, Savings and Cost Avoidance

For each benefit item, the total benefit flowing is one or a combination of three distinct elements: 'income', 'savings' and 'cost avoidance'. The definitions provided are:

- **Income:** actual monies received in report year (the only significant item is recycling revenues, of $5.6 million)
- **Savings:** the reduction in actual cost between report year and prior year
- **Cost avoidance:** additional costs other than the report year's savings, that were not incurred but would have been incurred if the waste reduction had not taken place.

'Income' is straightforward: monies received from recyclers for wastes collected and sent to them. However, the main source of benefit from the environmental programme over time is the opportunity offered to reduce costs.

The first part of this is 'savings', which is also relatively straightforward, in both concept and calculation. The amount calculated as 'savings' for each item is the difference between the actual amount spent in one year (the report year) and the amount spent in the previous year. This amount is therefore the *absolute* amount of any saving, and is irrespective of any changes in the underlying rates of business activity that produce the wastes. For a corporation such as Baxter, whose activity is growing steadily over time (though not necessarily evenly across all its different business units), the amounts calculated in this way will tend to underestimate the real value of the environmental initiatives.

However, the full benefit of cost reductions is measured not merely by year-on-year reductions in the total amounts incurred, but from a comparison with the hypothetical amounts that *might* have been incurred if the improvement had not taken place. This latter element is termed 'cost avoidance'. The amount of cost avoidance for each line-item is calculated as the difference between:

● The amount actually spent in the previous year

● The hypothetical amount that would have been spent this year, if the quantity of waste created were assumed to have increased at the same rate as production and distribution activities, which grew by 5.7% in 1996—i.e. if there had been no change in the rate of waste creation per unit of activity. (The current prices for materials and waste disposal costs are used in this calculation.) Whereas 'savings' are the *actual* improvement in profits, 'cost avoidance' therefore measures the *potential* deterioration in profits which might otherwise have occurred but has been prevented.

This calculates the amount of 'cost avoidance' for the current year (i.e. both realised in the current year, and arising from initiatives made in that year). In addition, a further calculation is made of the total cost avoidance realised in the current year that has arisen from initiatives made in earlier years (see also the following section).

This represents a more realistic measure of the value of the environmental initiative than would be obtained if only savings on actual costs were counted, since the measure has now been effectively normalised relative to business activity levels. Since one element in the calculation is by definition a hypothetical amount that cannot be extracted from conventional accounting records, some degree of estimation is inevitably required.

To demonstrate the method involved, Figure 2 shows the application of this to CFC reductions over the period 1989 to 1994. As the cost of CFCs has escalated rapidly, achieving reductions in their use has been important for both business and environmental reasons. The scale of the increase in 1990 was such that, despite successful efforts to reduce usage, the total cost increased by $2.7 million ($6.6 million – $3.9 million). However, the true benefit is measured only by considering also the cost avoidance of $4.2 million (= $8.1 million – $3.9 million), with the improvement in underlying performance being worth the net amount of $4.2 million less $2.7 million = $1.5 million. In 1991, an actual saving of $3.6 million was realised (= $6.6 million – $3.0 million), plus cost avoidance of $0.8 million (= $7.4 million – $6.6 million) attributable to efforts

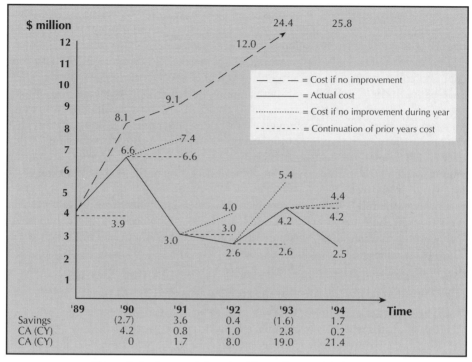

**Figure 2:** Savings and Cost Avoidance for CFC Reductions

initiated in that year. In addition, the efforts initiated in 1990 produced further benefits in 1991 of $1.7 million (= $9.1 million – $7.4 million). By 1994, the pace of price increases meant that, if the rate of usage of CFCs per unit of output had been the same as in 1990, a total cost of $25.8 million would have been incurred. The improvements made (a 90% reduction in CFCs used per unit of output) meant that the actual cost, at $2.5 million, was only one-tenth of this.

## 4. Current Benefits Arising from Efforts Initiated in Prior Years

There is little problem in determining what time-period the costs in the statement should represent—the amounts actually incurred in each year.

For benefits, this is less obvious, since most of the environmental initiatives produce benefits of continuing value. In principle, there are several possible ways in which the statement could reflect this.

- To ignore the problem, by stating only the benefits realised within the report year from initiatives implemented during that year. This would mean a systemic understatement of the value of the environmental programme, and would not distinguish between initiatives with a one-off impact and those of continuing value. The benefits measured would also be somewhat arbitrary, as an initiative implemented towards the end of a year would be generating little benefit within that year.

- This last point can be dealt with by measuring instead the annualised amounts of benefit, i.e. the benefits expected in a full twelve months. This would require some estimation, but would mean that the statement would show a full twelve months of both costs and benefits. However, if the period over which benefits are measured is still restricted to this twelve months, this method would still underestimate the total value of the initiative.

- In principle, some calculation of the present value of each initiative could be made, using an appropriate discount rate (though this would be contentious and complex in practice).

- In addition to stating in the main body of the statement the benefits realised in the report year (calculated under either of the first two methods above), a separate line-item could disclose also the amount of benefit realised during the report year from initiatives implemented in previous years.

This last is the method used by Baxter in its report. The 'cost avoidance in 1996 from efforts initiated in prior years back to 1989' of $93.5 million is the total improvement to, and prevented deterioration of, profits (including in this both 'savings' and 'cost avoidance') in 1996 that is attributable to environmental improvements that have been made in the period 1989–95. The only exception to this principle is that the continuing benefits arising since 1994 from past ODS reductions are no longer being included (although the benefits realised up to 1994 are still included in the $93.5 million). Since market prices of ODSs have increased rapidly in recent years as a result of the environmental legislation limiting their manufacture and use, and Baxter have now totally eliminated all but a minimal unavoidable amount from use in their operations, it is considered that it would not be legitimate to continue to claim credit for the *continuing* benefit from past reductions in those CFCs that Baxter no longer uses and which in any case are not available in the marketplace.

In principle, the cumulative amounts could increase indefinitely, since at present no limit is set on the maximum length of period over which benefits can be measured. This may in fact be correct for some initiatives; however, it could result in amounts that would lose credibility within the corporation as they increase indefinitely. Baxter has considered how best to determine an appropriate time-period. The most appropriate method, at least for product-related improvements such as packaging, would be to define a time-period equal to the length of the relevant product life-cycles. However, with a wide product range this would impractical. To deal with this pragmatically, Baxter has set an arbitrary cap of seven years on the period over which benefits are accumulated, since this is, on average, an adequate approximation of the useful life of a particular version, or generation, of a product.

A separate issue here is that this calculation makes an implicit assumption on the base-case against which current performance is being compared. The formula used is:

Current rate of business activity (production, distribution, etc.)
(by estimate, based on an annual growth rate of 5.7% in 1996)

×

Previous year's rate of waste generation per unit of activity

×

Current prices

Basing on the previous year's rate of waste generation implicitly assumes that, but for the environmental initiative, no improvement would have occurred otherwise. This may or may not be justified in each case: it could be that, for items where the rates being paid by the corporation are increasing rapidly (e.g. for ODS emissions), savings would have been prompted on purely commercial grounds, without any environmental considerations being needed. However, this is not an issue for Baxter, as the purpose of the statement is to reflect all economic gains generated by environment-related improvements, irrespective of which function in the corporation has developed or implemented the initiative, or whether their motive was primarily commercial or to improve environmental performance for its own sake. The key is to show that environmentally beneficial activity, regardless of the motive for undertaking it, often has financially beneficial results.

## 5. Data Collection

Collection of data for the Corporate Environmental Report, including the financial data on costs and savings, is done annually. The SOEP reporting form is the main vehicle. This is circulated by the corporate unit to divisions and then by them to facilities, specifying the data that is required. The facilities complete these forms, then their data is collated and aggregated up at divisional level and reported to the corporate centre.

This process is being developed over time, in order to simplify it and reduce the costs of the process for both providers and collectors of the data. In particular, the paper questionnaires used in the early years have been replaced by specifically designed computer templates, incorporating user-friendly features such as screen prompts to attract the data-inputter's attention if the data being entered exceeds the previous year's figure by more than a pre-set percentage, in case this indicates an error in input. Any changes made by the corporate unit to the data that they require must be planned well in advance so that the providers can set up systems for the capture of new types of data, which will often require the co-operation of facility operational staff.

The main quality control exercised on the process in the past has been a series of reasonableness checks carried out by corporate staff on the data submitted to them, using as benchmarks both the quantities reported in previous years and those reported by other divisions/facilities to identify any submissions that appear to be significantly out of line. These would then be queried with the division/facility to check whether there had been any mistakes or misunderstandings of what was required, and corrected if necessary. These checks continue, but are now being dealt with by teams that have been set up comprising both divisional and corporate staff to manage both the processes of the preparation of the divisional reports and their compilation at the corporate centre.

In principle, all the financial data required are also contained in the corporation's main accounting system. However, due to the way in which this is structured, the data are not in a form in which the corporate staff preparing the report can extract them directly; therefore, the SOEP survey is required (although this will be completed by divisions/facilities in part on the basis of data extracted from the accounting systems at those levels). Development is being done in partnership with the Finance function of one of the divisions (Cardiovascular) to identify how they can adapt their procedures in order to facilitate the collection of environmental data.

The EFS was not designed from the outset from a single clear concept of its function; rather it has evolved over time. The joint determinants of what is reported have been the needs of the environmental programme, as new issues and projects have developed, and the availability of data. Given the cost of collecting data, and sensitivity in the divisions and facilities to information requirements imposed by the corporate centre, it is rarely possible to have the data easily available in an ideal form, and pragmatism has been needed. Some figures have been estimated where actual amounts were not available (or only at an unacceptable cost), though efforts are made to restrict this to less material items, and to evolve over time from estimated to actual figures as data collection and reporting systems can be adapted. However, since Baxter estimates that there has to be a two-year lead period between the company's environmental leaders deciding that an additional item of information is needed and facilities being able to set up systems to generate the data that can be reported, this is a gradual process.

Inevitably, in a process in which information requirements from the centre have evolved over time and have required interpretation in over 100 facilities providing the data, inconsistencies in definitions have developed. As these have come to light they have been corrected. Where choices have had to be made between different possible definitions, these have usually been standardised on the more conservative basis, to maintain the credibility of the statement. Experience has shown the importance of keeping the figures 'clean and consistent'. At least up to a point, provided that the definition used is made clear, the particular choice of definition may be less important than users' ability to have confidence in the reliability of the figures.

## 6. Future Plans

Baxter continues to develop its EFS. Some of the main areas of current development are:

- Since the corporate reconstruction, the Health and Safety (H&S) function has been combined with Environmental Affairs. Bill Blackburn is keen either to include H&S in the financial statement together with environment, or to produce two separate statements. The aim is to prove the business benefit of good H&S performance—for example, by demonstrating the true cost to the company of accidents at work and employee ill-health, to show that spending on H&S can be an investment with positive net returns. This would provide a measure at a 'macro' level (at the level of the company as a whole) that would complement separate 'micro' measures of the costs associated with specific individual accident or work-related illness.

  However, it has not been possible to do this yet. Some of the cost elements to be included have been identified, but the question is how to measure them. Baxter has carried out wide benchmarking to ascertain other companies' practice in this area, but has found that, even where such exercises have been done, frequently they have not been continued and it has been impossible to substantiate their results. In many cases, these analyses depended on assumptions that were too broad to be acceptable. Work is continuing on this.

- It has for several years been Bill Blackburn's aim to 'cascade' the preparation of EFSs down the organisation by encouraging facility-level and division-level environmental managers to draw up their own statements, to communicate and publicise the value of environmental initiatives more locally. Recently, the Irish Manufacturing Operations facility (IMO) has trialed this and in 1997 produced its own EFS for 1996, based on the methods and format of the corporate version.[2] Although it is too early to make a full evaluation, the project proved that this is possible at a level below that of the corporation as a whole, and it was found to be a valuable experience by the environmental people involved, by providing a forum for them to work with financial and operational colleagues in a different situation than would otherwise have been possible. A revision of the SOEP form will re-sort the information at facility level in order to provide a classification that will support the preparation of facility-level EFSs.

It is hoped that, now that IMO have proved that this is possible, other facilities and divisions will also adopt the practice.

- To continue to improve the reliability of present methods of data collection— for example, as described in Section 5 above, to work with Cardiovascular divisional finance staff to identify ways in which the process can be improved—and to develop the computerised SOEP system so that facilities will be able to collect and record data over the course of a year, and then at the year-end to download them directly into the SOEP report.

- To improve some of the technical aspects of measurement—for example, to revise the measure of business growth rate that is used to calculate 'cost avoidance'. At present, a single rate (5.7% in 1996) is used across the whole of the corporation, based on the cost of goods sold as adjusted by an inflation factor. However, this may disguise varying rates of growth in different sectors. One approach being considered is to develop separate rates for each division, based on their own key measures of business activity, then to compute a weighted average for the company based on the divisions' revenues. There is also to be a re-evaluation of how packaging improvements are measured.

## 7. Evaluation and Conclusions

The central issue for any exercise such as Baxter's is the definition of environmental costs and benefits. One approach is to include only those benefits that are directly attributable to the environmental programme, and the costs related to them. However, this would exclude environment-related benefits created by mainstream business actions. The alternative approach—and the one adopted by Baxter—is to include the benefits of environment-related action from anywhere and any person in the organisation, even if this has occurred independently of the corporate environmental programme. While this makes it more difficult to justify the EFS as a decision-support tool, it makes it easier to achieve the objectives of raising awareness of environmental factors in business and generating motivation. Hence, the main use of the EFS is as the basis for an

---

2.   The project to develop a facility-level Environmental Financial Statement at Baxter's IMO facility was carried out as part of the EU 'Eco-Management Accounting as a Tool of Environmental Management' (ECOMAC) Project, sponsored by the Commission of the European Union (DGXII, Environment and Climate; contract no. ENV4-CT96-0267). A case study on the IMO EFS has been written by the authors of this case and is published by Wolverhampton Business School, UK, as a Working Paper (Bennett and James 1998b).

agenda item at meetings in Baxter's 100 facilities between operational management and environmental management.

To enhance credibility among a non-environmental audience, Baxter has found it essential to be open about the bases of calculation of the figures reported, and to ensure that these have been consistently calculated. It has also found that reliability is a higher priority than strict accuracy. It can be preferable that all divisions' and facilities' inputs into a corporate total are consistently calculated, even if this means standardising in the short term at the level of those able to provide only limited data, rather than to aggregate quantities computed on differing bases. The only exception to Baxter's general conservatism is the absence of a cut-off on continuing cost avoidance from previous initiatives. Although this does not appear to have affected internal perceptions of the statement, it is a weakness that Baxter is now taking action to remedy.

Data collection has also been an important area of learning. If the data are not already being collected in the facilities, new systems will be needed at local level. These will take time to establish before the information is available for reporting at a corporate level and, in the meantime, the best that is possible is to make estimates. However, as long as these are reasonable and open, they provide a better platform for judgement and decision than delaying until more reliable figures are available.

In large and decentralised organisations there can also be opposition at local level to demands from the centre for information, unless these are perceived to be reasonable. This is a further reason for an evolutionary approach. As the value of the statement is demonstrated through the use made of early versions, the additional information required to improve its relevance and reliability will be more readily accepted. In order to minimise the load on those providing the data, wherever possible Baxter have looked for short-cuts and approximations until more formal systems could be established at reasonable cost.

It is not difficult to criticise the EFS on grounds of the definitions of terms and some of the approximations behind several calculations. However, its stated purpose is not as a problem-solving tool, or even for direct decision support, but for attention-directing, to arouse interest and raise the profile within the company of environmental management. A frequent complaint from environmental managers is either that their colleagues see them as, at best, a necessary overhead burden, or that they have difficulty in justifying proposals by criteria that business colleagues would find convincing. Providing a means by which this can be addressed—and to position environment as a legitimate element of management generally—may be more important than getting the content right in an objective rational sense. Certainly Baxter believes the exercise has been a success. As Bill Blackburn puts it:

> When Baxter adopted its strong upgraded environmental policy in 1990, we were concerned that it not be a 'trophy policy'—one that hangs in the lobby like a moose-head merely to impress visitors. We wanted a living animal. Goal-setting and measurement, in helping with quality principles, has helped bring our policy to life. Our environmental balance sheet is one area of measurement that has strengthened our programme by bringing together our environmental and business professionals. It enables these professionals to focus on common opportunities using a common language, the language of business: money. It has been the ultimate tool for integrating our environmental programme into our business.

# 16 Full-Cost Accounting for Decision-Making at Ontario Hydro[1]

US Environmental Protection Agency

## ∎ Introduction

ONTARIO HYDRO is the largest utility in North America in terms of installed generating capacity. It was created in 1906 by provincial statute and operates today under the power corporation of Ontario. Its customers include 307 municipal electric utilities serving more than 2,800,000 customers, 103 large industrial customers serviced directly by Ontario Hydro, and almost 1 million rural customers serviced by 13 retail

---

1.  This chapter is an edited version of a longer report of the same title published by the Agency in 1996 (US EPA 1996c). It was prepared by ICF, Inc. for EPA's Environmental Accounting Project. The full report can be read from the Project's website at: *http://www.epa.gov/opptintr/acctg* or can be obtained free of charge from EPA's Pollution Prevention Clearinghouse by telephoning +1 202 260 1023.

    *Disclaimer.* This case study describes Ontario Hydro's approach to environmental accounting, which Ontario Hydro terms 'full-cost accounting', and implementation activities through February 1996. The case study focuses on the use of 'full-cost accounting' in planning and decision-making; it does not address external financial reporting issues. The case study intentionally uses Ontario Hydro's language and definitions in explaining its activities to incorporate environmental costs and impacts into its planning and decision-making. For example, Ontario Hydro uses the term 'monetise' to refer to the process of developing appropriate monetary (i.e. dollar) values for the impacts of emissions/pollutants on the environment. The concepts, terms and approach presented in this case study represent Ontario Hydro's view and not necessarily the position or views of the US Environmental Protection Agency (EPA). By publication of this case study, the EPA is not specifically endorsing Ontario Hydro's definitions or approach, but is offering this case study as an example of an approach to accounting for environmental costs and impacts.

    *Editors' note.* Ontario Hydro has undergone considerable change since this case was prepared and now appears to place less emphasis on full-cost accounting. However, the details of the case remain of considerable interest.

utilities wholly owned by Ontario Hydro. Its revenue for 1994 was approximately $8.7 billion with a net income of C$587 million. Ontario Hydro's supply system includes nuclear, fossil-fuelled and hydroelectric energy stations. Total system capacity is approximately 34,000 megawatts transmitted across 29,000 kilometres of transmission lines and 109,000 kilometres of distribution line. Ontario Hydro is a self-sustaining, government-owned utility without share capital, whose bonds and notes are guaranteed by the Province of Ontario.

Ontario Hydro is in a period of great change. As is true for many utilities, since 1990 Ontario Hydro has faced declining load demand due to economic conditions and has excess generating capacity. With an estimated 92% market share, Ontario Hydro traditionally has not been subject to competitive pressures. However, throughout North America the energy business is being redefined and competition is increasing. A new chairperson, Maurice Strong, was appointed in November 1992 to restructure Ontario Hydro and make it more competitive and customer-oriented. In 1993, Ontario Hydro underwent major restructuring in order to better meet the competitive challenges of the 1990s and beyond. Much of the restructuring was designed to contain costs, stabilise electricity rates and gain greater efficiency. The changes also involved dividing the company into separate business units, each with clear accountability for its activities, costs and environmental performance.

As this case study documents, Ontario Hydro has been considering internal and external environmental costs and impacts for many years. Ontario Hydro was the first Canadian company to publish an annual environmental performance report. This case study focuses on its more recent commitment to full-cost accounting (FCA) and extensive efforts to develop and apply environmental accounting under the FCA framework.

## ▮ How does Ontario Hydro define full-cost accounting?

Ontario Hydro calls its approach to integrating environmental considerations into business decisions 'full-cost accounting' (FCA). It defines FCA as follows:

> Full Cost Accounting (FCA) is a means by which environmental considerations can be integrated into business decisions. FCA incorporates environmental and other internal costs, with external impacts and costs/benefits of Ontario Hydro's activities on the environment and on human health. In cases where the external impacts cannot be monetised, qualitative evaluations are used (Ontario Hydro 1995b).[2]

Ontario Hydro recognises that some definitions of full-cost accounting include only 'internal costs' (also termed 'private costs'), which are the costs that affect a firm's bottom line, and exclude 'external costs' (also termed 'societal costs'), which is a term used to describe monetised impacts on human health and the environment that currently are not reflected in a firm's bottom line. Ontario Hydro's approach explicitly encompasses both internal costs and external impacts (both positive and adverse), even

---

2. Ontario Hydro's guidelines appear in full in Attachment A of the original case study. They have been endorsed by Ontario Hydro's Management Committee and discussed at Ontario Hydro's Board of Directors in October 1995 by the Sustainable Development Committee. These guidelines were tested with a number of stakeholders, including environmental, financial institutions, customers and government representatives.

if the latter cannot be quantified or expressed as external costs (i.e. fully monetised in dollars). In developing their FCA corporate guidelines (Ontario Hydro 1995), Ontario Hydro defined the following key terms.

- **Internal costs** can be thought of as the costs Ontario Hydro incurs in doing business. However, in some corporations, including Ontario Hydro, there are often less tangible, hidden or indirect internal costs, including environmental costs, which are often not identified separately or are misallocated to corporate or business unit overheads (e.g. contingent costs, community relations costs). If a business unit is not considering these costs, then the business may not understand the true costs of its products and services, and may, as a result, be making inappropriate business decisions.

- **External impacts** or externalities are effects on the environment and on human health that result from Ontario Hydro's activities, but are not included in the costs of its products and services. These impacts are therefore borne by society.

- **Monetised external impacts** are external impacts for which Ontario Hydro has developed monetary values. To date, Ontario Hydro has developed preliminary external cost estimates for the operation of its fossil stations and external cost estimates for fuel extraction through to decommissioning for its nuclear generating stations.

- **Non-monetised external impacts** are external impacts that can be described only qualitatively because there are scientific limitations in describing the full range of environmental and human health impacts. In other cases, the impact can be quantified (in physical units) but there are limitations in developing appropriate monetised values.

Figure 1 illustrates how these concepts relate to each other. Ontario Hydro has explicitly acknowledged that the dividing line between internal and external costs is not static. For example, a cost that Ontario Hydro considers external today may be internalised tomorrow because of new environmental regulations or corporate standards. Ontario Hydro's long-term goal is to better incorporate environmental impacts and costs into planning and decision-making.

For Ontario Hydro, FCA is

- Not *the* decision-making process
- Not full-cost pricing,
- Not an accounting system
- Does not require absolute or complete monetisation of all internal and external impacts.

All four points are important. Ontario Hydro sees FCA as providing information necessary but not sufficient for decision-making. It uses full-cost information as an input to its decision-making, not as the sole basis for making decisions. At this time, Ontario Hydro has no plans to include external costs in electricity prices; FCA does not require the corporation to adopt full-cost pricing. Because Ontario Hydro's goal is to use FCA in planning and decision-making, its focus is on changing management behaviour, not accounting systems. It feels that its information systems are only as good as the

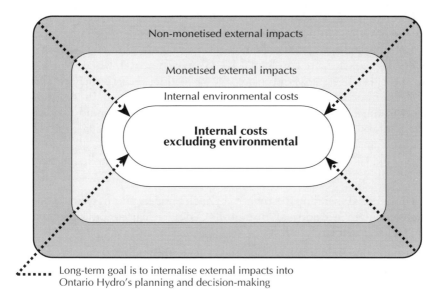

Long-term goal is to internalise external impacts into
Ontario Hydro's planning and decision-making

**Figure 1:** Ontario Hydro's Approach to Full-Cost Accounting

information put into them. It notes that internal and external environmental costs
must be calculated *before* they can be put into an accounting system. (For example,
an accounting system will not have the capability to quantify and monetise externalities.)
Finally, while quantification and monetisation of externalities is desirable whenever
possible, the key for FCA at Ontario Hydro is that environmental impacts be considered in planning and decision-making whether or not the impacts can be quantified or monetised.

## ▌ How did Ontario Hydro account for environmental costs before committing to full-cost accounting?

This section describes how Ontario Hydro approached internal and external costs in
the years prior to 1993.

### Process for Measuring Internal Environmental Expenditures

Although most major companies in Canada, including utilities, do not collect and
report overall environmental spending data,[3] Ontario Hydro has been estimating its
environmental expenditures since 1989. Revised in 1991, the guidelines developed
by Ontario Hydro's environmental staff and business managers, entitled 'Environmental Cost Concepts, Principles and Accounting Guidelines', serve as the basis for

---

3. '**Environmental spending** is any monetary expenditure, revenue, or revenue foregone, whether
capitalised or charged to current operating expenses, made by Ontario Hydro for the primary
reason of sustaining or protecting the environment. This definition includes any cost incurred for
control, reduction, prevention or abatement of discharges or releases to the environment of
gaseous, liquid or solid substances, heat, noise or unacceptable appearance' (1993 *Environmental
Performance Report*).

identifying environmental outlays and estimating environmental spending levels that indicate the environmental component of over 130 individual spending categories.[4] For example, outlays associated with monitoring groundwater conditions at ash and solid waste disposal sites are considered 100% environmental, while solid waste disposal site preparation expenses associated with construction activities are considered only 25% environmental. For some activities, *incremental* expenditures incurred to reduce environmental impacts are treated as environmental spending; for example, the incremental cost of right-of-way maintenance to reduce herbicide use is considered 100% environmental.[5]

Ontario Hydro's estimates of environmental spending are compiled in a couple of ways:

1. Environmental spending is estimated by each business unit in terms of operations, maintenance and administration (OM&A); major capital initiatives; and fuel and related.

2. Environmental spending is categorised by: material and waste management, water management, air management, land use management, environmental approvals and energy efficiency.

Because Ontario Hydro's environmental expenditures have not routinely been identified through its accounting system, the data have been manually collected and judgement applied to define the percentage of expenditures classified as environmental. As a result, the spending estimates represent a 'best judgement' and are considered gross estimates at best. Moreover, Ontario Hydro notes that the figures for capital expenditures include only major project initiatives and may not represent all outlays on capital. Since 1989, the results have been provided to Ontario Hydro's Board of Directors and summarised in the company's *Annual Sustainable Development and Environmental Performance Report*. One of Ontario Hydro's goals is to better define and allocate internal environmental costs to enable it to make better decisions and ensure value from environmental expenditures.

### Externalities Research

This section describes the history and status of externality research at Ontario Hydro prior to 1993, and explains Ontario Hydro's approach to quantifying and monetising externalities.

Ontario Hydro has been investigating externalities for many years. For the past twenty years, as part of its licence application process to export electrical energy to the United States, Ontario Hydro has been required to submit external cost studies to the National Energy Board of Canada to demonstrate that Ontarians are not being adversely affected by incremental generation for such sales. This led Ontario Hydro to the development of a methodology for identifying, quantifying and monetising external impacts and costs for its fossil and nuclear electricity generation system.

In 1991, Ontario Hydro established a Steering Committee on Environmental Costs (SCEC) with a mandate to co-ordinate and oversee all of Ontario Hydro's work on

---

4. The environmental spending guidelines were originally developed in 1982 and were updated in 1991. See Attachment B of the original case study.

5. Attachment B of the original case study offers more examples of environmental expenses identified by Ontario Hydro.

external environmental costs and benefits. In 1992, Ontario Hydro adopted a corporate position to adopt the damage function approach for quantification and *monetisation*[6] of external environmental impacts.

*Ontario Hydro's Approach for Quantifying and Monetising Externalities.* There are two main approaches currently being used by industry and government to place monetary value on externalities: (1) the cost-of-control approach; and (2) the damage function approach. The cost-of-control approach uses the cost of installing and operating environmental control technologies as a proxy for the dollar value of actual damages. The damage function approach uses site-specific data and modelling techniques combined with economic methods to estimate external impacts and costs.

Ontario Hydro supports the damage function approach to quantifying and monetising externalities and has used this approach since 1974. Although the cost-of-control approach is the simpler of the two approaches to calculate, Ontario Hydro does not support its use because it bears little relationship to environmental impacts and costs. Because the cost-of-control approach does not account for site-specific environmental factors or impacts, the external cost estimates derived for two similar power stations would be the same even if one station was located close to an urban centre while the other station was in a rural area.

The cost-of-control approach also is limited to pollutants for which control technology is available. The damage function approach, on the other hand, attempts to place a dollar value on the actual impacts to human health and the environment by considering site-specific impacts. Ontario Hydro advocates using market prices to estimate monetary values for those impacts (e.g. crop losses) that are traded in the market. For impacts that are not explicitly traded in markets (e.g. human health and mortality), Ontario Hydro believes that a number of valuation techniques can be used to derive estimates of willingness to pay (WTP) or willingness to accept (WTA) for changes in environmental quality.[7]

Ontario Hydro has acknowledged that even the most accurate externality estimates can be extremely sensitive to site-specific factors. For example, some pollutants create problems only when combined with other pollutants whose presence varies considerably from site to site. As a result, transferring damage cost estimates from one site to another can be very controversial. In addition, because damage estimates are based on scientific evidence regarding the relationship between pollution and human health, crop production, natural resources, materials, visibility, etc., impact estimates are limited by the nature of the scientific data available. Acknowledging these uncertainties, Ontario Hydro believes that the real benefit of the damage function approach is its focus on potential site-specific damages to receptors.

## ▌ Why did Ontario Hydro address full-cost accounting?

The activities described in the previous section were brought under the FCA framework in 1993 as part of Ontario Hydro's Task Force on Sustainable Energy Development

6. Ontario Hydro defines 'monetisation' to mean the process of developing appropriate monetary (i.e. dollar) values for the impacts of emissions/pollutants on the environment.

7. For more information, see Attachment C of the original case study.

initiative. This section describes the context for FCA as a key component of Ontario Hydro's commitment to sustainable development.

## Commitment to Sustainable Development

One of the major catalysts for Ontario Hydro's commitment to sustainable development was the appointment of a new Chairperson, Maurice Strong, in late 1992, who, in addition to his mandate to restructure the corporation, also had a strong sustainable development focus.

Maurice Strong came to Ontario Hydro with a personal commitment to sustainable development;[8] he recognised that movement towards sustainable energy development (SED) would be a key priority for the future. As a result, in 1993, Ontario Hydro incorporated sustainable development into its mission statement as follows:

> Ontario Hydro's mission is 'to make Ontario Hydro a leader in energy efficiency and sustainable development, and to provide its customers with safe and reliable energy services at competitive prices' (Ontario Hydro 1995a).

Ontario Hydro views sustainable development as a long-term strategy for achieving business success within environmental limits. It defines sustainable development as 'development which meets the needs of present generations without compromising the ability of future generations to meet their own needs' (WCED 1987). Ontario Hydro believes that moving towards sustainable development will enable it simultaneously to make progress on environmental goals and cost reduction, job creation and competitiveness. It also believes that business competitiveness cannot be achieved separately from environmental sustainability.

The SED concept applies the principles of sustainable development to the energy sector. A fundamental tenet of SED is the efficient use of energy, human, financial and natural resources. In this view, business success, ecological limits and inter-generational equity are related and should be managed together to drive decisions that are 'ecologically efficient' under the framework of SED. To emphasise that SED should not be seen as an add-on, Dr Al Kupcis, President and CEO of Ontario Hydro, indicated to business unit leaders that he did not expect to see specific SED action plans, but, rather, wanted SED to become the business norm throughout business units' planning processes.

Ontario Hydro acknowledges SED as a long-term goal. To move towards this goal, it recognises that its economic activities must be balanced with the capability of the earth's ecosystems to respond to the stresses or changes caused by those activities. To do this, it may need to make investments in the near term that do not meet its normal 'payback period' requirement. The evaluation of such investments will need to take into consideration the possibility for longer-term benefits for both the environment and business. Ontario Hydro believes that, by taking some actions now in order

---

8. Prior to coming to Ontario Hydro, Maurice Strong played a major role in the United Nations Conference on Environmental and Sustainable Development in Rio de Janeiro in June 1992. According to Strong: 'Sustainable development is a matter of economic survival in a world of finite resources and unlimited desire for growth. For present and future generations to enjoy a good quality of life, government, industry and individuals need to become ever more efficient in the use of materials and energy, minimise wastes through recycling and re-use, and develop new disposal methods.'

to reduce resource consumption, promote pollution prevention, minimise wastes and reduce environmental damage, it can contribute to the long-term objective of creating a healthier environment and saving resources available for future use.

### Relationship between Sustainable Energy Development and Full-Cost Accounting

What is the relationship between Ontario Hydro's commitment to SED and its sponsorship of FCA? Ontario Hydro sees FCA as one of the cornerstones of its sustainable development strategy. Of the ten central elements Ontario Hydro identified for SED actions, two were as follows:

- Integrate environment and economics in decision-making
- Adopt full-cost accounting

FCA can support sustainable development by helping to ensure that internal and external environmental impacts and costs are factored into business decisions. By better understanding the internal and external environmental costs associated with its activities, including quantifying and, where possible, monetising externalities, and incorporating this information into planning and decision-making, Ontario Hydro expects to be in a better position to fulfil its sustainable development mission and enhance its competitiveness. Ontario Hydro articulated in 1993 the expected benefits from introducing FCA (see Fig. 2). Notably, the FCA framework encompasses Ontario Hydro's ongoing efforts to estimate both internal environmental spending and external environmental impacts and costs.

## ▌ How did Ontario Hydro address full-cost accounting?

In 1993 two important and related events occurred at Ontario Hydro that catalysed its commitment to FCA and served as an impetus for action in 1994 and beyond. These events included:

1. The formation of the Sustainable Energy Development (SED) Task Force and the completion of its report, *A Strategy for Sustainable Energy Development and Use for Ontario Hydro* (Ontario Hydro 1993b)

2. In conjunction with the SED Task Force, the establishment of a full-cost accounting (FCA) team (as part of the SED Task Force initiative) and the

- Provides a powerful incentive to search for the most economic ways of reducing environmental damage
- Leads to choices that include explicit consideration of the present and future environmental impacts of alternative options
- Should lead to a more efficient and effective use of resources
- Should help in 'levelling the playing field' when evaluating demand and supply options (e.g. demand-side management, alternative power generation technologies, conventional supply options)*

* Energy utilities can meet their mandates to serve the needs of their customers by using a combination of two different strategies: (1) delivering power to meet energy requirements through various conventional and alternative supply options; and (2) helping customers use energy more efficiently, termed 'demand-side management' or 'demand management'.

**Figure 2:** The Benefits of Introducing FCA

completion of its report, *Full Cost Accounting for Decision Making* (Ontario Hydro 1993c)

The FCA team report served as the background document for the SED Task Force recommendations on FCA discussed below.

### Established SED Task Force

Having commissioned the SED Task Force in June 1993, Chairperson Maurice Strong launched it by stating: 'We must examine ways and means to incorporate full-cost accounting in our financial planning and controls and to monetise externalities and incorporate them in our planning.' On 3 June 1993, the twelve-member task force (assisted by over 150 Ontario Hydro staff members) held its first meeting, organising itself into ten teams for gathering data, identifying and analysing issues, and formulating recommendations. In September, the task force met to review and finalise its report and recommendations. The Sustainable Energy Development Strategy, including FCA recommendations, was formally submitted to the board in October 1993. Strong subsequently characterised the strategy as follows: 'The Sustainable Energy Development (SED) Strategy developed in 1993 reinforced and will build on many effective environmental initiatives already in place but, more than this, it provides the strategic vision and direction for further progress and new initiatives.'

### Established Full-Cost Accounting Team

The FCA team was one of the ten teams formed by the SED Task Force. The team consisted of eight members representing environmental economics, corporate finance, management, financial accounting, environmental and planning functions. When necessary, other contributors were called on for expertise in environmental science, engineering and strategic planning. The team's mandate was to:

- Define full-cost accounting and examine how it relates to Ontario Hydro's internal accounting and decision-making systems.
- Based on availability of data, provide estimates of internal and external costs of Ontario Hydro activities where possible. Identify data requirements and propose a research programme to expand on existing estimates of internal and external costs and develop external environmental cost estimates of the full range of Ontario Hydro's activities.
- Determine how internal and external costs can be integrated into a full-cost accounting framework for Ontario Hydro.
- Examine the potential applications of full-cost accounting and assess the implications of its implementation at Ontario Hydro.

Building on past research, the FCA team worked on an accelerated schedule established by the SED Task Force to analyse issues and develop recommendations. Team efforts entailed conducting substantial research, and discussion to explore internal costs and externality quantification and monetisation. The FCA team held a 'Full-Cost Accounting Workshop' in June 1993 (Ontario Hydro 1993a) and invited other Canadian and US environmental economists and accountants to share their knowledge of FCA, comment on Ontario Hydro's work, and offer guidance on next steps. The FCA team presented its recommendations to the SED Task Force and later, in December, issued a detailed report entitled *Full Cost Accounting for Decision Making* (Ontario Hydro 1993c). The report defined the concept of FCA, discussed the incorporation of full costs into Ontario Hydro's accounting and decision-making frameworks, presented a preliminary and partial assessment of external impacts and costs associated with Ontario Hydro's activities, and documented the FCA team's recommendations for developing and implementing FCA.

## What did Ontario Hydro's full-cost accounting team recommend?

The recommendations in Figure 3 were created by the FCA team and represent a detailed set of suggested next steps that were grouped into these umbrella recommendations.[9]

The following section summarises Ontario Hydro's implementation activities in response to these recommendations.

## What has Ontario Hydro done to implement full-cost accounting?

The 1993 FCA report represented a 'wish list' for FCA at Ontario Hydro. In moving ahead, Ontario Hydro has taken a more practical approach, and, in doing so, can report many concrete accomplishments. This section describes the many initiatives undertaken by Ontario Hydro to develop and implement FCA since 1993.

The first step was to establish an institutional foundation responsible for managing the implementation of the SED strategy. To do this, Ontario Hydro created a division called the Environment and Sustainable Development Division (ESDD). As part of ESDD, the Business–Environment Integration (BEI) department was established, which is responsible for FCA. The mandate of this department is to identify and implement means to better integrate environmental considerations into business decisions. Seven full-time staff are involved in these activities; developing and implementing FCA is a significant part of this work.

### Established FCA Corporate Guidelines
In 1995, ESDD developed corporate guidelines for FCA (Ontario Hydro 1995b). The guidelines define key terms, state the goal of FCA at Ontario Hydro, articulate Ontario Hydro's rationale for FCA, describe how Ontario Hydro plans to use FCA, delineate roles and responsibilities, and lays out an implementation plan through 1997.[10]

---

1. Modify the current accounting system into a full-cost accounting system
2. Augment the current financial evaluation framework
3. Support a research programme on full-cost accounting
4. Initiate a training programme on full-cost accounting
5. Take full-cost accounting beyond Ontario Hydro
6. Establish a fund for decommissioning, waste disposal, etc.

---

**Figure 3:** Ontario Hydro 1993 Full-Cost Accounting Recommendations

---

9. For more detailed information on the FCA report, the recommendations and the reasons the team made these recommendations, see Attachment D of the original case study.

10. These guidelines appear in full as Attachment A of the original case study.

Ontario Hydro's corporate guidelines articulate several reasons for supporting FCA:

- **Improved environmental cost management:** improve identification, allocation, tracking and management of environmental costs in each business unit
- **Cost avoidance:** improve ability of business units to anticipate future environmental liabilities and costs, so that corrective action can be implemented earlier
- **Revenue enhancement:** improve ability of business units to identify revenue enhancement opportunities, either through environmental technology innovations spurred by cost-cutting initiatives or by strategic alliances with companies that use waste products as material inputs in their own manufacturing
- **Improved decision-making:** aid business units to better integrate environment into decision analyses
- **Environmental quality improvement:** establish an optimal level for reducing emissions/effluents/wastes with consideration for least cost to society
- **Contribution to environmental policy:** contribute effectively to the development of environmental regulations/standards and emissions trading markets
- **Sustainable development:** assist in the transition to a more sustainable energy future

Ontario Hydro's adoption of FCA guidelines represents a fundamental change in the way it expects to do business. According to its *FCA Corporate Guidelines*:

> Managing resources wisely and minimising environmental damage will also contribute to Ontario Hydro's competitiveness, particularly in the longer term. By better understanding the environmental impacts of its activities and by making better resource allocation decisions based on this information, Ontario Hydro can save money, become more competitive and move towards the goal of sustainable development (Ontario Hydro 1995b).

Ontario Hydro has conducted stakeholdering of its corporate FCA guidelines in order to communicate its approach to interested parties and respond to their questions and comments. As part of this process, Ontario Hydro convened a full-day, professionally facilitated workshop in September 1995. Participants included representatives of the energy sector, consumers, environmentalists, university researchers and government agencies. Stakeholders were encouraged to raise issues and air any concerns. The facilitator sought their perceptions of the merits and constraints of the draft *Corporate Guidelines* and the proposed FCA research programme. Overall, the majority of workshop participants supported Ontario Hydro's efforts to develop and implement FCA and viewed the corporate guidelines as a reasonable step in that process.

### Applied FCA to Decision-Making

Ontario Hydro anticipates that FCA will evolve over time. Incorporating FCA into decision-making will take place on a step-by-step, pragmatic basis. In response to the FCA recommendations, Ontario Hydro has already taken concrete steps such as adding environmental considerations into investment decisions by implementing SED decision criteria in 1994, as described below.

Traditionally, environmental analysis and evaluation at Ontario Hydro have focused on compliance with environmental regulations. In the past, a generic set of questions was used to ensure consideration of the environmental implications of proposed projects or plans going to the board of directors for approval. These questions were:

- What are the environmental implications of this proposal? What environmental approvals are required?
- Does this proposal comply with existing environmental regulations? Is there sufficient flexibility to respond to more stringent, future environmental regulations?
- Is this proposal consistent with existing corporate environmental initiatives?
- Will this proposal contribute to a policy of sustainable development; for example: will waste products be recycled? Has energy-efficient equipment been incorporated?
- Will this proposal create a significant public concern—real or perceived? If so, then what measures are being considered to offset this effect?
- What are the environmental alternatives for/to this proposal? What are the relative merits of these alternatives?

Although this checklist of environmental considerations may have been effective in eliciting the general environmental implications of a proposal, this approach was quite limited. It relied mainly on qualitative and often subjective data. Because information on environmental impacts was not explicitly incorporated and monetised into cost information, Ontario Hydro had limited ability to rank investment alternatives using a common denominator.

**SED Decision Criteria.** Effective in September 1994, Ontario Hydro introduced new SED decision criteria as part of its Business Case Analysis Guidelines for evaluating investment decisions requiring senior management approval. The SED criteria represent a framework for Ontario Hydro to integrate environmental and economic information into decision-making. In addition, the SED criteria, notably the environmental impact sub-criterion, reflect an FCA approach. The SED criteria are intended to help Ontario Hydro's business units describe and evaluate the SED implications of expenditure decisions going to senior management or the board of directors for approval. Ontario Hydro believes that the SED decision criteria reflect its commitment to sustainable energy development and the movement towards FCA.

The SED criteria require that Ontario Hydro consider a project's (1) resource and energy use efficiencies; (2) environmental impacts; (3) social impacts; (4) employment of renewable energy sources; and (5) financial integrity. The five criteria were chosen because they are the macro SED indicators that Ontario Hydro uses to gauge its SED performance. According to the criteria, the evaluation should consider:

- Full life-cycle impacts,[11] where possible, but at a minimum, design, construction, operation, maintenance, decommissioning and disposal

---

11. Ontario Hydro has considered concepts of life-cycle costing (LCC) in developing its strategy for FCA. For internal costs, Ontario Hydro considers the full fuel cycle, inventorying energy requirements and generation of wastes/pollution. For external costs, involving the consideration of damages to human health and the environment, it aims to consider the full life-cycle but expects to emphasise at a minimum the stages of the life-cycle over which it has direct control and responsibility: design, construction, operation and maintenance, and decommissioning/disposal.

- *Expected* damage to ecosystems, community and human health (i.e. versus ability to meet existing or proposed environmental regulations)
- *Potential* positive and negative environmental impacts, including impacts that may be common to all the project alternatives being compared
- *Quantification and monetisation* of the potential impacts, where possible; but, at a minimum, a qualitative description
- *Trade-offs* made in selecting the preferred alternative

Ontario Hydro expects that this analysis will uncover relationships between competitiveness and sustainability that might otherwise go unnoticed and, as a result, lead to better investment decisions.

President and CEO Al Kupcis has charged Ontario Hydro's ESDD staff with providing senior management with an independent review of the SED component of business case summaries. ESDD staff are also available to work with the business units to advise on SED during the development of business case summaries (BCAs). Since the SED criteria were implemented in 1994, nineteen BCAs have been reviewed. The majority of these BCAs addressed the criteria appropriately and were recommended for senior management approval. In some cases, the SED implications analysis was effective in the development of alternatives that incorporated the principles of sustainable development. The SED analysis also exposed business unit staff outside of the environmental functions (i.e. financial staff) to sustainable development issues.

As an example, in one case, a proposed investment decision for a $24 million transmission line refurbishment, the SED implications were:

- 20% reduction in energy loss in transmission lines through the use of energy-efficient conductors
- $0.5 million annual increase in revenues through the re-use and recycling of removed line components
- Initiation of a programme to improve the biodiversity of rights-of-way by restoring and replacing natural habitats
- Provision of employment and economic benefits to local communities.

This investment decision was approved.

### Applied FCA to Planning
In addition to major investment decisions, Ontario Hydro has used FCA for planning activities such as the following:

- **Corporate Integrated Resource Plan (CIRP)**[12] is a business-wide, strategic exercise to evaluate different supply generation and demand management plans for the future. One of the criteria used to assess the plans was environmental impact. The assessment was performed on an environmental damage basis (using the damage function approach), consistent with Ontario Hydro's corporate guidelines for FCA. Impacts were either quantified, and monetised where possible, or qualitatively described, depending on the data

---

12. 'The objective of integrated resource planning is to ensure that all available options are considered in determining how best to meet customer energy needs' (Ontario Hydro 1994).

available. Additional SED considerations were included in the form of 'committed impacts', that is, impacts that would have to be managed by future generations (e.g. used nuclear fuel in storage; consumption of non-renewable resources; greenhouse gas emissions). The analysis was performed on a life-cycle basis.

- **Local Integrated Resource Plans (LIRPs)** address trade-offs in supply and demand management options for specific geographic areas with potential supply shortfalls. Ontario Hydro initiated six LIRP studies in 1993 and carried out nine LIRP studies in 1994. LIRP studies examine a wide range of options, including demand-side management (DSM) strategies, to meet customer needs. These studies offer customer participation in decision-making and aim to provide solutions that harmonise with environmental and social objectives. In 1994, Ontario Hydro evaluated environmental and other plan attributes such as cost and reliability within one LIRP process. Other LIRP studies have shown that DSM programmes could defer the need for constructing major new capacity.

In time, Ontario Hydro also plans to incorporate FCA into procurement decisions; it procures about $1 billion each year in goods and services.

*Use of Multi-Criteria Analysis in Planning.* Because Ontario Hydro has not yet developed monetised environmental impact estimates for all available supply, DSM and transmission options, an evaluation method is required to facilitate comparison of environmental impact information expressed in different units (qualitative, quantitative and, where available, monetised) and to integrate such data into Ontario Hydro's decision-making and planning processes. A similar evaluation method is also required to compare and make trade-offs between environmental and other plan attributes (cost, reliability, risk, etc.) in the planning process. Ontario Hydro uses multi-criteria analysis (MCA) for these purposes. MCA has been used in both Ontario Hydro's CIRP and LIRP processes to evaluate and compare these environmental 'unlikes', evaluate and compare environmental and other (e.g. cost, reliability) plan attributes, and make trade-offs. In 1995, MCA was used to select the key environmental indicators for evaluating the CIRP plans. In addition, Ontario Hydro is currently using MCA to evaluate plan attributes within its ongoing LIRP processes. Ontario Hydro believes that approaches such as MCA, combined with FCA, are necessary to evaluate trade-offs in decision-making and planning.[13]

## Undertook Full-Cost Accounting Research
Ontario Hydro has undertaken recent research on internal environmental cost accounting and external impact and costs issues. The following is a brief description of the results.

### Internal Environmental Cost Research.
*Environmental expenditures and overhead accounts.* Ontario Hydro believes that to implement full-cost accounting, it must be able to isolate (i.e. distinguish from other

---

13. Ontario Hydro's application of MCA is described in Boone, Howes and Reuber 1995.

types of expenditures) environmental expenditures, particularly from overhead accounts. For example, payments pursuant to compensation agreements with aboriginal peoples have traditionally been allocated to corporate overhead rather than to a business unit. Ontario Hydro is minimising the practice of charging expenses to overhead accounts, and has implemented the following procedures to ensure that each business unit is accountable for its own costs:

- All costs are incurred by or allocated to business units.
- Overhead charges for corporate services are limited only to those costs for which fees cannot be reasonably charged.

Making each business unit responsible for its own expenditures and costs helps Ontario Hydro achieve better internal environmental cost information, thereby minimising cross-business-unit subsidisation and the amount of money charged to general overhead accounts. Some business units are in the process of evaluating and implementing activity-based costing (ABC) systems, which will further aid Ontario Hydro in identifying and managing environmental costs.

*Allocation of energy-efficiency expenditures as internal environmental costs*. In 1994, Ontario Hydro expanded its definition of environmental expenditures to include costs associated with improving internal and customer energy efficiency.

Ontario Hydro is currently investigating methods to obtain more precise information on its internal environmental expenditures at the project/process level, to track and allocate these expenses on a life-cycle basis, and to accomplish this more explicitly than in the past. As the first step in this process, Ontario Hydro is undertaking a pilot study within one of its retail utilities. The pilot project is described below.

*Internal environmental cost pilot*. This pilot project is currently under way at Southwest Hydro, one of the thirteen retail utilities owned and operated by Ontario Hydro, and located in south-western Ontario. The Southwest Hydro utility territory includes approximately 75,000 customers and had a net income of $19 million in 1995. The goal of the pilot project is to identify and collect all internal environmental costs associated with Southwest Hydro's activities, identify and prioritise processes or products with higher environmental costs and liabilities, and develop recommendations leading to cost savings, cost avoidance, revenue generation, waste reduction and improved image in the community for the utility. Results from the pilot project are expected to benefit other management of the business.

Since environmental expenditures at present are not identified and recorded separately throughout Ontario Hydro's accounting system, the process of collecting environmental costs involved estimation based on physical data available or obtained through interviews with utility personnel, use of data from other utilities as proxy, and various other sources of information. A list of all major environmental activities and associated costs was then prepared by manually collecting data by separating environmental costs from other operating and capital costs, using environmental expenditure guidelines and allocation methods based on work practices and employees' experiences. Costs that were incurred and recorded as a one-time expenditure were annualised.[14]

---

14. Spread over frequency of occurrence, e.g. recurring every 4–5 years.

Internal environmental costs were defined as expenditures on both external and social environmental initiatives, whether capitalised or charged to operations for equipment, labour, fuel and programme to protect and restore the environment. The scope of this project covered costs incurred by the utility and did not include estimation of external costs relating to environmental impacts from its operations.

The total environmental costs were approximately 15% of the utility's operating, maintenance and administrative (OM&A) costs and 8% of the total annual expenditures. The top five environmental activities and associated costs were related to fuel consumption, transformer management, polychlorinated biphenyls (PCBs), energy efficiency and forestry work. While PCBs and energy-efficiency-enhancement-related expenditures were classified as being 100% environmental, others, such as fuel consumption, transformer management and forestry, did not attribute entirely to environmental costs, as they were incurred to meet operating requirements.

It is expected that the analysis of the results will lead to identification of cost drivers, fixed versus variable costs, regulatory versus non-regulatory costs, high-risk versus low-risk costs and future liabilities. Opportunities for managing these environmental costs and risks will also be identified for further analyses, such as evaluation of low-cost waste management/recycling options, green procurement (steel poles and pole extensions), alternate fuels for fleet (ethanol, gas), moving to PCB-free operations, adopting natural landscaping, investigating line loss-reduction options (shunt capacitors and transformer sizing), possible outsourcing of fuelling to reduce risk from underground tank leakages, optimising of tree-trimming cycles, future partnerships with other service providers (Bell, Cable TV, Parks for renewable energy technology [RET] applications). The results from the pilot study are also expected to help in benchmarking environmental expenditures and in the utility's business planning and budgeting activities.

***Externalities Research.*** As mentioned in the definition of externalities contained at the beginning of this case study, even after existing environmental regulations have been met, there are still residual emissions with associated environmental damage. It is Ontario Hydro's view that, by better understanding these 'residual' environmental impacts, the corporation will be in a better position to reduce future environmental liabilities and enhance its competitive position in the future. It is for this reason that Ontario Hydro is pursuing research on its external impacts and their associated costs. By understanding the external impacts and costs of its operations, Ontario Hydro can be better positioned to respond to tighter future regulations by developing process changes now to reduce its externalities, as well as better managing future environmental liabilities.

*Quantification and monetisation of externalities.* Ontario Hydro has developed monetised externality estimates for the operation of its fossil stations located in southern Ontario and for the full life-cycle of its nuclear stations. These are preliminary estimates and, certainly in the case of fossil, underestimate the health impacts associated with the operation of the fossil stations. Monetised externality estimates have yet not been developed for its hydroelectric stations, transmission or distribution line systems, renewable energy technologies or DSM initiatives.

In the summer of 1993, the FCA Working Group expanded on previous work completed within the corporation to identify, quantify and monetise external impacts and costs associated with Ontario Hydro's activities. Preliminary estimates were derived for external impacts associated with the operation of Ontario Hydro's fossil stations located in southern Ontario. Estimates were also developed for the nuclear system on a full life-cycle basis.[15]

Monetised estimates of physical impacts (i.e. statistically estimated impacts in terms of human mortality, morbidity, crop losses and building material damages) were developed based on the use of per-unit dollar values which had been previously developed (see original case study). As an example, Figure 4 presents one of the resulting tables; it summarises the FCA Team's preliminary estimates of the system's average external costs due to the generation of electricity in Ontario using fossil fuels.[16] As shown, the external costs associated with statistical premature mortality were estimated to be about $21.4 million (in 1992 Canadian dollars) or 0.088 cents per kilowatt. For all the impacts considered, the average monetised estimate was $95.79 million or 0.395 cents per kilowatt.

Since December 1993, ESDD has focused on developing research priorities to improve its externality impact and cost estimates and to broaden the range of environmental impacts for which externality cost estimates are developed. This is being done by working with the business units to better define and understand their external impacts and costs.

*Working groups.* During the period from 1994–95, a number of 'externalities working groups' were established at the business unit level to address issues relating to the development and implementation of FCA within Ontario Hydro and to define and, where possible, monetise external impacts. The working groups were initiated by the business units to examine externalities and assess how business case analysis could be undertaken with full-cost accounting-based information. Examples of three such working groups include:

1. **Energy Services Working Group.** This working group involved examining the implications of FCA for the evaluation of DSM technologies and programmes. The group examined environmental impact issues associated with DSM technologies and programmes and provided recommendations on how to incorporate environmental externalities into future decisions.

| Receptor | Pollutants of concern | Unit values | Monetised impacts | |
|---|---|---|---|---|
| | | | $ million 1992 | cents/kW |
| Mortality (statistical deaths) | $SO_2$, $SO_4$, $O_3$, $NO_3$ | $4,725,600 | 21.40 | 0.088 |
| Morbidity (admissions) | $SO_2$, $SO_4$, $O_3$, $NO_3$, TSP | $44,700 | 50.83 | 0.210 |
| Cancer cases | Trace metals | $408,397 | 9.53 | 0.039 |
| Crops | $O_3$ | n/a | 8.32 | 0.034 |
| Building materials | $SO_2$ | n/a | 5.7 | 0.024 |
| Total | | | 95.79 | 0.395 |

**Figure 4:** Monetised External Impacts of Fossil Generation in Ontario

15. These preliminary externality estimates are contained in the FCA team's report (Ontario Hydro 1993c) and are expected to change as research progresses.

16. These monetised impacts are based on the use of a specific methodology—termed the 'damage function approach'—that was described earlier in this chapter.

2. **Ontario Hydro Nuclear Working Group.** This working group examined the implications of FCA on business decisions relating to nuclear generation of electricity, focusing on environmental impacts and costs (i.e. external costs). The study revisited several recently approved projects and attempted to include FCA considerations. The study identified many items that should be included for a proper treatment of FCA; only some of these items could be included with information currently available. In some cases, FCA would not change the decision; in others, environmental impacts could be the deciding factor. The study addressed such issues as how to supply data, data consistency, cost-effectiveness of using FCA, necessary infrastructure, training needs, and required corporate guidelines.

3. **Transmission Working Group.** A GRID Externalities Team was established to undertake an examination of potential externalities due to activities associated with the transmission and distribution of electricity. The team consists of members from Grid System Strategies and Plans, Grid Operations, Grid Transmission Projects (Environment), Corporate Health & Safety, Corporate Strategic Planning, Aboriginal and Northern Affairs, and ESDD.

In 1995, ESDD developed its FCA research programme in consultation with each of the business units and with review and input from stakeholders. The research programme is undergoing review by Ontario Hydro's Business Planning process, as of February 1996.

Ontario Hydro is also monitoring trends in externality-related research in North America and Europe. It believes that the methodologies used, the issues identified and the estimates of external costs produced by recent studies supported by the European Commission and the US Department of Energy were consistent with those produced by Ontario Hydro. In developing its research programme, Ontario Hydro hopes to address some of the issues raised at the conference by other researchers.

### Internal and External Communication

ESDD devoted considerable resources to this, including a one-day seminar for over sixty environmental, financial and business unit staff within Ontario Hydro.

## ▌ What has Ontario Hydro learned about full-cost accounting?

Below are some of the lessons learned by Ontario Hydro to date in its effort to develop and implement FCA (Boone and Howes 1996). Ontario Hydro hopes that this information can be useful for other companies interested in or working on FCA issues.

Ontario Hydro has learned that:

- Full-cost accounting (FCA) must be positioned as an approach that makes 'good business sense' in order to promote integration of environment and business issues. Steps must be taken to demonstrate the benefits of understanding the environmental impacts and costs (internal and external) associated with business activities (i.e. the potential for reductions in future environmental costs and liabilities). If this is done, FCA will be considered to make 'good business sense'.

- Case studies and projects where FCA has been applied and have contributed to a better business decision provide concrete examples that may facilitate change in acceptance.

- FCA, for internal environmental costs and externalities, is not yet mainstream thinking. It is often difficult to get a 'foot in the door'. A way to overcome this barrier is to highlight the potential to avoid potential future environmental liabilities. In addition, if a company understands the environmental implications of its business activities, it can sometimes influence regulation.

- FCA needs an executive member of the organisation to champion its value and use for business decisions.

- FCA should be developed and implemented as part of a larger context; for Ontario Hydro, sustainable development is that context.

- FCA is only one of the elements that go into making business decisions; **it is not the decision-making process**. It is very important to communicate this point. Building on this, it is important to highlight that FCA can contribute to more informed decision-making that highlights a greater variety of the trade-offs involved in all decisions; Ontario Hydro has found that its approach to FCA, which includes Multi-Criteria Assessment, provides an effective tool for this.

- It is important to implement FCA as a central component of a corporation's overall environmental management system (EMS). In this regard, it is important to develop some high-level FCA guidelines and link them to the EMS.[17]

- Ontario Hydro also stresses that FCA does not mean 'full-blown monetisation' of all internal environmental costs and external impacts and costs. In this regard, Ontario Hydro stresses that it is essential to have a methodology for considering externalities that allows for the consideration of monetised (economic value of environmental damages) and non-monetised (i.e. qualitative description of damages or emission levels) environmental information. Ontario Hydro's use of the damage function approach to consider externalities and its use of Multi-Criteria Assessment have facilitated this.

- Developing and implementing FCA is a gradual process (for internal environmental costs and external costs). It will not happen overnight. However, just because it takes time and may be difficult does not mean that it should not be done. It is best to focus on those areas where it is possible to exert the most influence and obtain positive results. There are many environmental, economic and competitiveness benefits that will be realised by those companies that explicitly integrate externality concerns into the way they do business now. Ontario Hydro believes that it will become more competitive

---

17. According to Ontario Hydro, an environmental management system (EMS) is a management system designed to achieve organisational directives and policies regarding environmental impacts of an organisation's activities. Key issues include: full-cost accounting, sensitivity to issues of due diligence, an ability to monitor and respond to effects of ongoing activities and a commitment to continuous improvement through self-evaluation, correction and a capacity for learning and creation of economic incentives and instruments. ISO 14001, released in 1997, provides a framework for EMSs.

by knowing and integrating these considerations into its business practices through methods such as FCA.[18]

- The process of changing corporate culture and attitudes are key to fostering support and commitment to FCA; however, this is often a long, slow process. The challenge is to develop an appreciation for the business case for FCA and sustainable development. FCA is multi-disciplinary by its very nature. The successful development and implementation of FCA requires a team approach with input from a wide variety of professionals in the organisation such as: scientists and planners, environmental economists, and accounting-based disciplines. Full-cost accounting is **not solely** an accounting system issue. Rather, it is a framework that can be used to consider the broader financial and environmental implications of doing business.

- Terminology causes **many** problems; in part, because of the multi-disciplinary nature of FCA. There is a need to develop an agreed-upon set of terminology to address FCA. For example, terminology such as 'environmental cost accounting', 'full-cost accounting', 'total cost assessment', 'true-cost accounting', 'total social costing' and 'full-cost pricing' are often used interchangeably and are sometimes assumed to mean different things. In addition, some practitioners use FCA to describe only internal environmental costs; others refer to it when discussing externalities (for further discussion on terminology, see Chapter 2).

- There is a need to draw the links between internal and external environmental costs (see Fig. 1). It is important to understand that the boundaries between internal and external environmental costs are not static, but rather are dynamic because both regulations and company policies change over time. For example, a system-wide cap on greenhouse gas emissions, or new regulations on air toxins that may be either certain or possible, would lead to an expansion of the internal environmental cost domain and a reduction of the external cost domain. Whether voluntary or mandatory, it is certain that the external cost domain will contract over time. Corporations with a serious commitment to sustainable development will be at the forefront of this evolution.

- The process of identifying, quantifying and, where possible, monetising environmental impacts and costs (internal environmental costs **and** externalities) and integrating information into decision-making processes is data-intensive. Data must be analysed consistently if they are to be meaningful in decision-making and the promotion of sustainable development.

- Training and communication on what FCA means, the rationale for, the benefits of and the methods for implementation, should be a priority in order to drive the right behaviour of managers and decision-makers. However, it is sometimes difficult to provide broad-based training in an era of corporate

---

18. In a recent survey of Ontario electricity customers, environmental performance and environmental leadership were considered to be important by over 90% of respondents in their potential future selection of supplier. Ontario Hydro believes that it can strengthen its environmental performance and environmental leadership through initiatives such as FCA.

'right-sizing' because individuals and departments are usually only interested in training that is 'directly' relevant to their job. This is a barrier that must be overcome.

- There is a need to build bridges between environmental and financial staff in the organisation. Many of the capital investment decisions are made within the financial area of the organisation. If investment proposals are to be considered on more than just private costs, there must be communication and collaboration between the financial and environmental decision-makers in the organisation.

- Hydro clearly distinguishes between full-cost accounting for decision-making and full-cost pricing. As stated throughout this chapter, Ontario Hydro's approach to full-cost accounting focuses on planning and decision-making, not pricing. Full-cost pricing occurs when external costs are incorporated into the price of the product or service (i.e. they are explicitly accounted for in market transactions). It is important to recognise that consideration of internal environmental costs and external environmental impacts and costs in decision-making can facilitate better decisions without being explicitly incorporated into the price of a given product or service. While Ontario Hydro believes that, theoretically, prices should reflect all internal and external costs and benefits associated with production and consumption, the corporation does not intend to pursue full-cost pricing at this time due to competitiveness reasons and other issues. However, the development and use of FCA (internal environmental costs **and** externalities) in business decisions can help to move in a direction in which corporations make decisions that are least cost to society.

## ▌ Looking Ahead

In looking ahead, Ontario Hydro has identified several important challenges. For example, it believes that there should be greater support for its definition of FCA in its business sector. It has also found that some of their business sector's major customer groups are questioning the need for FCA. While believing that the practice of SED can enhance its competitiveness, Ontario Hydro has recognised a need to demonstrate results. It plans to address these challenges through better defining its externalities and costs, and through further communication, education and training. This section lists the major elements of the FCA corporate programme that Ontario Hydro is in the process of implementing.

### Use of FCA in Operating, Planning and Decision-Making Processes
Ontario Hydro plans to incorporate FCA into evaluations of:

- Major local integrated resource plans
- Operation and dispatch of Ontario Hydro's system
- Investment decisions
- Environmental externalities associated with imports and exports of electricity

- Decisions about retiring or rehabilitating existing stations
- Procurement decisions
- Benefits and costs of additional pollution control equipment
- Monitoring of environmental performance improvements

Ontario Hydro also believes that FCA will assist the corporation to:

- Provide input to the establishment of reference starting points for emission reduction trading
- Evaluate benefits and costs of new proposed environmental regulations
- Evaluate environmental externalities associated with private generation
- Contribute to decisions about demand-side management programmes to address societal issues (i.e. greenhouse gas reduction)

### Research Programme on FCA
Ontario Hydro has designed its research programme to focus on the following.

*Internal Environmental Costs.* To better understand its internal environmental costs and to determine if Ontario Hydro is getting value for its environmental dollars, this programme element will focus on:

- Continuing to estimate environmental expenditure for reporting in annual environment and sustainable development reports
- Developing methods to track, allocate and report on internal environmental costs
- Linking pollution prevention initiatives and internal environmental cost accounting to drive better pollution management decisions
- Completing the GRID and retail pilot studies; initiating pilot studies in fossil, nuclear and other business units.

*External Environmental Costs.* This programme element will focus on:

- Enhancing evaluation methods to ensure that qualitative, quantitative and, where possible, monetised environmental impact data are appropriately considered and integrated into decisions.
- Developing ecosystem approaches to assess environmental impacts
- Improving the current externality impact and cost data for the full life-cycle of fossil-fired stations and nuclear stations
- Developing full life-cycle externality impact and cost data for transmission and distribution systems, hydroelectric stations, renewable energy technologies and demand management
- Working with Canadian and Provincial governments, academics, businesses, professional associations and stakeholders to undertake research on environmental externalities
- Considering the development of an integrated externality impact and cost computer framework

***Expand FCA Communication/Education Programme.*** In order to develop internal awareness and understanding of FCA, this programme element focuses on:

- Developing communication materials on Ontario Hydro's approach to FCA
- Designing and delivering internal training programmes/workshops on FCA

***Promote FCA beyond Ontario Hydro.*** To promote the understanding and application of FCA beyond Ontario Hydro, this programme element focuses on:

- Working with the government, academics, businesses, professional associations and stakeholders to promote a better understanding and application of FCA
- Establishing business partnerships for environmental costing to identify and establish a network of Canadian and other experts engaged in FCA work, to educate others about FCA, and to identify opportunities to initiate or collaborate on FCA research
- Seeking opportunities to present papers on FCA

The development and implementation of FCA at Ontario Hydro is an ongoing process. While much progress has been made and much has been learned, Ontario Hydro looks forward to the next several years as it advances its research and use of this important management tool.

Ontario Hydro is facing some significant changes as the electricity sector moves forward with restructuring, which in turn facilitates movement towards a more competitive electric utility industry. One of the key challenges for Ontario Hydro relates to ensuring that key elements of sustainability are maintained in a more competitive electricity sector. The corporation believes that the electricity sector's move to an increasingly competitive market highlights the need for a regulatory framework that will promote sustainability in the energy sector in Ontario. In addition, Ontario Hydro believes that mechanisms/options will be required to ensure that environment/sustainability are addressed in a restructured and competitive electrical utility industry in North America.

Ontario Hydro firmly believes that FCA has a key role in enhancing the corporation's competitive position in a new open electricity market. Ontario Hydro also firmly believes that the energy utilities that prosper in the 21st-century competitive marketplace will be those that exhibit strong environmental leadership and sustainability qualities. Ontario Hydro realises that most companies already operate in a competitive market place and believes that the future for such companies will be equally linked to environmental leadership and sustainability.

# 17 Environmental Accounting at Sulzer Technology Corporation

Georg Schroeder and Matthias Winter

FEW, if any, companies are as yet fully tracing their environmental costs directly to the source. This oversight reduces the transparency of economic benefits to decision-makers who might be considering pollution prevention measures. In order to overcome these obstacles and to test the practical usefulness of environmental accounting, Sulzer and the MIBE project ('Managing the Industrial Business Environment') at IMD ran a pilot project at Sulzer Hydro in Kriens, Switzerland.

The Sulzer Technology Corporation is a leading Swiss-based technology company with a turnover of about Sfr6 billion. Sulzer is active in a large number of markets, including weaving machines, plant and building services, medical technology, process engineering, reciprocating compressors, locomotives, surface technology, thermal power systems, hydraulics, thermal turbomachinery and pumps. Sulzer Hydro belongs to the Sulzer Wintherthur division and produces hydroelectric machinery and engineering. Sulzer Hydro's turnover was SFr330 million in 1995; more than one-third of this turnover is produced by 430 employees at Kriens (near Luzern). It is in this facility that the MIBE-Sulzer environmental accounting pilot project took place. At Kriens, 50% of the turnover comes from new machinery and 50% from service contracts, i.e. the repair and upgrading of old turbines.

The objective of the project was to develop and test new methods of environmental cost management in order to identify opportunities for environmental protection and potential cost cuts. The following chapter describes the details of this environmental accounting project and also explains the experience gained in the pilot project at Sulzer.

The project was essentially based on six steps:

1. Definition of the boundaries and choice of team
2. Process definition
3. Definition of environmental costs
4. Physical and financial data collection
5. Identification of environmental costs
6. Identification of possible cost savings

## ▐ Definition of the Boundaries and Choice of Team

The first essential step in an environmental accounting project is to define the boundaries of the environmental accounting project very precisely and to maintain the definition strictly throughout the project. These boundaries have two dimensions. The first dimension is the physical boundaries which define the relevant physical and financial data. The importance of these boundaries lies in the fact that all physical and financial data collection will be limited within them, while everything outside the boundaries will be ignored. The second dimension is the time during which the environmental costs will be identified. As physical and financial data gathered has to be historical, this can be done only by reference to a particular time-period.

The physical boundaries for an environmental accounting project can be manifold, and examples include:

- An individual process or group of processes (e.g. production line)
- A system (e.g. lighting, waste water treatment, packaging)
- A facility, department or location
- Regional groups of departments or facilities
- A corporate division, affiliate, or the entire company

It is important that the physical boundaries of an environmental accounting project be defined in such a way that the identification of physical and financial data within these boundaries does not become too difficult. Therefore, they should be chosen according to the convenience with which current information systems can provide data on physical processes and financial costs. Data collection on single facilities, for example, is especially convenient.

The time boundaries should be chosen so that the data gathered within the time-frame is representative. In most cases, a time-frame of one year can serve this purpose. This choice has the additional advantage that conventional information systems support the data gathering.

At Sulzer, it was decided to begin the environmental accounting project at one of the Sulzer Hydro facilities in Kriens. This plant was seen as appropriate because it is largely independent of in-company deliveries from other Swiss Sulzer plants. The physical boundaries of the pilot project were defined as being the borders of the Sulzer Hydro plant in Kriens, meaning that all processes within the facility would be examined. The time boundary was defined as the year 1995. It was further decided that the environmental accounting project would not consider changes in stock, as they were only minor.

The choice of the project team should be made so that together the team members are able to gather all the necessary physical and financial data. At Sulzer Hydro, the team consisted of an engineer, a financial analyst, the Vice Director of Quality and Environment, an environmental manager at group level and a MIBE research associate.

## ▮ Process Definition

As a first step, the project team had to identify the most important physical processes within the boundaries, as they are the sources of environmental costs. Typically, the amounts of waste, waste heat, etc. differ considerably when comparing the processes within a company. It is also typical to find little data available regarding which processes incur which exact environmental costs.

The identified processes can usually be split into main processes and auxiliary processes; the definition of these is as follows:

- **Main processes** are those processes that directly produce products from material by treatment.

- **Auxiliary processes** support these main processes. They may be involved in the treatment of products, but they are not modifying them (e.g. testing). Generally, the auxiliary processes are necessary for the company infrastructure to function.

Process definition is just as important as choosing appropriate boundaries for the environmental accounting project as a whole. Only if the processes are properly identified and, as a consequence, the physical and financial data for the processes meaningfully collected, can the environmental costs be properly traced to their sources. The main and auxiliary processes identified at Sulzer Hydro are listed in Figure 1.

| Main processes | |
|---|---|
| **Manufacturing processes** | **Auxiliary processes** |
| Welding | Stock receipt |
| MicroGuss™ | Testing, examination |
| Ceramic coating | Heating |
| Heat treatment | Maintenance |
| Mechanical machining | Engineering, administration |
| Machine assembly | Packaging, dispatch |
| Vapour cleaning | Model test laboratory |
| Sandblasting | |
| Corrosion protection | |

**Figure 1:** Process Definition at Sulzer Hydro

## ▌ Definition of Environmental Costs

In order to carry out environmental accounting, it is important to define which costs will be considered environmental costs. The decisions taken here are of central importance for the project, as they act as a filter for further calculations. By defining the scope of environmental costs, the direction of possible improvements in decision-making is already determined.

From our experience, it is recommended to define the environmentally relevant costs using a two-level separation method. The first level includes environmental costs, which depend directly on the data from an energy and material balance sheet, while the second-level costs do not require such data. The two levels can be separated into further cost categories as illustrated in Figure 2.

Usually, it is recommended that environmental costs of waste reduction are not considered. The reason for this is that experiences with eco-balances have proven that a comparison between different options according to their environmental friendliness is very difficult. In most cases, every product or production process has, at the same

---

### Level 1: Direct environmental costs (on the production level)

**Environmental costs from waste**
These are all costs resulting from the disposal of waste and waste water (collection costs, disposal costs, storage costs, transport costs), but without the costs of the end-of-pipe-technology for waste treatment itself (cleaning equipment). They also include labour costs and the cost of the inputs (e.g. metals, chemicals, cleaning oil) which leave the company again as waste.

**Environmental costs from the end-of-pipe-treatment of waste and waste-water**
These costs result from the use of end-of-pipe-technologies (machines that have their sole purpose in reducing environmental damage). Costs in this category include all costs of inputs resulting from the usage of end-of-pipe-technologies (e.g. filters), wages in connection with these technologies, depreciation, etc.

**Environmental costs from recycling**
These are costs/income through recycling programmes (material costs). An example may be the recycling of iron waste where a company pays US$3 per kg. However, the original material cost was US$20 per kg, so the costs per kg of recycled material is US$17, as this is exactly the amount that will be saved when reducing waste. The costs in this cost category include also labour costs for the time spent on recycling programmes.

**Environmental costs from waste reduction**
In this cost category, there is a comparison between different options (products or processes) according to their environmental friendliness. Examples of these costs include higher material costs resulting from the use of environmentally friendly input (e.g. water-based solvents, lead-free gas) and labour and material costs resulting from the use of integrated technologies.

### Level 2: Indirect environmental costs (in overheads)

These costs are not based on the material balance sheet. Instead, they occur in general administration and can be split into pure environmental cost centres (e.g. environmental department) and mixed cost centres where environmental and other costs occur.

**Figure 2:** Definition of Environmental Costs

time, both environmental advantages and disadvantages compared to other options. For example, one production process can produce more waste-water, while the alternative process produces more waste air. It is then very difficult to define which of the two processes can be considered more environmentally friendly. In order to avoid endless discussions and thus casting doubt on the value of the entire project, it is recommended that these costs be considered only at a later stage.

## ▌ Physical and Financial Data Collection

The most important step in an environmental accounting project is the collection of physical and financial data. In many company projects on environmental accounting, it has been proved that a company should not try to review absolutely all physical and financial data in a first attempt: a reasonably fair overview of all inputs and outputs should be sufficient for a first try. Even so, physical and financial data collection is probably the biggest challenge in all pilot projects: it can be very difficult to collect complete and accurate information.[1]

Before beginning detailed physical data collection, different material categories for inputs and outputs must be defined. These definitions are essential, as they identify which material belongs to which category. The definition of material categories used at Sulzer Hydro is shown in Figure 3. Within each of these categories, different forms of material/energy were identified, which can be seen in Figure 4.

The material input–output sheets are much easier to develop than the often-cited life-cycle analysis, as inputs and outputs are listed only if they occur within the defined boundaries, e.g. the production process or the facility. Besides the input and output categories, this data sheet should have columns for the physical amount of every material category (plus the energy equivalent for all energy categories), the material costs, the data source for future references, the data quality and the person responsible for the data. Defining the material balance sheet in such a comprehensive way gives a good overview for future reference. This can be seen in Figure 5.

Usually, in order to acquire the necessary physical and financial data, a number of different data sources have to be used. Nevertheless, during the project, it became obvious that, very often, certain data do not exist in information systems, and a time-consuming process of manually identifying them has to be carried out. In this case possible data sources might be:

- Usage statistics
- Security papers
- Material flow statistics or material flow diagrams in the electronic data processing systems
- Delivery invoices
- Disposal proofs
- Invoices from the disposal company
- Statistics on dangerous substances

1. Probably the most comprehensive eco-balance was made at the Kunert AG, Immenstadt, Germany (Kunert AG 1995).

| Inputs | |
|---|---|
| **Energy** | Energy or sources of energy comprise electrical power and material that is used exclusively for the production of energy. |
| **Material** | Material is any input except energy or customer products. |
| **Semi-finished material** | Semi-finished material is treated by a main process to become part of a product. Semi-finished material used for maintenance of the infrastructure and equipment is auxiliary material. |
| **Finished materials** | Finished materials are products that are assembled together with other components and are generally assigned to become part of the product. Finished material used for maintenance of the infrastructure and equipment is auxiliary material. |
| **Process material** | Process material is used and necessary to run a main process. Process materials are consumed in proportion to the process running. Process material does not become part of the product. |
| **Auxiliary material** | Auxiliary material is used within the auxiliary processes. It comprises all material not defined under the definitions above. |
| **Customer product** | Customer products are machines or components that are the property of the customer. They are provided to the company for a specified treatment such as repair, revision or assembly. |
| **Outputs** | |
| **Products** | Products are machines or components manufactured in accordance with the customer's and the company's specifications. Finished products become the property of the customer. |
| **Recycling material** | Recycling material is waste, which is used as raw material for the production of the same or another product, normally by specialised companies. |
| **Waste** | Waste is any waste for disposal. |
| **Energy** | Energy (i.e. heat). |
| **Waste-water** | Waste-water comprises all waste water including sewage and rain water collected and transferred to the community's sewerage. |
| **Waste air** | Waste air from the heating chimney and end-of-pipe equipment or buildings. |

**Figure 3:** Definition of Material Categories

- Rights to dispose of waste-water
- Waste water invoices
- Emission statistics

When dealing with energy in the physical data collection, a useful definition was that all energy entering a plant is, in the final stage, transformed into heat and thus leaves the plant as waste. Only in certain cases can this definition not be applied: for example, when chemical processes powered by energy transform raw material into products with a higher energy value (e.g. transformation of bauxite into aluminium). So the decision about the way energy is dealt with has to be made according to each company's specific needs.

| Energy and material categories | | | |
| --- | --- | --- | --- |
| **Input** | | **Output** | |
| **1.1 Energy** Electricity Oil Propane Petrol | **1.4 Semi-finished material** Steel Non-ferrous metals Pipes and tubes Welding material | **2.1 Products** Large Hydro Compact Hydro Service FEA (shop job) | **2.6 Energy** Heat |
| **1.2 Process materials** Water Coolants (emulsion) Corundum sand Steel sand Inert gas | Ceramics Paints Synthetics Adhesives **1.5 Finished material** Machine elements | **2.2 Recycling material** Ferrous and non-ferrous metal Wood Glass | **2.4 Waste water** Sewage |
| **1.3 Auxiliary material** Water Spare parts (filters) Tools Replacement material Lubricants Solvents Detergents Packing material (general) Packing material (wood) Office material Paper (office) General commodities | Motors Lubricants (products) **1.6 Customer products** Hydro Service FEA: shop job 1 FEA: shop job 2 | Sandblasting material Paper and cardboard Synthetics **2.3 Waste** Garbage Emulsions Pickle Rubble Fluorescent tubes Gas discharge lamps Toxic waste | **2.5 Waste air** Heating Manufacturing process |

**Figure 4:** Material and Energy Categories at Sulzer Hydro

| 1. Input 1995 | Amount (physical) | Amount (energy) | Amount (financial) (1,000Sfr) | Data source | Data quality | Data provided by |
| --- | --- | --- | --- | --- | --- | --- |
| **1.1 Energy category** | | | | | | |
| **Electricity** **Oil** **Petrol** **Propane** | 468 tonnes 1,618 litres | 3,342 MWh 392 MWh 4.2 MWh | 578 140 1.9 | Meter POs Invoices | Measured Calculated Measured | Heer Wy MMu |
| **1.2 Process material** | | | | | | |
| **Water** **Coolants** **Corundum** **Steel sand** **Inert gas** | | | | | | |

**Figure 5:** System for Data Tracing in a Material Flow Sheet

| Energy source | kWh/kg |
| --- | --- |
| Electricity | – |
| Heating oil light | 11.86 |
| Propane | 12.83 |
| Diesel | 10.75 |
| Acetylene | 12.50 |
| Natural gas | 10.11 |

**Figure 6:** Relationship between Physical Amounts of Energy and their Energy Equivalent

| Category | Physical amount | Unit | Financial amount (Sfr1,000) | Category | HA (Hydro-manufacturing) machines | HA (Hydro-manufacturing) electronics | HS (Hydro-servicing) machines | HS (Hydro-servicing) electronics | FEA (shop job) | FEA (shop job) assembly | Steel and non-ferrous metals | Wood | Glass | Paper | Synthetics | Batteries | Garbage | Emulsion | Oils | Pickle | General waste | Electronic waste | |
|---|---|---|---|---|---|---|---|---|---|---|---|---|---|---|---|---|---|---|---|---|---|---|---|
| | | | | | | | 2.1 Products | | | | | 2.2 Recycling material | | | | | | 2.3 Waste | | | | | |
| **1 Input** | | | | Physical amount | 2,829 | 1.6 | 2,225 | 208 | 1,592 | 467.4 | 692 | 32.4 | 0.8 | 21.19 | 0.40 | 0.02 | 78 | 25.4 | 3.5 | 8.3 | 24 | 1.8 | 1, |
| Energy and physical substances to produce energy | | | | Unit | t | t | t | t | t | t | t | t | t | t | t | t | t | t | t | t | t | t | Nur |
| **1.1 Energy** | | | | | | | | | | | | | | | | | | | | | | | |
| Electricity | 3,341,590 | kWh | | | | | | | | | | | | | | | | | | | | | |
| Oil | 468.2 | t | | | | | | | | | | | | | | | | | | | | | |
| Propane for heating | 2.1 | t | | | | | | | | | | | | | | | | | | | | | |
| Petrol | 1.46 | t | | | | | | | | | | | | | | | | | | | | | |
| Propane for oven | 16.8 | t | | | | | | | | | | | | | | | | | | | | | |
| Acetylene, H, O2 | 1.0 | t | | | | | | | | | | | | | | | | | | | | | |
| Air | 187.3 | t | | | | | | | | | | | | | | | | | | | | | |
| Material (except customer material) | | | | | | | | | | | | | | | | | | | | | | | |
| **1.2 Process material** | | | | | | | | | | | | | | | | | | | | | | | |
| Coolants | 3.6 | t | | | | | | | | | | | | | | | 0.2 | 3.4 | | | | | |
| Beizpaste | 0.05 | t | | | | | | | | | | | | | | | | | | 0.05 | | | |
| Corundum sand | 7 | t | | | | | | | | | | | | | | | | | | | 6.95 | | |
| Steel sand | 6 | t | | | | | | | | | | | | | | | | | | | 5.95 | | |
| Protection gas Argon, CO2, N | 20 | t | | | | | | | | | | | | | | | | | | | | | |
| **1.3 Auxiliary material** | | | | | | | | | | | | | | | | | | | | | | | |
| Rain water | 43,507 | t | | | | | | | | | | | | | | | | | | | | | 6.5 |
| Water | 22,511 | t | | | | | | | | | | | | | | | | 22 | 8.25 | | | | |
| Lubricants | 4.69 | t | | | | | | | | | | | | | | | 1.79 | | 2.9 | | | | |
| Tectyl (valvoline) | 0.105 | t | | | 0.05 | | 0.05 | | 0.005 | | | | | | | | | | | | | | |
| Petrol | 0.48 | t | | | | | | | | | | | | | | | 0.18 | | 0.3 | | | | |
| Solvents VOC | 5 | t | | | | | | | | | | | | | | | | | | | | | |
| Testing material | | | | | | | | | | | | | | | | | | | | | | | |
| Filter (infra) | 0.1 | t | | | | | | | | | | | | | | | 0.1 | | | | | | |
| Filter (runner grinding) | 1 | t | | | | | | | | | | | | | | | 1 | | | | | | |
| Tools | 160.5 | t | | | | | | | 160.5 | | | | | | | | | | | | | | |
| Lamps | 0 | Stk | | | | | | | | | | | | | | | | | | | | | |
| Packaging material general | 2 | t | | | 0.6 | 0.1 | 0.55 | 0.05 | 0.3 | 0.1 | | | | | | | 0.3 | | | | | | |
| Packaging material wood | 25 | t | | | | | | | | | | 24 | | | | | | | | | | 1 | |
| Wood from packaging from suppliers | 9 | t | | | | | | | | | | 8.4 | | | | | | | | | | 0.6 | |
| Office material | 5 | t | | | 0.5 | 0.1 | 0.5 | 0.1 | 0.2 | 0.1 | | | | | 0.2 | | 3.3 | | | | | | |
| Paper (office) | 12.4 | t | | | 1 | 0.2 | 1 | 0.2 | 0.1 | 0.1 | | | | 9.8 | | | | | | | | | |
| Paper from mail | 51.3 | t | | | | | | | | | | | | 11.29 | | | 40 | | | | | | |
| Detergents | 0.73 | t | | | | | | | | | | | | | | | 0.63 | | 0.1 | | | | |
| General commodities | 6 | 0 | | | | | | | | | 2 | 0 | | 0.1 | 0.2 | | 1.2 | | 0.2 | | 0.5 | 1.8 | |
| Glass | 1 | t | | | | | | | | | | | 0.8 | | | | 0.2 | | | | | | |
| **1.4 Semi-finished material** | | | | | | | | | | | | | | | | | | | | | | | |
| Steel | 4,000.00 | t | | | 1,800 | | 308.7 | 100.0 | 1,070.0 | 210.8 | 509.0 | | | | | | | | | | | | |
| Non-ferrous metal | 50.00 | t | | | 14.0 | 0.1 | 14.0 | 1.0 | 13.9 | 2.0 | 5.0 | | | | | | | | | | | | |
| Welding material | 100.00 | t | | | 40.0 | | 31.0 | 2.0 | 20.0 | 2.0 | 5.0 | | | | | | | | | | | | |
| Electrodes | 100.00 | t | | | 40.0 | | 32.0 | 2.0 | 19.0 | 2.0 | 5.0 | | | | | | | | | | | | |
| Additives for welding | 100.00 | t | | | 40.0 | | 30.0 | 2.0 | 23.0 | | 5.0 | | | | | | | | | | | | |
| Synthetics | 10.00 | t | | | 2.0 | | 2.0 | | 1.0 | | | | | | | | 5 | | | | | | |
| Ceramics | 0.98 | t | | | 0.45 | | 0.45 | | | | | | | | | | | | | | | | |
| Paints | 8.75 | t | | | 3.2 | 0.1 | 3.5 | 0.6 | 0.5 | 0.3 | 0.5 | | | | | | | | | | | | |
| Adhesives | 0.0005 | t | | | 0.0002 | | 0.0002 | | | | | | | | | | | | | | | | |
| **1.5 Finished material** | | | | | | | | | | | | | | | | | | | | | | | |
| Machine elements | 645.50 | t | | | 300 | | 215 | 25 | 100 | | | | | | | | 5 | | | | 0.5 | | |
| Motors | 587.50 | t | | | 300 | | 200 | 22 | 60 | | | | | | | | 5 | | | | 0.5 | | |
| Smaller parts | 263.90 | t | | | 100 | | 100 | 10 | 45.595 | | | | | | | | 7.8 | | | | 0.5 | | |
| Components | 166.00 | t | | | 100 | 0.5 | 40 | 25 | | | | | | | | | | | | | 0.5 | | |
| Electrical components | 164.30 | t | | | 100 | 0.5 | 40 | 18 | | | | | | | | | 5.3 | | | | 0.5 | | |
| Lubricants (products) | 14.55 | t | | | 7.2 | | 6.3 | 0.1 | | | | | | | | | 1 | | | | | | |
| **1.6–1.7 Customer products** | | | | | | | | | | | | | | | | | | | | | | | |
| **1.6 Parts for maintenance** | | | | | | | | | | | | | | | | | | | | | | | |
| HS (Hydro-servicing) Switzerland | 1,100 | t | | | | | 1,100 | | | | | | | | | | | | | | | | |
| HS (Hydro-servicing) Export | 100 | t | | | | | 100 | | | | | | | | | | | | | | | | |
| **1.7 New components (for assembly)** | | | | | | | | | | | | | | | | | | | | | | | |
| FEA Hatebur (Shop Job 2) (including oils) | 238.4 | t | | | | | | | 238.4 | | | | | | | | | | | | | | |
| FEA (Shop Jobs) others | 250 | t | | | | | | | | 250 | | | | | | | | | | | | | |
| Total | | | | | 2,849.0 | 1.6 | 2,225 | 208 | 1,592.0 | 467.4 | 692 | 32.4 | 0.8 | 21.19 | 0.4 | | 78 | 25.4 | 3.5 | 8.3 | 24.0 | 1.8 | 1 |
| | | | | | 2,829.0 | 1.6 | 2,225 | 208.0 | 1,592.0 | 467.4 | 692.0 | 32.4 | 0.8 | 21.2 | 0.4 | | 78.0 | 25.4 | 3.5 | 8.3 | 24.0 | 1.8 | 1 |
| Deviation in % | | | | | 0.7 | 0.0 | 0.0 | 0.0 | 0.0 | 0.0 | 0.0 | 0.0 | 0.0 | 0.0 | 0.0 | 0.0 | 0.0 | 0.0 | 0.0 | 0.0 | 0.0 | 0.0 | 0.0 |

| | 2.3 Waste (continued) | | | 2.4 Waste water | | | 2.5 waste air | | 2.6 Energy | | | | | | | | | | | | | | | | | |
|---|---|---|---|---|---|---|---|---|---|---|---|---|---|---|---|---|---|---|---|---|---|---|---|---|---|---|
| Gas discharge lamps | Waste from ceramics (filter) | Waste from ceramics (from blasting) | Toxic waste | Sewage | Water evaporation | Rain water | Waste air from processes | Waste air from heating | Ventilation | Heating | Water | Lighting | Other uses in production | Offices | EDP | Other uses | Compressed air production | Cantina | Noise | Heating production site | Heating offices | Heating stock | Welding | Heat treatment | Total (t; kWh) | Deviation in % |
| 106 | 0.45 | 1.25 | 0.01 | 21,981 | 9,000 | 25,000 | 88.15 | 99.2 | 275,400 | 250,000 | 12,400 | 437,830 | 1,377,130 | 769.030 | 122,750 | 16,700 | 71,280 | 9,070 | 0 | 374.6 | 93.6 | 2.1 | 21.0 | 16.8 | | |
| Number | t | t | t | t | t | t | million m³ | million m³ | kWh | kWh | kWh | kWh | kWh | kWh | kWh | kWh | kWh | kWh | dBA | t | t | t | t | t | | |
| | | | | | | | | | 275,400 | 250,000 | 12,400 | 437,830 | 1,377,130 | 769.030 | 122,750 | 16,700 | 71,280 | 9,070 | | 374.6 | 93.6 | | | | 3,341,590 | 0.0 |
| | | | | | | | | | | | | | | | | | | | | | | | | | 468.2 | 0.0 |
| | | | | | | | | | | | | | | | | | | | | | | 2.1 | | | 2.1 | 0.0 |
| | | | | | | | | | | | | | | 1.46 | | | | | | | | | | | 1.46 | 0.0 |
| | | | | | | | | | | | | | | | | | | | | | | | | 16.8 | 16.8 | 0.0 |
| | | | | | | | 88.15 | 99.2 | | | | | | | | | | | | | | | | | 1 | 0.0 |
| | | | | | | | | | | | | | | | | | | | | | | | | | 187.35 | 0.0 |
| | | | | | | | | | | | | | | | | | | | | | | | | | 3.6 | 0.0 |
| | | 0.05 | | | | | | | | | | | | | | | | | | | | | | | 0.05 | 0.0 |
| | | 0.05 | | | | | | | | | | | | | | | | | | | | | | | 7 | 0.0 |
| | | | | | | | | | | | | | | | | | | | | | | | | | 6 | 0.0 |
| | | | | | | | | | | | | | | | | | | | 20.0 | | | | | | 20 | 0.0 |
| | | | | | 18,500 | 25,000 | | | | | | | | | | | | | | | | | | | 43,507 | 0.0 |
| | | | | 21,980.8 | 500 | | | | | | | | | | | | | | | | | | | | 22,511 | 0.0 |
| | | | | | | | | | | | | | | | | | | | | | | | | | 4.69 | 0.0 |
| | | | | | | | | | | | | | | | | | | | | | | | | | 0.105 | 0.0 |
| | | | | | | | 5 | | | | | | | | | | | | | | | | | | 0.48 | 0.0 |
| | | | | | | | | | | | | | | | | | | | | | | | | | 5 | 0.0 |
| | | | | | | | | | | | | | | | | | | | | | | | | | 0.1 | 0.0 |
| | | | | | | | | | | | | | | | | | | | | | | | | | 1 | 0.0 |
| 06 | | | | | | | | | | | | | | | | | | | | | | | | | 160.5 | 0.0 |
| | | | | | | | | | | | | | | | | | | | | | | | | | 2 | 0.0 |
| | | | | | | | | | | | | | | | | | | | | | | | | | 25 | 0.0 |
| | | | | | | | | | | | | | | | | | | | | | | | | | 9 | 0.0 |
| | | | | | | | | | | | | | | | | | | | | | | | | | 5 | 0.0 |
| | | | | | | | | | | | | | | | | | | | | | | | | | 12.4 | 0.0 |
| | | | | | | | | | | | | | | | | | | | | | | | | | 51.3 | 0.0 |
| | | | | | | | | | | | | | | | | | | | | | | | | | 0.73 | 0.0 |
| | | | | | | | | | | | | | | | | | | | | | | | | | 6 | 0.0 |
| | | | | | | | | | | | | | | | | | | | | | | | | | 1 | 0.0 |
| | 0.45 | 1.07 | | | | | | | | | | | | | | | | | | | | | | | 4000 | 0.0 |
| | | | | | | | | | | | | | | | | | | | | | | | | | 50.00 | 0.0 |
| | | | | | | | | | | | | | | | | | | | | | | | | | 100.00 | 0.0 |
| | | | | | | | | | | | | | | | | | | | | | | | | | 100.00 | 0.0 |
| | | | | | | | | | | | | | | | | | | | | | | | | | 100.00 | 0.0 |
| | | 0.08 | | | | | | | | | | | | | | | | | | | | | | | 10.00 | 0.0 |
| | | | 0.0005 | | | | | | | | | | | | | | | | | | | | | | 0.98 | 0.0 |
| | | | 0.0049 | | | | | | | | | | | | | | | | | | | | | | 8.75 | 0.0 |
| | | | 0.0001 | | | | | | | | | | | | | | | | | | | | | | 0.0005 | 0.0 |
| | | | | | | | | | | | | | | | | | | | | | | | | | 645.5 | 0.0 |
| | | | | | | | | | | | | | | | | | | | | | | | | | 587.5 | 0.0 |
| | | | | | | | | | | | | | | | | | | | | | | | | | 263.895 | 0.0 |
| | | | | | | | | | | | | | | | | | | | | | | | | | 166 | 0.0 |
| | | | | | | | | | | | | | | | | | | | | | | | | | 164.3 | 0.0 |
| | | | | | | | | | | | | | | | | | | | | | | | | | 14.55 | 0.0 |
| | | | | | | | | | | | | | | | | | | | | | | | | | 1100 | 0.0 |
| | | | | | | | | | | | | | | | | | | | | | | | | | 100 | 0.0 |
| | | | | | | | | | | | | | | | | | | | | | | | | | 238.4 | 0.0 |
| | | | | | | | | | | | | | | | | | | | | | | | | | 250 | 0.0 |
| 06 | 0.45 | 1.25 | 0.0055 | 21,981 | 19,000 | 25,000 | 93.1 | 0 | 0 | 0 | 0 | 0 | 0 | 0 | 0 | 0 | 0 | 0 | 0 | 0 | 0 | 0 | 21 | 16.8 | | |
| .0 | 0.5 | 1.3 | 0 | 21,981 | 19,000 | 25,000 | 88.1 | 99.2 | 275,400 | 250,000 | 12,400 | 437,830 | 1,377,130 | 769.030 | 122,750 | 16,700 | 71,280 | 9,070 | 0.0 | 374.6 | 93.6 | 2.1 | 21.0 | 16.8 | | |
| .0 | 0.0 | 0.0 | -81.8 | 0.0 | 0.0 | 0.0 | 5.4 | 0.0 | 0.0 | 0.0 | 0.0 | 0.0 | 0.0 | 0.0 | 0.0 | 0.0 | 0.0 | 0.0 | | 0.0 | 0.0 | 0.0 | 0.0 | | | |

**Figure 7:** Sulzer Hydro's Complete Material Balance Sheet

There is a second problem with energy: it can exist in two different forms; namely, as power (kWh) and as physical material (kg). Some energy inputs exist in both forms (e.g. heating oil), while others have no physical mass (e.g. electricity). We decided therefore to identify energy in both forms where possible and to put both values into the material flow sheet. The relationship between the physical amount of energy and the energy equivalent can be seen in Figure 6, which lists some of the energy categories.

The complete material balance sheet for Sulzer Hydro can be seen in Figure 7 (with the exception of financial figures). Here we can see that the total weight of the inputs in all material categories equals the total weight of the outputs.[2] This means there is an easy way of checking the accuracy of the data. In order to arrive at plausible data, we continued to change the weights of the inputs and outputs until the difference between the total input and output of one material category was less than 1%. The difference in percentage values can be seen in the row marked 'Deviation'.

## ▌ Identification of Environmental Costs

In order to trace environmental costs to their sources, the next step is to identify the scope of environmental costs based on the acquired data. In order to do so, a first intermediate step is to link the energy and material data that has been collected on a plant level to the identified processes. It is especially important when building up these linkages to identify the 'non-product outputs', as they are a major source of environmental costs. As it is very easy to get lost in too much detailed data, it is recommended to concentrate on the most important inputs and non-product related outputs (waste) in each of the processes. An example for such a list of inputs into single processes and non-product-related outputs can be seen in Figure 8.

| | Processes | Inputs | Non-product outputs/waste |
|---|---|---|---|
| **Manufacturing and assembly** | Machining and machine assembly | Energy, lubricants, coolants, detergents, tools | Metal scrap, oil, toxic waste, heat |
| | Welding | Energy, welding material | Metal scrap, heat |
| | Micro-cast | Energy, welding material | Heat |
| | Runner grinding | Energy, tools | Filters, air |
| | Ceramic coating | Energy, ceramics | Ceramic waste |
| | Heat treatment | Energy | Heat |
| | Corrosion protection | Energy, paint, solvents, air | Waste air, detergents, paint |
| | Sandblasting | Energy, corundum, steel sand | Corundum, steel sand |
| | Vapour cleaning | Energy, water | Waste water, toxic waste |
| **Auxiliary processes** | Stock receipt | (Semi-)finished material | Wood, cardboard, paper |
| | Testing, examination | Testing material, water | Air, heat |
| | Heating, maintenance | Energy, water, lubricants | Toxic waste, rubble, waste water |
| | Engineering, administration | Energy, paper | Paper, garbage, electrical scrap |
| | Packing, dispatch | Energy, packing material | Wood, cardboard |
| | Model test laboratory | Energy, water | Waste water |

**Figure 8:** Sample Linkage of Processes, Inputs and Non-Product Outputs

2.  Note that this statement is not necessarily true if chemical processes transform one material category into another.

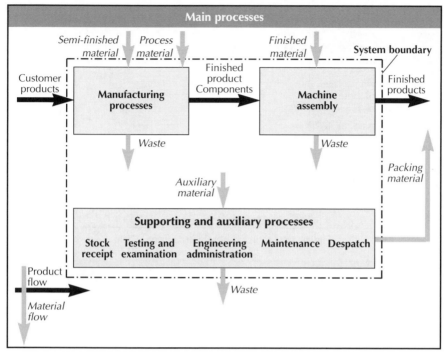

**Figure 9:** System of Linkage between Processes, Inputs and Non-Product Outputs

In order to identify the amounts of inputs and non-product outputs, a chart showing the company-specific production process can be of help. Figure 9 is an example of such a chart. Here, the horizontal axis shows the flow of products and the vertical axis the flow of all materials except the products themselves within the defined boundaries.

The next step is to calculate the costs according to the cost categories defined in Figure 2. Here, the data from the material and energy balance sheet and the additionally collected financial data must be combined. The best method is to use tables in which the data is aggregated, and examples of these are shown in Figures 10–14.

An analysis of the environmental costs can be made based on these aggregated tables. This analysis may include a detailed review of the most important environmental costs, an identification of processes where most environmental costs occur and an analysis of relations between different categories of environmental costs. At Sulzer Hydro, the values shown in Figure 15 were identified for the material costs of all inputs and outputs (as a percentage of the overall material costs). Based on this data, the environmental costs can be calculated as shown in Figure 16 (as a percentage of total turnover). We can see that, at Sulzer Hydro, the non-value-adding costs amount to 4.8% of the total turnover.

## ■ Identification of Savings

As a next step, the team identified the economically most important waste streams, which were metal waste and energy usage. The dominating factor at Sulzer Hydro was not disposal costs, but the purchase values of the residual materials. Thus, avoidance of residual

| | Waste | | | | | |
|---|---|---|---|---|---|---|
| | Waste stream A | Waste stream B | ... | Waste water | Waste air | Waste heat |
| Disposal costs (costs for waste collection, storage, transport and disposal, fees, taxes) | | | | | | |
| Company's wages, related to waste disposal (without administration) | | | | | | |
| Costs of input (material, which leaves the factory as waste) | | | | | | |
| **Total** | | | | | | |

**Figure 10:** Environmental Costs from Waste

| | End-of-pipe technology 1 | End-of-pipe technology 2 | End-of-pipe technology 3 | ... |
|---|---|---|---|---|
| Purchase value (SFr) | | | | |
| Usage time (years) | | | | |
| Yearly depreciation | | | | |
| Interest rate on unit | | | | |
| Space requirement (m$^2$) | | | | |
| Costs of space (SFr per m$^2$) | | | | |
| Costs for space (SFr per year) | | | | |
| Costs related to expected future repairs | | | | |
| Wages in connection with facility | | | | |
| Special costs | | | | |
| **Annual costs** | | | | |

**Figure 11:** Environmental Costs from the End-of-Pipe Treatment of Waste

| | Recycling stream 1 | Recycling stream 2 | ... |
|---|---|---|---|
| Input material costs for waste recycled | | | |
| Costs/income from recycling programmes | | | |
| Wages in connection with recycling programmes | | | |
| **Total** | | | |

**Figure 12:** Environmental Costs from Recycling

| | Area 1 | Area 2 | Area 3 |
|---|---|---|---|
| Costs for future compliance | | | |
| Costs related to take-back guarantees | | | |

**Figure 13:** Other Environmental Costs

| Level 2: Indirect environmental costs (overhead) | | | |
|---|---|---|---|
| | **Department or manager 1** | **Department or manager 2** | **...** |
| Salaries | | | |
| Training | | | |
| Costs for environmental managers in headquarters | | | |
| Costs for identification of liabilities | | | |
| ... | | | |
| **Total** | | | |

**Figure 14:** Indirect Environmental Costs

| Balance in financial terms (material value) | |
|---|---|
| **Input categories** | **Percentage of Input** |
| I-1 Energy | 1.20% |
| I-2 Process material | 0.26% |
| I-3 Auxiliary material | 3.32% |
| I-4 Semi-finished material | 37.58% |
| I-5 Finished material | 57.65% |
| **Total input** | **100.00%** |
| **Output categories** | **Percentage of output** |
| O-1 Products | 91.52% |
| O-2 Recycling material | 5.66% |
| O-3 Waste | 1.30% |
| O-4 Waste water | 0.02% |
| O-5 Waste air | 0.30% |
| O-6 Energy | 1.20% |
| **Total output** | **100.00%** |

**Figure 15:**
Total Direct Environmental Costs

| Cost category | | Total |
|---|---|---|
| **1** | **Direct environmental costs** | |
| 1.1 | Environmental costs from waste | 0.67% |
| 1.2 | Environmental costs from end-of-pipe-treatment of waste and waste water | 0.16% |
| 1.3 | Environmental costs from recycling | 3.28% |
| 1.4 | Other environmental costs | 0.36% |
| | **Total direct environmental costs** | **4.47%** |
| **2** | **Indirect environmental costs** | |
| 2.1 | Pure environmental cost centres | 0.12% |
| 2.2 | Mixed cost centres | 0.21% |
| | **Total indirect environmental costs** | **0.33%** |
| | **Total environmental costs** | **4.8%** |

**Figure 16:**
Total Environmental Costs at Sulzer Hydro

materials is considerably more profitable than was assumed before the project. Traditional cost accounting showed only the 'tip of the iceberg' of the environmental cost.

Basically, two areas for savings were identified:

1. MicroGuss™, with its high consumption of electrical energy

2. Ceramic coating, with the high value of ceramic waste

The concrete proposals for improvement were:

1. Reduction of recycling material related costs by 44%

2. Reduction of waste-related costs by 10%

3. Cost reduction related to electrical energy by 13%

The total savings are estimated to be approximately 33% of the environmental costs and about 1.6% related to turnover. The project proved that environmental management does not necessarily increase costs, but can rather lead to savings and trigger innovation. In order to realise these savings, companies must view environmental issues from a different perspective.

Apart from the identification of possible savings, the general findings in the pilot project in environmental accounting were that a precise definition of energy, material, product and process categories is helpful and, in fact, a requirement. Generally speaking, it was observed, that the company's existing material classifications and categories do not reflect the needs for an environmental classification. Means to improve this situation are efficient periodical update of data and adaptations of the data processing system and related education of the users. Information and expertise gained can also be useful for other environmentally-related projects such as eco-balances, the introduction of an environmental management system and ISO 14001 or EMAS certifications.

# 18 Life-Cycle Costing and Packaging at Xerox Ltd[1]

Martin Bennett and Peter James

## ▌ Introduction

XEROX LTD is a subsidiary of Xerox Corporation, selling its products to markets in Europe, the Middle East and Asia—some eighty countries in all. It employs 21,000 people and has a turnover of approximately £3 billion. It began as a 50–50 joint venture between Xerox and the British company, Rank plc, and was known as Rank Xerox, but is now a wholly owned subsidiary of Xerox Corporation.

Xerox Ltd defines its purpose in this way:

> The primary mission of Xerox is to develop, manufacture, market and service a range of document processing products. The principal goals of the company are: customer satisfaction; employee satisfaction and motivation; market share; return on assets.

Its core business is manufacturing photocopiers, most of which are leased rather than sold to customers. However, it now defines itself as 'The Document Company' and provides a wide range of products and services that allow business customers to capture, create, store, print and distribute documents. The company has undergone con-

1. This case study was prepared for the 'Eco-Management Accounting as a Tool of Environmental Management' (ECOMAC) Project, sponsored by the Commission of the European Union (DGXII, Environment and Climate; contract no. ENV4-CT96-0267). Further information on the ECOMAC Project is available from the authors of this chapter (e-mail: m.bennett@wlv.ac.uk) or from the project co-ordinator, Dr Teun Wolters (e-mail: two@eim.nl).

siderable rationalisation over the last decade. It now operates manufacturing plants in Egypt, India, the Netherlands, Spain and the UK, and research and development centres in France and the UK.

This chapter reports on a project to reduce the packaging used within the European logistics function. In Europe, Xerox manufactures copiers from two factories in Mitcheldean (UK) and Venray (Netherlands) respectively, which then ship the bulk of their output (apart from a small proportion of shipments direct to customers) to the European Logistics Centre (ELC), also at Venray. The ELC then ships to Xerox's operating company in each of eighteen countries. These operate a total of 68 'platforms', i.e. delivery points, which are run for Xerox by contractors who also handle other companies' goods. Each contractor has trained delivery crews with specialised vehicles for installing Xerox equipment. The change four years ago to the present system of 68 platforms has increased the demands on the company to reduce (or, ideally, to eliminate) packaging.

Xerox has an enviable reputation as one of the leaders in Western industry in quality management and continuous improvement. One driver of this is the embedding of benchmarking in the organisational culture—Xerox managers refer to 'the benchmark' as a regular term in their corporate language.

Another process is a commitment to multidisciplinary teams (Quality Improvement Teams [QITs]) as a way of cutting across the clearly defined functional boundaries in order to identify opportunities for improvement. A QIT will be set up to address a problem—usually, though not invariably, cost-related—which is often based on a prior benchmarking process. Brian Rolfe, an accountant and strategist in corporate logistics who was responsible for setting up their financial system (see below) identifies the value of QITs as their ability to:

- look across functions
- involve people from all areas, including those who will have to implement any changes in practice.

## ▌ Environmental Systems and Tools

Xerox is often seen as one of Britain's best-managed companies. It was the first winner of the European Quality Award in 1992. In doing so, it drew upon the experience of Xerox Corporation, which was a pioneer of total quality management (TQM). Like the Xerox Corporation, Xerox Ltd tries to apply this TQM approach to environmental management. Each of its main business units, sites, etc. undertakes an annual self-assessment and validation exercise, based on the European Quality Award template. 'Impact on society', which incorporates both environment and community affairs, forms one component of this assessment.

Xerox operates on the basis of five environmental principles:

1. Protection of the environment and the health and safety of our employees, customers and neighbours from unacceptable risks takes priority over economic considerations and will not be compromised.

2. Xerox operations must be conducted in a manner that safeguards health, protects the environment, conserves valuable materials and resources, and minimises risk of asset losses.

3. Xerox is committed to designing, manufacturing, distributing and marketing products and processes to optimise resource utilisation and minimise environmental impact.

4. All Xerox operations and products will be, at a minimum, in full compliance with applicable governmental requirements and Xerox standards.

5. Xerox is dedicated to continuous improvement of its performance in Environment, Health and Safety.

Xerox has a long history of 'unintentional environmental improvement' as a result of its leasing of its copiers. By retaining ownership, the company was encouraged to re-use either whole machines or components and materials within them. This has resulted in a highly successful copier reprocessing centre at Venray, Netherlands. Specifically, environmental initiatives were introduced in the 1970s and 1980s—including a programme of environmental auditing in 1985—but the major impetus came in 1990, when the company established an Environmental Leadership programme. The goal of this programme has been summarised as 'waste-free products from waste-free factories', and it has resulted in a number of initiatives:

- Introduction of environmental management systems at all major sites, with the aim of achieving certification to ISO 14000 and EMAS

- Elimination of hazardous materials from manufacturing

- A design-for-environment initiative to minimise environmental impacts over the life-cycle and to maximise the durability and re-usability of components, for example by making them easy to disassemble, and interchangeable and compatible across a wide range of products

- A number of recycling, waste minimisation and packaging reduction initiatives. As a result, landfill from the company's own activities was reduced by 75% between 1993 and 1997.

## ▌ The Environmental Challenge and Issues

The experience of reprocessing and recycling copiers has taught Xerox the importance of understanding the full costs and benefits of environment-related activities. Environmental and packaging staff decided to apply this logic to the packaging area, which was both a major cost and the target of proposed European and national legislation in the late 1980s and early 1990s. As a result, there have been a number of success stories:

- Standardisation of re-usable transport packaging from over 8,000 variants down to ten has generated savings of $2.1 million per annum and greatly reduced waste.

- Increased use of 100% recycled packaging materials, reducing weight by an average of 5 kilos and cost by $20 per unit

- Introduction of two standard re-usable totes for internal transport, saving $1.2 million per annum (the subject of this chapter)

A key part of these initiatives has been to go beyond the company's conventional unit costing procedures and to encompass full acquisition cost, based on life-cycle costing. This is discussed in more detail in the following sections.

## ▮ Management Accounting Systems

At a corporate level, Logistics is organised into four 'channels'. These channels differ from the basis on which the operational side is organised, which would be inappropriate in Logistics where the economies of scale are cross-functional. These channels are:

- **Direct:** from the ELC in Venray to each operating company, or (increasingly) direct to customer
- **Indirect:** a relatively new channel, set up to deliver small machines, in bulk, to dealers. The aim in setting this up is to help to develop a low-cost, no-frills logistics service for small machines.
- **Service:** spares such as photo-receptors and toners, distributed to Xerox's service engineers
- **Returns:** carcasses of machines returned at the end of their lives, which are then either remanufactured or stripped down in order that their component parts can be re-used.

Although the costs of logistics are incurred by operating companies and reflected in their reported profits, the Logistics function is managed from the centre. The reason is to combine two potentially conflicting aims.

First, Xerox's philosophy is to devolve responsibility to operating companies, which is reflected in evaluation of their performance against profit and revenue targets. On the other hand, in logistics, economies of scale are obtained by being holistic in looking at the whole of the supply chain. The ability to achieve this is one of the ways in which the corporate centre is able to add value to the business operations. One simple example is the ability to identify when a delivery from Venray (in the Netherlands) to a distant destination (e.g. in France) can also carry, and drop off en route, a separate delivery to a nearer destination, e.g. in Belgium; if logistics planning were left entirely to the operating companies, the outcome could be two separate lorries used instead of only one. Although the operating companies' managements are evaluated primarily on profit performance, a close corporate control is still maintained on the procedures followed.

At the same time, the operating companies are allowed to purchase from outside (e.g. from external parcels carriers) if they find this more cost-effective than using centrally provided logistics services, which makes it imperative that the centre is able to understand its own costs. For this reason, activity-based costing (ABC) has become increasingly more important as a way of allocating costs. The aim is that all cost eventually ends up back with the business (i.e. the operating company) which means that a better basis of cost allocation, even if not full ABC, is needed. These analyses have meant that there have been some areas where the centre has deliberately chosen not to compete with alternative outside suppliers—this is what stimulated the establishment of the new 'Indirect' channel, whose function is to offer a low-cost no-frills service that central Logistics does not consider itself able to offer economically.

Understandably, operating company managers tend to be cautious in adopting radically new policies, so that a part of the role of the central Logistics function is to develop new initiatives and encourage their adoption by the operating companies. However, if the potential success of changes can be established through pilot projects, the potential benefits realisable are then incorporated into the profit targets against which they are evaluated, which are based on:

- Anticipated volumes
- An expected regular rate of improvement in productivity of 3%–5% per annum
- The potential benefit of any centrally driven improvements in practice

Up to a decade ago, the total cost of logistics was not apparent at the centre, since these were embedded in product costs rather than being calculated separately. Estimates of their total amount ranged widely, from $100 million per annum (based on the results of a benchmarking exercise) up to $600 million per annum. Those close to this area of the business were convinced that the benchmarking figure significantly underestimated the true position, which they felt was likely to be towards the higher end of this range, but in the absence of a systematic way of capturing and reporting costs were unable to substantiate this. Brian Rolfe was charged with setting up a data collection and reporting system. It took some time to achieve this—two years simply to agree definite boundaries—and at the start the necessary data had to be collected by questionnaires and direct inquiry rather than through a formal system. However, after three years, this had been integrated into the general ledger coding systems and was part of Xerox's regular systematic management information.

The existence of this system means that, although logistics costs are largely incurred and captured at operating-company level, all the figures are pulled through the centre, who are therefore enabled to identify opportunities for cost savings and operational improvements. This provided a key resource for the Quality Improvement Team that was charged with finding cost savings in logistics.

## ▌ Problem Definition

Much of the existing environment-related management accounting literature reports on projects to minimise wastes and maximise energy efficiency. These are clearly worthwhile for both environmental and economic reasons, but are limited to identifying clearly 'low-hanging fruit'. For further gains, more thorough analyses of business processes and costs will be needed.

There are frequent references in the literature to the theory of life-cycle costing as a way of identifying costs more broadly than conventional business accounting methods do, and of identifying long-term strategic opportunities. This project was based on an analysis of costs over the whole of the relevant section of Xerox's logistics chain ('whole-chain costing'), which, in the context of logistics, is equivalent to product life-cycle costing.

## ▮ Life-Cycle Costing and Packaging at Xerox[2]

### Quality Improvement Team on Logistics and Packaging

Based on the figures that were made visible by the new financial reporting system, logistics was targeted as an area where significant cost savings could be achieved by taking a holistic view of the logistics chain as a whole. Based on the figures generated by the system, the cost of logistics was estimated to have been approximately 12%–13% of revenues in 1989. A long-term target of 6% was set and, after taking a few 'low-hanging fruit' opportunities, such as contracting out rather than self-managing warehousing, this was quickly reduced to 9%.

To identify further improvements, a QIT was set up and tasked with the target of taking $16 million per annum out of the total delivery process. This covered eight distinct areas of activity, including packaging, where savings of $1 million were targeted. It was soon obvious that this could not be achieved by the packaging procurement team alone but needed a cross-functional response. Merely incremental improvements would not be sufficient to achieve this—they realised that the team would have to look 'outside the box'.

A large multi-functional project team looked at all parts of the logistics cycle, including end-delivery crews, manufacturing and the operating companies' platforms. According to Environmental Manager Irina Maslennikova:

> Nobody had ever looked at things in this way. Each platform did their own thing and there were a number of outside contractors. As soon as people saw what the flows were, they realised that not only could there be cost savings but that it was environmentally sub-optimal.

The initial focus of attention was the 'Direct' channel, which was not only the largest and most costly channel but had also reported some dissatisfaction from customers which the company was keen to address. The QIT utilised quality costing principles to undertake an initial cost analysis, which showed that 50% of the total distribution cost was in the final stage (delivery to customers). Understanding this required information from the contractors who handled this stage—Irina commented that 'it was sometimes easier to get information from contractors than internally'!

A careful examination of polystyrene packaging also revealed that, although its purchase cost was low, total acquisition costs were high because manufacturers would deliver styrene only in bulk—hence secure storage facilities were needed. In addition to this barrier of organisational fragmentation, a further barrier was a common feeling that contractors' packaging practices were not an issue for Xerox. For Irina, 'it was difficult getting a realisation that packaging costs are internalised into contractors' tariffs. But the more they can reduce them, the lower the costs can be for Xerox.'

---

2. The totes project has evolved over several years, with a series of iterative and mutually dependent processes of cost analysis, trial, and design and redesign of both the new type of package (the tote) and the system in which it was used. However, for simplicity of exposition, the next three subsections deal with each of these aspects separately, rather than attempt a chronological account of the project. In reading the following it should be borne in mind that each of these were moving in parallel over a period of several years.

## The 'Totes' Solution

Packaging is a crucial aspect of product quality for Xerox. With complex high-value products being shipped across Europe, it is essential that they are protected against possible damage in transit. The previous method of shipping products from Venray to end-customers used several different types of packaging, each specific to the particular product (for larger products, this would include a pallet on which the machine rested). This meant a total of 23 different pallet and pack sizes from the two manufacturing plants. Some were completely over-packed, while others were secured to a pallet with standard metal brackets and then stretch-wrapped.

When the product was received by the customer, this could sometimes be re-used, if they were also returning the carcass of an old machine, *and* if the packaging of the new machine also happened to fit the old one. However, this would be coincidental: more often it did not, and the packaging would be waste. This meant both the loss of its original cost, and increasingly a problem and cost for the customers who then had to dispose of it. They also had to incur additional cost of new packaging for the return of their carcasses.

Based on information provided on costs at various stages in the chain (see the following section), the solution devised by Bill Starkey, the packaging expert who led this project, was to standardise the packaging used at this stage of the logistics chain in a standard pack or 'tote'. A tote comprises a pallet with collapsible sides and a lid (the sides are collapsible so that on occasions when the customer has no carcass to return, space in transport can be saved). The design of the standard totes has evolved over the period of the trial studies. The customer requirements that Bill Starkey had to meet were:

1. To eliminate all packaging

2. To introduce a single robust, returnable pack that suits the whole product range

3. A pack that can be used for both new machine delivery and also to return carcass hulks to the Asset Recovery Operation

4. Design features incorporating quick-release mechanisms to reduce de-palletisation time

5. Packs must be stackable and collapsible.

One challenge was the range of different products, of different sizes, which the standard tote had to be able to cope with. As well as the current range of 23 product-lines, since Xerox products have a long working life (typically around fifteen years), it had to be able to cope with the return of the carcasses of up to forty old and discontinued lines. For the final design, an analysis was done of all the new product lines likely to be introduced in the next seven years in order to 'future-proof' the design. Now that the totes are an established part of Xerox's operations, new products are being designed in order to fit the tote, rather than vice versa.

It was quickly realised that a single tote design would not be adequate to cope economically with the whole product range, from small desktop copiers to large floor-standing machines. However, with careful design, two basic models of tote were found to be sufficient. For the smaller desktop products, following a benchmarking exercise with both competitors and suppliers, an off-the-shelf product of a timber design produced by a Swedish supplier was tested. Following these tests it was then modified in order to meet all Xerox's customer requirements, with:

- Quick-release straps to secure the product internally
- Quick-release security toggles for ease of assembly and disassembly
- Built-in base cushioning
- Total collapsibility
- Optimum size for Xerox's product range and transportation

For the larger products, a clean-sheet design steel tote was conceived, again incorporating a quick-release mechanism. Optional extensions can be added for greater flexibility so that operating companies can return copiers without first having to remove attachments such as feeders and document handlers. One result of using totes is that de-palletising times have been reduced from twelve minutes per product to thirty seconds.

### Cost Analysis

The totes project was based on a holistic analysis of the costs incurred over the whole of this part of the logistics chain, from:

- Manufacturing
- ELC, at Venray
- The platforms
- Asset Recovery Operations (ARO), at Venray

An example of this analysis is provided in Figure 1. This particular example was carried out at an intermediate stage in the project rather than at its start, but indicates the type of analysis that was also carried out in a more limited form from the outset, and the approach taken throughout. This example is for a particular combination of destination (France) and product (5317 + ADF), and was carried out during Phase 1 of the project (see the following section for a description of the progress of the project).

The main comparison in this analysis is between columns 1 and 2, which represent the costs, respectively, of the previously existing system and of the new totes system. In this case, the whole-chain costs of the tote system, at $47.71 per unit, are substantially less than the $59.79 incurred previously—this is typical of these analyses. Similar analyses were carried out for each possible combination of eighteen destinations and 23 product lines.

*Data Collection.* Completing this analysis required data from across the organisation. Generally, Bill Starkey found that this was not available from the accounting system, but that much was available from operational measures taken by cost engineers, such as their measurements of the time taken at each stage of the process (e.g. that de-palletising took approximately twelve minutes per product). These data had to be pulled together by Bill Starkey from different sources, and in some cases it was found that the data needed were not already being collected so that additional measures had to be made.

| **FRANCE: 5317 + ADF** | | | | |
|---|---|---|---|---|
| **Activity** | **Current pack (1995$/unit)** | **Proposed tote ($/unit)** | **Proposal (better/[worse])** | **Saving/(on-cost) Accounted for in:** |
| Packaging material cost | 14.02 | 6.54 | 7.48 | ELC/Manufacturing |
| Pack time: Mitcheldean | 1.70 | 0.80 | 0.90 | Mitcheldean Manufacturing |
| Transit Mitcheldean to ELC | 3.41 | 5.76 | (2.35) | ELC inbound freight |
| Transit ELC to platform | 3.04 | 5.13 | (2.09) | Operating company inbound freight |
| Unpack time platform | 1.70 | 0.80 | 0.90 | Contractor tariff |
| Platform pack disposal | 5.20 | – | 5.20 | Contractor tariff |
| Platform purchase new packs for carcass returns (pallet and case) | 10.27 | – | 10.27 | Contractor tariff |
| Platform carcass pack time | 3.33 | 0.80 | 2.53 | Contractor tariff |
| Transit platform to ARO (Oostrum) | 3.85 | 6.50 | (2.65) | ELC inbound freight |
| Transit ARO to Mitcheldean | 6.67 | 6.67 | – | Inter-plant freight |
| Platform storage space-saving | 4.00 | – | 4.00 | Contractor tariff |
| Pack disposal at ARO | 2.60 | – | 2.60 | ARO recharge (supply trading) |
| Consolidation/handling costs | – | 0.43 | (0.43) | Manufacturing |
| Tote losses/repair (15%) | – | 1.83 | (1.83) | Tote budget |
| Line balance contingency (10%) | – | 1.22 | (1.22) | Tote budget |
| ARO de-tote and repack activity | – | 6.23 | (6.23) | ARO labour (recharge to Supply Trading) |
| ARO purchase new packs | – | 5.00 | (5.00) | ARO (part of ARO recharge) |
| **Total** | **59.79** | **47.71** | **12.08** | |

**Figure 1:** Cost Analysis

Inevitably some trade-offs were required. For example, standardisation of the previous wide variety of packs into only two designs of tote meant that some over-specification was unavoidable, so that the totes required more space in storage and transport than the previous packs which had been designed specifically to fit a particular product line. The move to totes was therefore not universally profitable for all product lines, and in some areas it meant that costs were increased. However, overall the benefits of standardising and taking a consistent approach exceeded the costs, although finding these exceptions helped to inform the tote design and led to modifications in design.

*Realising the Benefit.* The final column indicates in which part of the Xerox organisation (or outside, with contractors) the costs and potential benefits are realised. This reflects the fact that, although the economics of the logistics operation need to be looked at holistically over the whole of the chain, in a normal organisation the responsibilities for different parts of the chain belong to different sub-units within the organisation. This is where costs are incurred, which are then recorded in their respective accounting systems and charged against the budgets of the respective managers, and used in evaluating their performance. It is because of this that a central holistic view is needed—no individual budget-holder at a single stage in the chain would be able to justify a major change in policy on the basis of only the benefits realised in their own area, even if they were in a position to identify the potential opportunity in the first place.

It is then necessary to ensure that the potential gains are actually realised in practice. This requires that budgets and performance targets are adjusted to take them into account, for internal sub-units. As described above, these targets are usually based on volume increases, an expected annual 3%–5% improvement in productivity, plus any specific centrally driven benefits such as those arising from the use of totes.

For external contractors, this would need to be taken into account in renegotiating contracts. This is now done at a corporate level rather than by each separate operating company (although the costs incurred are ultimately charged to the operating companies). This also provides an opportunity to identify further efficiencies, such as the example of a drop-off in Belgium en route from the Netherlands to France, which was described earlier.

## Three-Phase Experiment

The original QIT was set up, and produced its initial proposals, in 1993. Trials started in 1994. Full implementation of the system across Europe was achieved in early 1997.

Trialing and implementation was carried out in three phases (see Figs. 2 and 3). In Phases 1 and 2, the tote system covered only the distribution sections of the logistics chain, with manufacturing still using the old system based on pallets. Phase 1, for three months, covered only France and Germany, trialing 200 totes of each type. During this trial, close contact was kept with those operating the system, and their feedback was collected through questionnaires and meetings as well as through direct observation. This suggested some redesign, e.g. increasing the size of the smaller totes, and helped to identify which processes had to be changed. Phase 2 extended this to all operating companies in Europe, using 3,000 of each type of tote for an extended trial. Phase 3 brought in manufacturing, at the two factories at Mitcheldean and Venray respectively.

Figures 2 and 3 describe the system, in flowcharts. Figure 2 refers to Phases 1 and 2, when the system covered only distribution, before manufacturing was brought in. The main procedures in the system were:

- Manufacturing: as previously, products were built on pallets and shipped in this form to the ELC.
- ELC: products were de-palletised and transferred onto totes.
- Shipment from ELC to the operating companies' platforms on totes.
- Shipment from operating companies to customers, on totes. The totes are immediately brought back, with the carcasses of the old machines (if any) being returned, and taken to the Asset Recovery Operation (ARO).

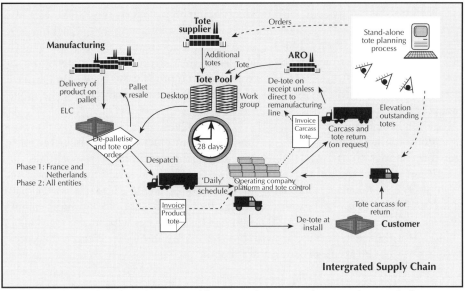

**Figure 2:** Flowchart for Returnable Totes, Phases 1 and 2 (Distribution Only)

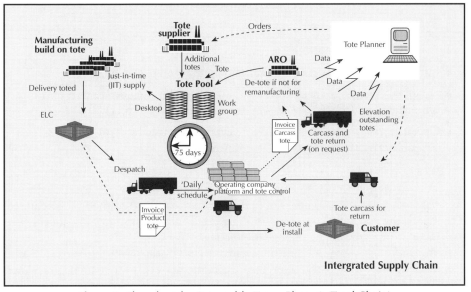

**Figure 3:** Flowchart for Returnable Totes, Phase 3 (Total Chain)

- ARO: remanufacture the products, or disassemble and re-use their parts (depending on the condition and degree of obsolescence of the particular machine). The empty totes are returned to the Tote Pool (at Venray), which has a repair facility for any damaged totes.

- The Tote Pool returns totes to the ELC, where the cycle recommences.

The total cycle time in Phases 1 and 2 was 28 days.

Phase 3 (Fig. 3) added two main features to this process. First, manufacturing was brought into the tote system, with products being built directly onto the totes. This meant that the Phase 1/2 procedure, of the ELC de-palletising and repacking onto totes, is no longer necessary. This required some modest investment at Mitcheldean, of $140,000, in order to widen the roller base of the conveyors, plus some automatic equipment to handle totes. Second, and in part prompted by the need to manage the expanded system, a Tote Planner was appointed to monitor and control the movements of totes.

The total cycle time of the Phase 3 system was planned to be 75 days, to allow for the additional holdings of stocks of totes, though in practice, with close control by the Tote Planner, shorter cycle times have been achieved—on average, 31 days.

The last stage of the full implementation of the tote system was finally completed in early 1997. At full operation, the system now requires in total 28,000 plywood (small) and 20,000 steel (large) totes, representing a total investment of $4–5 million at an average cost of approximately $100 per tote. This compares with an average cost of only $20 for the packs previously used, but this extra capital cost is more than compensated for by the savings in use and re-usability.

**Administrative Procedures.** Within this part of the logistics chain, the totes project has created a totally closed-loop system—the first in Xerox. This has required changes in procedures in several different areas of the company:

- All logistics processes for final delivery
- Asset Recovery Operations
- Introduction of a new tote pool function and its management
- Financial changes, as packaging costs are no longer included as part of the Unit Manufacturing Cost (i.e. the standard cost per product)
- Operating companies are charged for each tote received by them, and receive a credit of the same amount on its return.

On this last point, to maintain accountability for totes in use, each internal invoice includes—as well as the cost of the product—a charge of $130 per tote. This is re-credited to the operating companies, in full, when the tote is returned. This required a change to the company's invoicing systems which, although simple in concept, was a major and time-consuming change in practice.

In the early stages, some administrative investment by the operating companies was also needed, since they had to convince the corporate centre that their procedures were adequate to control the return of totes. Setting up adequate systems took some time for a few operating companies, sometimes reflecting and bringing to light a broader problem of internal control generally.

As part of the centralised tote planning and control process, all totes now have an individual identifying barcode, which is scanned on despatch and at subsequent stages of the process, to monitor their movement and identify at any time who is accountable for them. Tote planning and control is far from a straightforward function, since it is essential that sufficient quantities of totes are available to cope with fluctuations in manufacturing volumes, while at the same time minimising the company's total investment in stocks. To support this, the Tote Planner is linked into the company's manufacturing planning systems.

Bill Starkey considers:

> The Tote Pool is the key to the whole process, since it acts as a preferred supplier to the manufacturing operation. Manufacturing demands fluctuate on a weekly basis, and the Tote Planner has to calculate this demand and also understand what old carcasses are being returned from the field. If a shortfall occurs, it is his responsibility to top up the quantities.

## ▌ Results

The main benefits of the system are quantified and reflected in the whole-chain cost analysis sheets. In total, the investment of $4–5 million in totes is estimated to have generated directly quantifiable annual savings of $1.2 million. There are also some further, more intangible benefits, such as the easier handling of standard totes rather than the previous variety of different pack sizes. Another intangible benefit, related to quality, is that previously, when the larger machines were delivered to customers, these frequently had to be set up by the riggers on-site, from modular packages (from eighteen separate boxes, in the case of one product). This can now be done in the manufacturing stage, in a controlled factory environment. There is also the benefit of reduced administration workload, since the number of different packs to be purchased, stored and handled has reduced. However, it has not been feasible to put a definite monetary value on this.

In addition to the financial benefits, Xerox also sees customer service benefits. According to Irina, 'end-customers are impressed with the reduction in packaging and have more contact with our delivery crews—there's an overall impression of slickness.' Following the initial success, the tote scheme has now been introduced into Xerox's manufacturing operations to move components, etc. between plants.

All the participants were also positive about the process of developing the totes. According to Bill Starkey, 'it was totally participative and generated great motivation and atmosphere. People could see that there were benefits and that they applied to every part of the chain—there were no winners or losers.' A final perceived benefit from Brian Rolfe's perspective was that the packaging innovations have made Xerox's Integrated Supply Chain initiative more visible—this was initially 'lost in accounting in its first two years and didn't have the visibility it needed'.

## ▌ Conclusions

The main conclusions to be drawn from this case are:

- The value of whole-chain costing
- Mixed messages on the value of conventional accounting systems
- The balance between central control and local autonomy
- The importance to the process of a supportive corporate culture

## Whole-Chain Costing

The value of the change to a tote system was established by a holistic analysis of costs over the whole of the chain. These are costs that are actually being incurred by the business, but which in normal financial methods of reporting and analysis are not made visible to management.

## Conventional Accounting Systems

Alternative views were expressed on the usefulness to this project of the existing mainstream accounting systems. On the one hand, Bill Starkey, as leader of the totes project, found that the data that he needed for the cost analyses were not available from the accounting systems, but had to be collected directly from operational sources (although the company's use of standard costing in its accounting systems, and at least limited introduction of activity-based costing, probably meant that there was a higher likelihood that supporting data—for example on process times—were already being captured for other purposes). He attributed this in part to the attitude of those responsible: 'I find that accountants are OK if they have a straightforward problem, but not for anything totally new—you have to go and develop your own data.'

However, the initial identification of logistics, including packaging, as an area worth review was enabled only due to the systems previously set up by Brian Rolfe to collect and report logistics-related costs across the chain as a whole. Prior to this, there had been widely differing perceptions of the total cost being driven by logistics. Accurate measurement and tracking of this meant that the scale of the challenge, and the potential benefits of improvements, could be assessed, and quantitative targets defined (e.g. to reduce logistics costs from 12%–13% of revenue to 6%).

It may be fair to conclude that the outputs from the more obvious, 'back-end' stages of the accounting process will not necessarily be immediately in the most appropriate form for studies such as the totes project, which by definition are looking at the business in new ways which have not (yet) been reflected in formal ongoing reporting processes. However, this is only part of the total system, which rests on a sub-structure of data capture and collection subsystems which may serve several purposes (for example, the reporting and analysis of labour costs is based on data that are generated in the first place for operational reasons, i.e. meeting the regular weekly or monthly payroll). Companies that have developed more sophisticated internal financial reporting systems (such as Xerox's corporate logistics system) are more likely also to have developed support systems whose data can then be used for further purposes.

## Balance between Central Control and Local Autonomy

The totes project raised the endemic management issue of the appropriate balance between central planning and control on the one hand, and local autonomy and accountability on the other. The opportunity was apparent in the first place only because the information was visible at the centre, which was then able to design and establish a system which was then implemented throughout the organisation. Bill Starkey observed that it might have been more difficult to achieve this in the Xerox Corporation in the US, which manages itself in a more devolved style and is therefore less able to control its logistics from the centre.

At the same time, centrally driven changes need to be introduced and implemented with care and tact in order to avoid appearing as corporate intrusion, which might disempower and demotivate local managements. This is more likely if the changes can first be trialed with a limited number of co-operative operating companies before being implemented more widely, as with this project. It is also necessary to ensure that the expected benefits are captured in the budgets and targets for operating companies which are agreed annually between each operating company and the centre. The closer understanding by the centre of the details of operations, and the likely effects of changes that are generated by projects like this, means that they are more likely to be able to set realistically achievable budgets and targets that are credible to operational managers.

## Supportive Corporate Culture

The totes system was devised and developed by a cross-functional Quality Improvement Team. These are an established part of Xerox's culture, and the project benefited from being able to build on this.

# 19 The Cost of Waste at Zeneca[1]

Martin Bennett and Peter James

## ∎ Introduction

ZENECA WAS CREATED following a de-merger from ICI in 1993. It now has a turnover of £5 billion and is seen as one of Britain's most successful and most profitable companies. It has three autonomous businesses: pharmaceuticals (45% of turnover), agrochemicals (34%) and specialities (21%). It also has a separately run American healthcare subsidiary, Salick Healthcare.

The Pharmaceuticals Business produces products in several clinical areas such as cancer, cardiovascular and metabolism, and respiratory, its main customers being hospitals and medical practitioners. The Agrochemicals Business produces herbicides, insecticides, fungicides and seeds for farmers, distributors and retailers. The Specialities Business is a collection of businesses such as biocides, industrial colours and life-science molecules.

The main sponsor of the 'Cost of Waste' initiative has been the Agrochemicals Business which has its two main manufacturing sites in the UK at Grangemouth and Huddersfield. These sites are supported by a Process Technology department which owns the process technologies and is responsible for the support and development of existing products and the delivery of new products to manufacture.

1. This case study was prepared for the 'Eco-Management Accounting as a Tool of Environmental Management' (ECOMAC) Project, sponsored by the Commission of the European Union (DGXII, Environment and Climate; contract no. ENV4-CT96-0267). Further information on the ECOMAC Project is available from the authors of this chapter (e-mail: m.bennett@wlv.ac.uk) or from the project co-ordinator, Dr Teun Wolters (e-mail: two@eim.nl).

Overall, almost 40% of Zeneca's 30,000 employees work in the UK, though its extensive international operations mean that 94% of its annual sales are outside the UK, with the main market being the Americas, followed by Europe.

Zeneca's activities are based on bioscience and involve the creation of mainly high-value biologically active chemicals. Hence, it has high research and development expenditures and its manufacturing processes tend to be complex, multi-stage syntheses with low overall yields and high losses in process. Until recently, high profit margins have minimised financial pressures to reduce these, but this situation is changing as margins are now becoming tighter in several areas. One implication of the high rate of research and development in the industries in which Zeneca operates, together with the complexity of their production processes, is that product lines have limited lives and it may take a substantial portion of this time to move down the learning curve to full process efficiency.

The main corporate environmental issues are summarised thus in its annual report:

> A key element of Zeneca's SHE [Safety, Health and Environment] policy is to seek to reduce environmental impact. For existing manufacturing facilities this is achieved by modifying processes to reduce wastes, and introducing more comprehensive waste treatment programmes such as the new liquid effluent treatment plant at Grangemouth in Scotland, which will remove metals, acidity and organic compounds from the present discharge.
>
> When new and replacement parts are sanctioned, Zeneca's engineers and process technologists are able to apply their skills and knowledge to introduce more substantial environmental improvements. For example, the environmental protection measures built into the new paraquat plant at Huddersfield in the UK include impervious membranes under the working areas to minimise the risk of any contamination of soil or groundwater. These, and other improvements at production sites around the world, are aimed at reducing waste emissions. Zeneca is particularly concerned that it should curtail its hazardous wastes, and has made significant progress in limiting the volume produced. This is particularly encouraging since Group sales (and hence the volumes of products manufactured) have increased in recent years.
>
> All waste emissions, whether from vents, discharges to sewers and water courses, or disposals to landfill, are disposed of safely and in accordance with local regulations, and as such do not normally cause environmental problems. Operating procedures make sure that accidental leaks or spills are minimised and that swift action is taken to contain their effect. All such events are recorded and investigated to pinpoint the cause and apply the lessons learned to make improvements.
>
> Zeneca's concern for safety, health and environment extends to its products and their use by customers. For example, a new water-based formulation of the herbicide 'Fusilade' eliminates the use of organic solvents, improving safety for the user. Micro-encapsulated formulations, where crop protection products such as acetochlor (the active ingredient in 'Surpass') are enveloped in a benign biodegradable polymer, pose much less hazard to spray operators and are more easily degraded in the environment.

Peter Doyle, the corporate Executive Director with overall responsibility for research and development, manufacturing, safety, health and the environment, is also a member of the Royal Commission on Environmental Pollution.

## Zeneca Agrochemicals Business

Zeneca Agrochemicals accounts for approximately one-third of Zeneca's turnover. It is one of the world's largest producers of herbicides, insecticides, fungicides and seeds. Many of its products—paraquat, for example—and intermediate chemicals are highly toxic or otherwise hazardous. It therefore has several business-level safety, health and environment staff—notably the Environmental Manager, Peter Natkanski.

## Zeneca's Huddersfield Works

Huddersfield Works occupies a 100-hectare site and employs around 1,100 production and support staff. It is also the base of the Technical Development department (chemists and process engineers) of approximately 300. The site serves several Zeneca core businesses, as well as manufacturing chemicals for a number of external companies, including ICI. However, around 60% of total output by volume goes to Zeneca Agrochemicals.

The Works contains over thirty individual plants—and 24 processes registered under Integrated Pollution Control (IPC) legislation—producing around 300 different products, with annual volumes ranging from a few hundred kilos to several hundred thousand kilos. All types of chemical and process engineering unit operations are employed, and the multi-business influence ensures a high rate of change in products, chemical routes, product capacities and processing techniques. Most of the facilities are multi-product batch units, but there are also large single-product batch and continuous plants.

The Process Technology department, based at the Works, includes a large semi-technical facility for process and equipment development, and for producing initial small quantities of new products for toxicology assessment and field trials.

Huddersfield Works is divided into shared service and manufacturing cost centres. Shared-service cost centres include Human Resources, Production Planning, Boiler, Landfill Site and the on-site neutralisation and effluent treatment plants. Their costs are allocated to 24 manufacturing cost centres, based on discrete buildings and production processes. They recharge their costs (but without any internal profit add-on) to the businesses who use their services, on a range of bases: actual as far as possible, apportioned on a reasonable basis where not. (They are introducing activity-based costing [ABC] in some areas in order to be able to recharge on more appropriate bases than in the past, and also to help cost centre managers to understand and manage better their costs.) All costs have to be recharged between the businesses, so it is in each business's interest to try to minimise their own share of the total costs (so that a higher proportion would be borne by other businesses).

The boundaries of the manufacturing cost centres have been drawn so that, as far as possible, each manufacturing cost centre services a single Zeneca product and/or business—so that costs are much easier to allocate—but in many cases this has not been possible. In these instances a reasonable basis of apportionment has to be applied. However, all new facilities built at the site will be sole producers for an individual business. Each of these centres has a cost centre manager whose brief is to maximise efficiency and minimise cost while respecting SHE and other constraints.

The site produces its own steam and electricity from a CHP (combined heat and power) facility which incorporates gas turbine technology, and which uses a variety of fuels. Water, nitrogen and compressed air services are also provided. The effluent from individual plants after appropriate pre-treatment is sent to a common site drainage system, where it is mixed with other effluents. This site effluent is pre-treated on the site to neutralise and remove suspended solids, before going via a dedicated sewer to Yorkshire Water Services plc (YWS) for biological treatment. The effluent has to meet a number of legal consents applied by YWS before discharge, mainly on heavy metals, some organics, chemical oxygen demand (COD) and flow. Some liquid effluents are sent off-site for recovery, disposal or destruction. Solid wastes can be handled in a nearby wholly owned state-of-the-art landfill site, or can be sent off-site for landfill or incineration.

## ▌ Environmental Systems and Tools

Zeneca's Safety, Health and Environment policy states that:

> In pursuit of its business objectives, it is Zeneca's policy to manage its activities to give benefit to society ensuring that:
>
> - they meet all relevant laws, regulations and international agreements
> - they are conducted safely, protecting the health of all employees and all persons who may be affected
> - they are acceptable to the community at large
> - their environmental impact is reduced to a practicable minimum at an acceptable cost to the Group and Society.

The company implements this policy through nineteen Group SHE Standards and associated guidelines. The main components are an environmental management system at each site, regular assessment of environmental impacts, and regular auditing. At the end of each year, the chief executive officer of each business reviews SHE performance and delivers a letter of assurance to the Zeneca executive board outlining the extent of compliance with company standards and indicating areas, plans and timescales for improvement.

## ▌ The Environmental Challenge and Issues

The nature of Zeneca's processes means that it generates large quantities of wastes, some of which are hazardous. The Huddersfield Works generated wastes of 180,503 tonnes in 1995, split as follows:

- To air: 3%
- To water: 32%
- To land (including off-site disposal and incineration): 65%

This was a substantial increase over the 1992 level of 143,380 tonnes, largely due to increased production. Tightening environmental legislation is increasing the costs of dealing with these wastes. In addition, the newly formed (in 1996) UK Environment Agency has unified the regulation of emissions and wastes and is encouraging businesses to adopt a multi-media approach to environmental management, and to endeavour to minimise waste at source through clean technology and other measures.

The financial incentives to reduce waste generation have also been increased by changes in Zeneca's business. Market conditions are becoming more competitive, putting pressure on margins. Hence, improvements in manufacturing efficiency are now of greater strategic importance than in the past. When Zeneca was formed, it inherited an old ICI target of 50% 'across-the-board' waste reduction—interpreted as applying to every site. According to Peter Natkanski:

> The pluses of this were high profile, external publicity and creating a climate for change. However, there were a lot of downsides. It had no environmental logic as all wastes were treated the same. It wasn't cost-effective to have the same percentage reduction target for every site: the USA sites had practised waste minimisation since the early '80s and had little potential for a further 50% reduction. And no one was sure what was to be reduced—there was no definition of waste.

These problems led the Agrochemicals Business to redefine its targets in relation to waste minimisation. It was decided to structure waste reduction on a commercial basis, i.e. the new target was a financial saving rather than a percentage reduction figure. This was termed the 'Cost of Waste' programme. In the words of two members of the Process Technology department, Frank Bradburn and Mark Peacock:

> It was clear that the true cost of waste was unknown and that if this were displayed alongside the other business cost parameters, then there was an opportunity for the businesses themselves to target waste minimisation.

## ▌ Management Accounting Systems

### Product Costing at Huddersfield

The accounting system primarily relevant to this case study is the product costing system. This is based on a standard costing system, which aims to track all costs through to processes and products. Costs are collected primarily at cost centre level (i.e. at the level of the individual plant), then broken down between the businesses that use that plant. Costs incurred at a higher level, such as in providing and operating the infrastructure of the site, are tracked through to cost centres. David Hollingdrake, an accountant at the Huddersfield Works, explains:

> So far as possible the aim is to allocate costs to businesses as specifically as possible, so that each business is aware of its own costs. This includes environment-related costs which are allocated (so far as possible) on a 'polluter pays' basis.

The business system SAP is used to record and process operational costs, in a single central system. Business-owned SAP systems are used to list in detail the various materials inputs, to monitor actual usages and to compare these against standard quantities of yields and prices. There are separate headings in the process/product costing system for:

- Materials
- Energy
- Labour
- Capital
- Maintenance

Materials (defined as described below) is the largest single cost, representing approximately 60% of the total operating costs incurred for the Huddersfield Works as a whole. Labour costs represent around 25%, with slightly less than half of this being direct labour, the remainder being indirect labour. Energy costs represent approximately 10%, although this under-represents the total energy consumption since most of the site's requirements are generated by Zeneca's own on-site power-generating plant which produces energy at a substantially lower cost per unit than energy purchased from outside.

The term 'materials costs' is used to include raw materials, packaging and waste products that can be directly associated with the product. In connection with waste costs, for example, the direct cost of tankering out wastes from a particular process is included (and therefore directly tracked to the business); costs such as gypsum, waste sludge and boiler ash, which cannot be directly identified with a particular process or product, are treated as general costs. 'Materials costs' include a variable element of effluent-related costs such as the charges, based on the BOD (biological oxygen demand) and COD (chemical oxygen demand) content of effluents, which are levied by Yorkshire Water Authority. They also include packaging, but not energy costs, since these are not variable with production volumes, at least in the short run. Costs such as energy are tracked to products not through the materials accounting systems but through expenses costing systems.

The system calculates a cost for each product, which can be used to support the preparation of financial reports. Variances between actual performance and pre-set standards are regularly reviewed in order to monitor managerial performance, requiring an explanation for all major variances at every month-end. Where a variance has been persistent over a significant period of time (several months), this may prompt a closer scrutiny during the annual year-end review of standards, which could result in the standards in the recipes being adjusted. This is often needed because of learning improvements as new processes become mature over time and more efficient ways of operating them, including the reduction of wastes, are developed.

However, care is needed in interpreting the significance of variations, which can reflect several possible causes. The interpretation of reported variances—and the appropriate management response—is not always obvious in a highly complex business. For example, each process has an optimal level of volume, above (or below) which yields may be reduced, so that an adverse variance is reported. However, if the demand for a particular product has increased, it may be possible—up to a point—to achieve the extra volume within the existing plant capacity, although at the expense of a reduced process yield. This may, at least in the short term, be more cost-effective than to invest in additional fixed plant. Zeneca's experience is that working beyond optimal capacity also provides a strong practical stimulus to look for creative ways to expand production in the longer term within the existing capacity, and thereby to avoid the need for further investment.

## Allocating the Costs of Landfill

Cost analysis and control exercises such as the Cost of Waste programme depend on accurate cost identification and allocation. One particularly environment-relevant cost is the cost of disposing of solid wastes, which for most non-hazardous wastes occurs at Zeneca's own landfill site close to the Huddersfield Works. The recently introduced UK landfill tax has focused additional attention on this, since it represents a substantial extra cost. Other costs associated with the landfill site are its capital and operational costs.

The landfill site takes both ordinary wastes such as fly ash and demolition wastes, and also most of the wastes defined in the terminology of the landfill tax as 'special'. Some special wastes have to be dealt with off-site, in which case the outlaid cost is charged back to each product's variable cost. The site consists of not one but several separate compartments, which are created and used in sequence: as one compartment is filled (which on average is between three and five years after its opening), the next is started. The main capital cost incurred is the cost of creating compartments, for example in digging, lining and installing extra drains specific to that compartment. This is amortised over its life, though there are also some capital costs for infrastructure, such as drains that serve the site as a whole, and which are expected to last for its whole life.

Allocation of the operating costs of the landfill site, such as the digging equipment used (as for other waste disposal costs, these are incurred at the level of the site as a whole rather than at the level of an individual plant), is based on detailed and accurate records of the incoming wastes and their source (i.e. from which plant each batch of waste comes). This is applied not to recharge actual costs immediately as they are incurred, but for an annual review and (if necessary) adjustment of the apportionments. The 'allocation key' (i.e. basis of apportionment) used for this is weight, since this is the basis on which the landfill tax is charged. The capital costs of creating the compartment(s), such as for digging, lining and installing drainage, are apportioned back on a basis that reflects the volume of wastes as well as their weight, since volume is the main determinant of the life of the compartments.

The capital costs of the landfill site are depreciated and charged out on the basis of the historic cost spread over the useful life. Although Zeneca uses indexation as a method of allowing for the effects of changing prices in appraising potential investments in long-life assets, this is not done in reporting on past performance. In any case, the main capital cost associated with the landfill site is for the creation of compartments. Since these usually have a life of between three and five years, the likely scale of price changes within this period would not justify indexation here.

## Problems of Tracking and Allocation

Tracking and cross-charging costs of providing infrastructure, such as on-site roads and service plants (including the end-of-pipe treatment plants), is inherently problematic on a site such as Huddersfield, since a level of infrastructure that was previously adequate may become inadequate if new plants are built, posing additional demands on the infrastructure. With the significant increase in activity at Huddersfield over recent years (owing mainly to the building of new plants rather than to additional throughput in the existing plants), this is an ongoing issue. In these situations, who should pay for the extra capacity? Should this be only the new plant whose marginal demands have triggered the need for more infrastructure, or should it be shared between all plants, existing and new, in some proportion that reflects their relative demands on the infrastructure as a whole? The principle adopted is that specific additional requirements should be borne by the business unit that requires them; however, if there is a general increase in volume, this will be charged back across all the businesses, though it can be difficult in practice to keep the boundaries clear. Attempts are currently being made to develop a capital allocation system that is based on forward requirements rather than on past actual demand.

## ▌ Problem Definition

Zeneca was seeking a better picture of the financial implications of different waste disposal actions. This case focuses on the process of doing this and on the actual data generated at its Huddersfield Works, which was the focus of its activities. It provides evidence on opportunities for, and barriers to, more effective calculation of waste costs and their allocation to individual budgets in a multi-product environment, and also interesting evidence about the significance of organisational factors.

## ▌ Actions Taken and Results

The Huddersfield initiative began when Peter Natkanski put waste reduction proposals to the Zeneca Agrochemicals Business Team, based on the old ICI waste reduction targets. The proposals listed a number of projects that were designed to reduce the largest waste volumes from specific processes. However, there was insufficient resource to fund all the projects and no system to select the most economically viable ones. The Business Team was not prepared to sanction the projects as an environmental benefit case *per se*, and requested additional financial data in order to make the selection.

The material waste data were readily available and indicated that the Huddersfield Works was responsible for 50% of the waste. The next step, in Peter Natkanski's words, was:

> Using the existing waste data we said: let's focus on the biggest waste-generating site and look at it from a cost viewpoint. We'll find the waste reduction projects that will give a payback and see the percentage reduction they'll generate. We started with a pilot analysis of one product. We found its waste costs were unexpectedly high and dominated by the purchase cost of wasted raw materials.

The analysis of the pilot agrochemical herbicide was conducted at the Huddersfield Works and co-ordinated by Frank Bradburn of the Process Technology department. The starting point was a definition of costs of waste, which was:

> Waste is materials that have been paid for once and have to be paid for again in order to be disposed of. Therefore the 'cost of waste' comprises:
>
> **A.** The purchase cost of raw materials that are then discarded in the process
>
> **B.** Costs of treating discarded materials
>
> **C.** The capital costs of plant to treat materials.

The main elements in these three categories were:

A. **Materials costs:** all materials entering the process that do not form part of the final product (e.g. excess reactants, lost yield, auxiliary chemicals)

B. **Waste treatment costs:** materials (caustic, activated carbon etc.), labour, energy, landfill, incineration, effluent treatment plant, legal charges, transport

C. **Capital costs:** all capital allocation for plant and landfill expansion

The waste costs do not include 'occu-pacity' costs as Zeneca terms them—i.e. the expenses incurred in processing materials that end up as waste rather than as saleable product, since these are modest compared to the above.

The data to analyse the materials costs of the pilot product were taken from process mass balances—which was helped by the installation of a new effluent-monitoring database—and the product cost sheets. The treatment costs were taken from the existing charges from the Works, with apportioned calculations where no charge was levied. The capital costs were calculated by apportioning the cost of relevant plant and converting this to a discounted per-annum cost over a nominal fifteen-year plant life.

This revealed that wasted materials accounted for 35.7% of total product costs, waste treatment 15.1% and capital costs 5.9%. All the wasted materials costs and approximately a third of the treatment costs were calculated to be variable costs, indicating that over 40% of total product costs were variable costs that could theoretically be influenced by short-to-medium-term action. The figures are especially striking given that the pilot product has already been in existence for ten years and has been subject to frequent process development over that time to increase capacity and improve chemical efficiency.

Subsequent analyses of further products found that these figures are not atypical, neither in size nor in their domination by the costs of wasted materials. According to Peter Natkanski:

> At Huddersfield we found that waste costs were around £43 million a year—almost half of total manufacturing costs. Some of these can't be prevented but we calculated that there were £10 million per annum of potential annual savings. When people get this kind of data they get very excited—much more so than with corporate waste reduction targets.

In 1995 Zeneca achieved £1.5 million of these savings (on an annualised basis) and reduced hazardous waste by 6% and—at the time of the research—was on target for £5.4 million annualised savings and a 13% fall in hazardous waste in 1996. Its US operations—which are mature businesses with less potential for cost savings—have also achieved annualised reductions of £0.8 million in 1995 and £1.5 million in 1996.

One interesting feature of the work was that it was carried out by chemical engineers and manufacturing chemists—accountants played little part. However, they and site managers were impressed by the results, not having previously realised just how high the costs of waste were. Frank Bradburn and Paul Green (Environmental Adviser at Zeneca Specialties, who is located in the plant and worked on the pilot analyses) also believe that chemists had previously focused on increasing yields from core materials and expensive materials within processes, and aggregated all the others together. The costs of wastes approach makes losses of other materials more visible and therefore maps out improvement opportunities for them.

Frank Bradburn considers the involvement of the chemists to be a major reason for the initiative's success:

They can look at the numbers and have a feel for which materials will be technically difficult to alter and which are more promising. With their involvement it takes around half a day for each process stage to do a good mapping.

Peter Natkanski adds that 'we've got these savings just basically by looking at our position. People thought they knew all their areas of loss but they didn't. Now we're much better able to make a business case for environmental improvement.' He also stresses how useful the data are in setting priorities. Previously 'we had a gut feel that we should spend money on process improvement but we didn't have the hard data. Now we can show that it pays.'

The Cost of Waste approach now forms a part of Zeneca's six-monthly strategic review process. The costs of wasted materials and waste treatment are calculated for each process and product—initially in detail and subsequently as changes to the key parameters identified by the detailed analysis—and included within the review documentation. The analysis also contains data about development/improvement options with their likely environmental and financial returns, based on four key parameters: replacement of costly raw materials; removal of raw materials; reducing the quantities of raw materials used; and recovery/recycling of raw materials.

However, Frank Bradburn highlights one disincentive to further progress, which is the relatively low costs of waste disposal that are charged to businesses, due to the existence of an in-house landfill facility. As noted above, this is costed on an actual (historical) cost rather than a replacement cost basis, so that the costs with which business managers are confronted may be significantly lower than prevailing market prices for waste disposal.

## ▍ Conclusions

Some of the main conclusions to be drawn from this case are:

- Organisational factors are major determinants of environment-related management accounting.
- Accountants are not essential to environment-related management accounting.
- In businesses that are already operationally well managed, much of the data needed for environment-related management accounting is probably being collected already.

### Organisational factors are major determinants of environment-related management accounting.

It appears that a crucial aspect of Zeneca's experience was the location of the Process Technology department at the Huddersfield site. This gives it day-to-day involvement with site processes and good personal relations with site production staff. If it had been based at corporate level, and geographically at a distance, the improvements could have been more difficult—and perhaps impossible—to achieve.

### Accountants are not essential to environment-related management accounting.

The Cost of Waste project made little direct use of accounting data or the accounting systems. The data on production that were available from the accounting system were often not sufficiently accurate or detailed to be of use in exercises such as this.

Frank Bradburn identified one aspect in which the accounting function could help in future exercises (and/or updates of this exercise), though only at the fairly low level of first providing current costs for each raw material (i.e. maintaining a database of data from supplier invoices) and doing some of the basic computational work. However, the exercise was mainly outside the competence of the accountants:

> Most accountants don't always understand chemical processes—for example, the distinction between 'spine' chemicals (which end up in the final product) and 'crutches' (which enable reactions and separations)—so they're not in a position to suggest improvements. Their role is to check the proposals made by chemists and engineers.

Nor are the company's chemists disadvantaged by any lack of specific accounting skills, since they are regularly working with, and familiar with, product cost sheets. However, as in many businesses, there is a perception that, as operational champions, some chemists have a long-standing reputation for predicting possible cost savings which are subsequently found to have been over-optimistic when the benefits that they predicted then fail to materialise. This might be interpreted to indicate that, although the execution of a Cost of Waste study necessarily requires primarily scientific production-related skills, there remains a role for the accounting function to play in confirming and subsequently monitoring the accuracy of the results.

### Much data already exists.

Experience has shown Zeneca that a large proportion of the necessary data was around—but needed to be collected. According to Frank Bradburn:

> The main issue wasn't getting hold of the data but getting the methodology right. The key is going through existing cost-sheets and mass balances and knowing what percentage of each number to put in the box for individual products. Mass balances are notoriously difficult in batch processes, even to monitor yields, because of multiple syntheses. So we sometimes have to draw boundaries around two or more processes combined, since we cannot separate them.

This process has now been made easier by a new software package.

Peter Natkanski's overall conclusion is that

> there's nothing spectacular about what we've done. We just decided what information we needed, gathered it and then carried out a detailed analysis. It was initially problematic in terms of time and effort but now it's much easier as we've got systems and a common format. My advice to others is to gather all the existing information there is and look at that, rather than to set up new systems at the start.

# 20 The Road Not Taken

## Acting on 'Beyond Environmental Compliance' in Managerial Decision-Making[1]

### Timothy T. Greene

## ∎ Introduction

Environmental managerial accounting, the process by which companies actually use environmental costs to support decisions, has developed without explicit attention to the fact that firms both incur *internal* and cause *external* environmental costs. That is, there is a distinction between *internal*[2] costs which are on the firm's individual income statement and *external*[3] costs which are beyond the corporation's technical responsibility and off the corporation's books. Internal environmental costs are defined by statute, regulation, common law, consumer choice, investor demands and fashion. External costs are the costs of impacts, emissions, pollution, rubbish, waste, heat, accidents or scrap for which the firm is not legally responsible. The distinction between internal and external costs is depicted in Figure 1.

Accounting for external costs has received limited attention. Only two threads in the literature of the last 25 years have directly addressed external environmental costs from the firm's point of view. First, as reported by Shane (1982: 6), the American Accounting Association (AAA) charged standing committees from 1973 to 1976 with determining whether and how firms should accumulate and report external environmental costs. Second, research into the voluntary internalisation of external costs in

1. The author wishes to thank Germain Böer at the Owen Graduate School of Management at Vanderbilt University for the initial encouragement to engage this topic; also, the author thanks Clifford S. Russell, Director of the Vanderbilt Institute for Public Policy Studies, for his review of this work. Finally, the author wishes to thank the anonymous reviewers of the chapter.

2. Internal costs are also called 'private' environmental costs.
3. External costs are also called 'public' costs, 'societal' costs, 'social' costs or 'externalities'.

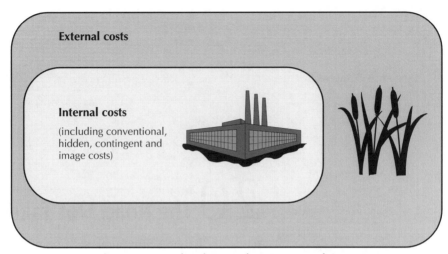

**Figure 1:** Internal and External Environmental Costs
*Source: US EPA 1995a: 15; White and Savage 1995: 49; Ditz, Ranganathan and Banks 1995*

managerial accounting is found in work pertaining to the US electricity utility industry. Jurewitz (1978), Freeman *et al.* (1992) and Agathen (1992) all outline how externalities might be included in electricity utility resource supply planning.

The arguments in the literature outlined above hinge on the idea that external costs should be examined and 'beyond-compliance' performance considered because firms have a fundamental measure of social responsibility. The AAA committees of the 1970s had social responsibility as part of their committee charters, and authors working on the electricity utility industry felt that the quasi-public nature of the industry required a greater consideration of social responsibility as a necessity. But there is a basic conflict in these literatures—one that is not easy to resolve—between making decisions based on social responsibility and making decisions based on profit and return on investment. This conflict, it may be noted, is at the heart of whole fields of economics because the welfare theory—the 'invisible hand' working to maximise social welfare—fails in the presence of external costs. This conflict may be the 'killer' of the efforts of the last 25 years to work on external cost accounting.

Epstein aptly summarises the state of the field today, noting that current environmental accounting efforts to manage internal environmental costs 'leave the question of how to include social costs in management decisions unresolved' (Epstein 1996c: 162). As Anderson points out, with these questions unresolved, '…if corporations do not bear the cost of preventing waste gases from being discharged, then the public must. Environmental economies must be allocated' (Anderson 1992: 64). This lack of work on external cost theory, applications or empirical research is not the result of a lack of interest in the topic. Recent work by Ditz, Ranganathan and Banks suggests that managers might be wise to address the issue:

> The wiser course of action is to truly internalize environmental costs—not merely to bear them, but to anticipate and manage them. This means identifying the activities that give rise to them. Firms must be able to generate, evaluate, and implement alternatives to the status quo. They must be able to motivate executives, managers and employees towards this common goal. In this endeavor, more effective, accurate accounting is crucial (Ditz, Ranganathan and Banks 1995: 13).

This business advice is a typical illustration of the state of thinking about external environmental costs today. The literature advises that managers manage the issue, but the focus of environmental accounting effort remains on studies of private environmental costs.

How should the manager deal with external environmental costs? Should the manager even try? In other words, how does 'beyond compliance' fit into an expenditure decision analysis system? The next section of the chapter addresses the question of why and the subsequent sections the question of how. Trying to find answers to these questions aligns with Owen's forecast that '…one can anticipate accountants becoming increasingly concerned with the theoretical and applied development of investment appraisal methodologies capable of incorporating environmental factors' (Owen 1993: 63). The question at hand is the extent to which these investment appraisal techniques should go beyond the status quo.

## ◼ Should Managers Care about External Environmental Costs?

External environmental costs, when assumed voluntarily by the firm, move the firm 'beyond environmental compliance'. If the firm starts to move towards further improvements in environmental performance—reductions in emissions, scrap, waste, heat and/or noise—then it begins to reduce its impact on society and thus reduce the level of external costs it causes. Offering empirical verification of the usefulness of 'beyond compliance' has turned out to be a tricky task indeed. Part of the problem lies in the need to know not only what firms actually decided to do but also the options they had available to them that they decided not to pursue. 'Beyond-compliance' investments which were not made—the road not taken—are seldom revisited within companies and only very rarely reported. Post-decision audits, one potential source of information for examining 'beyond-compliance' investment options, are still not widely performed. One would need more information about the road not taken in order to test for the dominance of 'beyond-compliance' alternatives over other possible paths. Only recently has empirical work begun in this area.[4]

With empirical proof out of reach, one cannot state (yet) that *always* selecting 'beyond-compliance' investments over 'compliance-only' ones is a practical way to run a company. 'Beyond compliance' remains, for most corporations, the road not taken, and external environmental cost accounting is not yet a necessity. Managers are not willing to move in favour of accepting 'beyond-compliance' investments simply because they achieve a higher level of environmental performance. However, *thinking about* 'beyond-compliance' options may add value to decisions even though the jury is still out on adopting the options unconditionally. That is, if 'beyond compliance' is *possibly* a dominant strategy, there is ample opportunity for managers to move beyond the status quo and include some external costs in their decision calculus. Two arguments can be made to further support this position that external environmental cost accounting is a valuable adjunct to today's decision-making processes and environmental accounting.

---

4. The author has a study under way of 53 chemical firms in the United States. The study is designed to test the hypothesis that firms pursuing 'beyond-compliance' hazardous waste disposal practices in the 1970s before the passage of the Superfund law and before final rule-making under Subtitle C of the Resource Conservation and Recovery Act continued to do so into the 1980s. The study is examining facility-level data for the waste management practices of the companies; the scope of the study is large, with over 1,600 facilities covered. Information about alternatives available to the firms is based on a comparison of facilities in the same four-digit SIC codes.

First, history shows today's legal emissions, which generate external environmental costs, may well become transformed through regulation into tomorrow's internal costs. At the heart of this position is the idea stated by White: 'With continuously evolving US environmental regulations, public expectations, and emerging international environmental management systems standards, cost boundaries are anything but static. External costs today may well become internal, less tangible costs tomorrow' (White and Savage 1995: 49). The inevitability of forced internalisations is not 'proof' of the need for external cost accounting, but it does suggest that managers would be wise to think about 'beyond compliance' in practice.

Second, there is a theoretical possibility that 'beyond compliance' is a fundamental guiding principle for environmental management in the same way that 'maximise expected utility' is a fundamental guiding principle for making decisions under uncertainty or risk. In the risk/uncertainty context, 'maximise expected utility' is accepted as a theoretical benchmark; in the environmental management context, one might vest 'beyond compliance' with the same deductive and definitional status. If 'beyond compliance' is a fundamental guiding principle like 'maximise expected utility', then accounting for external costs becomes a necessity for the complete examination of 'beyond-compliance' investment options. The analogy is strengthened by the status of the guiding principles within their respective fields. In the risk/uncertainty field, there is a large body of empirical evidence suggesting that individuals do not always, or even usually, follow the guiding principle and make decisions so as to maximise expected utility; likewise, in the environmental management context, firms follow many different fundamental rules from 'beyond compliance' to 'compliance only' to 'wait and see if the firm gets caught'. The possibility that a general theoretical agreement may emerge on 'beyond compliance' again points to the potential value of learning more about external environmental cost accounting.

Any of these three reasons—empirical proof, the reality of forced internalisations or the emergence of 'beyond compliance' as a fundamental guiding principle—give managers sufficient reason to care about external environmental cost accounting, at least to the extent that it becomes an adjunct to existing environmental accounting processes. To illustrate how a decision-maker might implement 'beyond-compliance' thinking as an adjunct to environmental accounting, let us consider a paper mill in the late 1960s. Suppose that the mill's executives and managers are considering purchasing a new paper-making machine. One of the dimensions of difference among the managers' choices are the raw water pollution loads the machines create. These water pollution loads have, in turn, implications for the size and type of water treatment systems that the paper mill would have to install, given changes in water pollution laws.

The management team at the paper mill begins by weighing at least the following two alternatives in considering the capital expenditures for the new paper-making machine—the net present value of costs of:

**D** The 'dirty' (traditional) machine which generates currently legal water pollution requiring no treatment under the law of the time

**C** A 'clean' machine which does not produce pollution loads

Comparing the above is the standard decision-making procedure at the paper mill. However, the paper mill's managers are also thinking about the possibility of regulatory changes which would require the addition of the waste-water treatment system during the useful life of the equipment. The late 1960s witnessed a groundswell of pub-lic sentiment for enhanced protection of the environment in the United States. With the possibility of forced internalisations (driven by law) in mind, the managers are considering going 'beyond compliance' by addressing situations in which external cost internalisations might occur. One such situation is a policy change that internalises the costs of the waste-water pollution load. This scenario produces the following additional cost comparison:

**D + T** The 'dirty' machine given some sort of policy change that shifts the external costs of the water pollution load to the firm—in this case translating into the need to install a waste-water treatment system (**T**)

**C** The 'clean' machine which does not produce pollution loads

The four options faced by the paper mill can be set up in a matrix as per Figure 2. There are two facts the managers know about the relationships among the costs in the matrix:

1. **C > D:** the clean machine is more expensive than the dirty one.

2. **D + T > D:** the waste-water treatment plant adds costs to the dirty machine.

With these fundamentals in mind, risk-averse paper mill managers apply the 'mini-max cost' decision criterion to the matrix.[5] In this example, making a choice between clean and dirty machines under minimax cost requires knowing the costs of **T**, the waste-water treatment plant. If **T** turns out to be larger than the price differential of the clean machine over the dirty machine (i.e. **T > C − D**), then the paper mill would decide to buy the clean machine. If the treatment costs turn out to be less than the additional expenditure required for the clean machine (i.e. **T < C − D**), then the paper mill would not go 'beyond compliance'.

Although seemingly straightforward, this simplified decision analysis requires a dif-ficult prediction in order to make the decision—figuring out **T** as dictated by possi-ble regulatory shifts in water pollution policy. Assigning values to **T** is a form of external cost accounting. Here, the paper mill's managers are making a forecast of the regula-tory future by defining option costs 'given some policy change'. Making this 'beyond-compliance' prediction is not an easy task:

| | No policy change | Policy change | *Row maximum* |
|---|---|---|---|
| **Dirty machine** | D | D+T | *D+T* |
| **Clean machine** | C | C | *C* |

**Figure 2:** Paper Mill Option Matrix

---

5. Minimax decision-making, a conservative or risk-averse strategy for selecting an option, has two basic steps. First, the managers identify the maximum for each machine option (each *row* of the matrix). Second, the managers select the smaller of these two—the minimum of the maxima or 'minimax'—as the preferred option (the smaller value in the *column*). This decision-making can be extended to more complex matrices of options.

> The problem…is that 'doing the right thing' [going 'beyond compliance'] gives no real guidance to the executives trying to make the complicated, subtle tradeoffs that environmental management involves…what do people mean by 'the right thing', anyway? That which science dictates? What regulators want? What environmentalists value? What the public desire? What consumers would like? These are often poles apart, leading down completely different paths (Colby, Kingsley and Whitehead 1995: 142).

The next section turns to putting some meat on the bones of the example above by introducing three candidate approaches to actually performing external environmental cost accounting as part of thinking about and evaluating 'beyond-compliance' investment options. The candidate approaches might help the paper mill 'do the right thing' as it reviews 'beyond-compliance' options for the paper-making machine.

## ▌ Candidate Approaches for External Environmental Costs in Environmental Managerial Accounting

Parker provides a succinct abstract of the problem the practitioner faces when looking for practical advice on how a manager might include external costs in decisions:

> Higher standards for air and water quality are examples of governmental actions that have pushed former external costs into the domain of internal costs. Management accountants and financial managers are familiar with noting liabilities associated with litigation and penalties, actual and potential, and the related [balance sheet] disclosure issues. But they are just beginning, on an ongoing basis, to identify and assign environmental costs effectively to support managerial decisions (Parker 1996: 52).

Three candidate approaches are considered below for identifying and assigning external costs for managerial decisions: (1) the firm's share of social damages; (2) forecasting the regulatory future; and (3) control, avoidance or liability cost forecasts. Following the introduction of these three approaches, an example demonstrates application of the methods to an expenditure decision in the tyre and rubber industry.

### Social Damages
First, the firm could include its share of social damages in the expenditure decision. In the example above, the option costs 'given some policy change' would be revised to 'option costs including social damages'. Instead of trying to predict $T$ (waste-water treatment costs), the water pollution loads produced by the dirty option would be transformed, based on the valuation literature in economics, into dollar amounts of social damage; call them $S$. Although simple in concept, the execution of this strategy is difficult because of the complexity of the valuation literature. As reported by Magness (1997: 16), utility company Ontario Hydro has developed dollar measures of external environmental costs—measures of $S$ (see also Chapter 16). Specifically, the company, a Canadian utility, has used a damage function approach—one of the three major ways of measuring social damages. The following is a very brief review of these three major techniques in use in economics (Cropper and Oates 1992; Russell 1996a, 1996b).

- **Indirect methods** attempt to infer from actual choices, such as choosing where to live, the value that people place on environmental goods. Indirect methods may examine:
  1. Averting behaviour/expenditures where people make purchases to ameliorate the effects of pollution
  2. Weak complementarity/travel cost where the value, say, of cleaner water is assumed to be connected somehow to visits to a lake
  3. Hedonic market methods where the price of a house or a job can be decomposed into attributes, one or some of which are environmental attributes
- **Direct methods** ask people to make trade-offs between environmental and other goods in a survey context. Direct methods involve direct questioning about willingness to pay (WTP) or willingness to accept compensation (WTA) for some change in health, recreation, visibility or other environmental attribute. These direct questioning methods have come to be known by the jargon name 'contingent valuation'.
- **Mixed methods** include the 'damage function' approach taken by Ontario Hydro. Mixed methods generally blend results from indirect and direct methods.

Measuring social damages is attractive for its 'fullness' in capturing the cost of production externalities. Valuation methods in economics have become an accepted standard for estimating the impacts of pollution, waste, scrap and spills; for example, contingent valuation was used extensively in the wake of the Exxon Valdez spill to document environmental impacts (Brown 1992; Carson 1992). However, no firm has adequate resources to perform economic valuation for every external cost it faces. At the same time, using existing valuation results requires 'carving up' aggregate damages so that they can be attributed to single firms, single production lines or individual pieces of capital equipment within a plant. In the paper mill example, this 'carving' activity would require knowing the pollution loads from the new machine as a single point source distinct from the loads generated by all other sources within the plant.

The firm must also address the extent to which regulators might force internalisations of social damages. There may be, in practice, a level beyond which the 'rational regulator' would not force additional internalisations based on economic assumptions about how 'optimal' pollution control is achieved. Even in practice, where the regulator is not rational, it is politically unlikely that a firm would be required to internalise all of its damages.

## ▌ The Regulatory Future

Second, the firm could try to predict the regulatory future—a sort of scenario analysis based on the policy changes that the firm's managers deem likely to occur. Forecasting the regulatory future requires predicting changes in public policy and then identifying the costs shifted onto the firm. This method relies on heuristic approaches to evaluating how control authorities might regulate in the future. Monty argues in favour of this type of approach; he states that business leaders need a good understanding of the 'regulatory dynamic' or the rate of policy change they face (Monty 1991: 4). The results of this process are internalised environmental costs specific to individual policies; Figure 3 captures a wide range of options for this type of analysis.

| Policy change | Policy mechanism | Costs |
|---|---|---|
| Emission limits/ performance standards Technical standards for pollution control | Limiting the production of externalities or mandating technologies to limit their production | Costs to acquire or modify equipment to meet a new waste emission standard, acquire or modify equipment to meet a new technology-based standard or design standard, meet stakeholder demands, hire, fire and/or train employees, or, modify product designs, formulations, production methods or production schedules |
| Liability, remediation, restoration | Requiring clean-up or repair of damage that has already occurred | Costs to clean up or remediate wastes, dispose of the wastes from the clean-up, assure that the now properly disposed wastes remain safe over a long planning horizon, repair natural resource damages or restore ecosystems, or compensate for damages to human health or property |
| Product prohibition/ limitation | Banning the activity producing the externality | Costs of process/product re-engineering; lost sales of product when no technology exists to replace; bankruptcy of company |
| Payment | Private parties pay the generating party to eliminate the externality | Negotiation costs, transaction costs, cost of contract, escalation |
| Merger/symbiosis | Applies in the special situation where two parties each deliver an externality to the other | Costs of negotiation transaction, contract, escalation, potential cost of acquiring other firm |
| Taxes/charges | Sources pay charge/tax equal to source's marginal social damages | Cost of tax, cost of assessing marginal damages |
| Subsidies | Government pays companies to abate pollution | Cost of higher taxes or loss of other government-provided services in order to pay |
| Technical assistance | Government gives company information to reduce pollution | Cost of higher taxes or loss of other government-provided services in order to provide; contingent loss of industry advantages |
| Permits/pollution rights | Licences give producer right to pollute | Costs of acquiring permits, options on future permits, costs of selling permits into the market, losses of expiring permits |
| Challenge regulation | Agency challenges sources to achieve a standard | Cost of achieving voluntary standard, contingent cost of choosing the 'wrong' standard, other costs arising from regulatory uncertainty |
| Information reporting | Company reports all emissions to EPA and public | Highly uncertain costs from financial and consumer market impacts |

**Figure 3:** Implications of Specific Possible Policies for Internal Costs

*Source: Belkaoui 1984; Russell and Powell 1994*

The single most common type of cost shift results from a reduction in allowable emission levels (row 1, Fig. 3). The reduction in allowable emissions translates into the need to spend more money to control or avoid the production of the pollutant. In the paper mill example above, the managers identified the possibility that the plant would not be able to emit pollutants in the water. This change in legal emission levels implied the need for greater control of the emission; the greater control generates new internal costs for the firm—for the paper mill those costs are driven by the installation and operation of a waste-water treatment facility.

Managers frequently examine myriad contingent situations in order to verify that an investment choice offers returns under a variety of sales levels, market conditions and production levels. Analysing potential regulatory changes that force external cost internalisations is an extension of this business forecasting discipline. Forecasting the regulatory future is straightforward in practice. Firms may employ or contract with regulatory specialists or lobbyists who have access to information and personal contacts that can shed light on the likelihood and direction of policy changes.

### Cost Forecasting

Third, the firm might conduct a simplified forward-looking exercise and make a forecast of control or avoidance costs based on past experience, much in the same way as managers use demand forecasting tools to predict sales. This forecast of pollution abatement and control (PAC) costs is a simplification of the regulatory future forecast method. Where the method above relied on making specific policy predictions, this forecast looks at general trends in avoidance, control and liability costs.

Operationalising PAC forecasts is done in two possible ways: (1) applying escalation factors to known PAC costs; or (2) assigning PAC costs where none is currently spent. Drawing the analogy to demand forecasting, the latter is like predicting sales of a new product while the former is similar to predicting next year's sales based on last year's performance. In the paper mill example, the former method would be applied by identifying the part of the costs of the dirty machine that are PAC expenditures (such as air pollution control or solid waste disposal costs) and applying growth rates or escalation factors to predict future compliance costs.

Data sets maintained by the US Government give some insight into the figures on which these forecasts might be based. Two such US data series are charted in Figure 4 in real and current dollar terms: aggregate PAC expenditures and replacement costs of installed PAC equipment. A risk-averse firm might predict that its internal environmental expenditures would grow by the highest amount known historically (8.3% for real dollar PAC expenditures 1983–84 or 19.5% for current dollar expenditures 1973–74), while a more risk-tolerant firm might adopt the 22-year average (3.6% for real dollar PAC expenditures or 9.5% for current dollar expenditures). However, it must be noted that using aggregate figures may not capture the mercurial nature of forced external cost internalisations. Consideration must be made of the potential for lump-sum costs arising from the whims of the courts, regulators and citizens' groups that shift external costs onto the firm unexpectedly.

The next section of the chapter turns to illustrating these candidate approaches to performing external environmental cost accounting as an adjunct to environmental accounting.

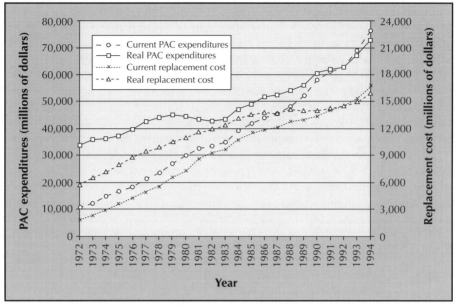

**Figure 4:** Growth in Expenditures and Replacement Costs for Control/
Avoidance Cost Forecasts

*Source: Vogan 1996: 56-63; Hahn 1994: 318*

# ▌ Application of Candidate Approaches:
## Tyre and Rubber Plant Hazardous Waste

Having considered the basic methods of these three candidate approaches, let us turn to illustrating their application to a real decision problem. In the United States in the 1970s, there was no systematic federal regulation of hazardous waste, and state laws created a patchwork of requirements for corporations. The management of hazardous waste was simply considered one dimension of the larger solid (non-hazardous) waste regulatory scheme at the federal level. Hazardous materials were generally disposed of in pits, ponds, lagoons or in sanitary (mixed waste) landfills. These disposal techniques generated very small internal environmental expenditures. Dumping a 55-gallon drum of still bottoms out the back door of the paint plant was essentially free for firms that selected this method of disposal. However, by 1977 problems such as Hooker Chemical's 'Love Canal' dumpsite in Niagara Falls, New York,[6] revealed to managers, the public and the US Congress that the external cost dimensions of waste disposal could be very large indeed.

---

6. 'Love Canal' is arguably the most famous (infamous?) dump site in the Americas. Discovered in the mid-1970s, the site consisted of numerous trenches filled with buried drums of mixed hazardous waste produced at the Hooker Chemical plant in Niagara Falls, NY. The Love Canal site had been developed as a residential housing area following Hooker's dumping activity. An elementary school was also built on this site. Health problems among the families living at Love Canal precipitated investigation of the site. The extent of the dumping received national media attention and intense scrutiny from the US Congress and environmental protection authorities. It is widely believed that Love Canal led to the passage of the Superfund hazardous waste site clean-up law in the United States.

To illustrate how managers might implement the three candidate approaches out-lined above, let us consider the case of a tyre-manufacturing plant in the 1970s. According to a US EPA report on the rubber and synthetic plastics industry (US EPA 1975: I-33), the typical tyre plant in 1974 produced 59.4 million kilogrammes of finished goods each year—about 75% passenger tyres and 25% truck and bus tyres. It also generated the eight hazardous or potentially hazardous waste streams listed in Figures 5 and 6. Figures 5 and 6 also detail the generation rate of the waste per 1,000 kilogrammes of production. For this example, let us consider the most hazardous of the three waste streams in Figure 5—Banbury 'oozings'—as we address an expenditure decision at the tyre plant. A Banbury mixer is a large heated and pressurised vessel in which rubber

| Potentially hazardous waste name Type I wastes | kg waste per 1,000 kg product | Description/source |
|---|---|---|
| Floor sweepings | 1.9 | Waste from receiving, compounding and mixing areas containing carbon black, pigments, accelerators, anti-oxidants, etc. |
| Air pollution control abatement equipment | 8.3 | Dusts containing carbon black, pigments, accelerators, anti-oxidants, etc. |
| Banbury mixers and other high-energy mechanical equipment | 0.5 | Oils from Banbury 'oozings' contaminated with materials in rubber mixture |
| **Total** | **10.7** | |

**Figure 5:** Hazardous or Potentially Hazardous Waste Streams from Tyre Production
*Source: US EPA 1975: III-63*

| Potentially hazardous waste name Type II wastes | kg waste per 1,000 kg product | Description/source |
|---|---|---|
| Scorched rubber | 11.3 | Waste rubber from mixing, stuck to sides or bottom of Banbury mixer |
| Uncured stock | 3.4 | Waste from calendering, tread manufacture and tyre-building |
| Coated and uncoated fabric | 15.9 | |
| 'Green' tyres | 2.7 | |
| Rubber, steel wire and rubber-coated fabric | 0.7 | Bead-manufacturing |
| Rubber dust or rubber crumb | 4.8 | Waste from tyre-finishing and trimming |
| Scrapped tyres | 21.9 | Defects from final inspection process |
| **Total** | **60.7** | |

**Figure 6:** Non-Hazardous Waste Streams from Tyre Production
*Source: US EPA 1975: III-63*

is mixed with various additives before being formed into a sheet for further production. Producing the mixed rubber is a batch process much like making biscuit dough. The 'oozings' from the mixer occur on the outside of the vessel where the mixed, uncured rubber pushes past the bearing seals for the mixer's rotor shafts. The result is a waste that has many components, including rubber, additives and lubricating oils and greases. Standard practice for the typical tyre plant in 1974 was to either dump the 'oozings' into a pit or to drum them in 55-gallon drums and store them on-site.

The tyre plant has a number of options to deal with the 29,700 kg of Banbury 'oozings' that it generates each year. The firm could, based on US EPA information (US EPA 1980: V-36):

$C_D$  Make no additional expenditures and continue to pay the costs of landfill disposal or drummed storage ($8.25 per 1,000 kg)

$C_L$  Go beyond compliance to send the wastes to a 'secure' landfill, a landfill specifically designed for hazardous wastes ($60.50 per 1,000 kg)

$C_I$  Go beyond compliance and have the wastes incinerated, either off-site or on-site ($550 per 1,000 kg), which reduces the need for land disposal to a small amount of residual ash which is disposed of in a secured landfill (2% of waste total at $60.50 per 1,000 kg)

$C_M$  Go beyond compliance to invest in waste minimisation equipment to retrofit the seals on the Banbury mixers in the first year ($100,000); this change results in a small amount of waste which can be recycled back into the rubber mix and 1,000 kg per year of waste from the regular maintenance of the new seals to be disposed in a secure landfill ($60.50 per 1,000 kg).[7]

The column of figures in Figure 7 labelled 'No changes' outlines the net present value of these options for the tyre plant given no change in the regulatory scenario. Clearly, the status quo option of land disposal or storage is the least expensive for all three discount rates. The costs are based on a ten-year decision span and real discount rates of 6%, 10% and 14%. Also, the waste generation amounts are assumed to grow at 3.5% per year in line with US EPA estimates (US EPA 1980: III-1).

Four illustrations of the candidate approaches follow: the social damage approach, two versions of the regulatory future approach, and the PAC expenditure cost forecast approach. The total waste management option costs for each illustration are summarised for each of the three discount rates in Figure 7.

### Social Damage Approach

There are a number of ways to approach the estimation of the firm's share of social damages: one simple way is to use the national average damage per kilogramme of waste. According to the US EPA, between 41.2 (1980: III-2) and 264 (1984: 135) million metric tonnes of hazardous waste were generated in 1980–81 and the growth rate in this amount was around 3.5% (1980: III-1). Based on these figures, we might conclude that about 125 million metric tonnes were produced in 1975. As to total damages, let us use a proxy measure to bypass the complexities of the valuation literature. To demonstrate

---

7.  The waste minimisation costs are higher than the range given in US EPA (1980: V-36) due to the essential non-substitutability of the Banbury mixer technology.

| | No changes | Social damage approach | Regulatory future: pre-treatment | Regulatory future: charges | PAC expenditure forecast | Row maximum |
|---|---|---|---|---|---|---|
| **6% discount rate** | | | | | | |
| **Compliance-only: land disposal or storage** | $2,100 | $14,700 | $91,400 | $27,300 | $2,400[1] | $91,400 |
| **Secure landfill disposal** | $15,300 | $27,900 | $103,900 | $40,500 | $18,300[2] | $103,900 |
| **Incineration** | $139,000 | $139,300 | $139,000 | $141,600 | $139,000 | $141,600 |
| **Waste minimisation** | $94,900 | $95,300 | $94,900 | $95,200 | $94,900 | $95,300 |
| **10% discount rate** | | | | | | |
| **Compliance-only: land disposal or storage** | $1,700 | $12,100 | $81,000 | $22,600 | $1,900[3] | $81,000 |
| **Secure landfill disposal** | $12,600 | $23,000 | $91,400 | $33,500 | $15,000[4] | $91,400 |
| **Incineration** | $115,000 | $115,100 | $114,900 | $117,000 | $115,000 | $117,000 |
| **Waste minimisation** | $91,300 | $91,700 | $91,300 | $91,600 | $91,300 | $91,700 |
| **14% discount rate** | | | | | | |
| **Compliance-only: land disposal or storage** | $1,400 | $10,200 | $72,300 | $19,000 | $1,600[5] | $72,300 |
| **Secure landfill disposal** | $10,600 | $19,400 | $81,000 | $28,100 | $12,600[6] | $81,000 |
| **Incineration** | $96,600 | $96,800 | $96,600 | $98,300 | $96,600 | $98,300 |
| **Waste minimisation** | $88,000 | $88,400 | $88,000 | $88,300 | $88,000 | $88,400 |

Note: Using lump sum of $250,000 in PAC forecast approach, the figures are as follows:
[1] $141,700; [2] $154,900; [3] $98,100; [4] $109,000; [5] $68,900; [6] $78,000

**Figure 7:** Banbury 'Oozings' Total Waste Management Option Costs

the working of the social damages estimator, let us take as a proxy for the impact of hazardous waste disposal the number of deaths[8] from leukaemia in the United States in 1975—about 15,000 (Statistical Abstract of the United States 1996: Table 129). Assume that, of this number, one-quarter can be linked to the disposal of hazardous waste. This assumption of a connection between waste disposal and leukaemia is borne out by the experiences of residents of Woburn, Massachusetts, site of a massive waste disposal problem (see, for example, Harr 1995 or NRC 1991: 131). Based on one economics survey article, these statistical lives are worth between '$1.6 million to $9 million ($1986) with most studies yielding mean estimates between $1.6 million and $4.0 million' (Cropper and Oates 1992: 713). To get a single 1986 value, let us double the minimum amount—about $3.2 million. Using the US consumer price index as a deflator (Statistical Abstract of the United States 1997: Table 751), this $3.2-million value would be $1.6 million in 1975 dollars. Multiplying lives lost by the statistical value estimate, the estimation of

---

8. If we observed the actual health effects of hazardous waste disposal, we would not see social damages as deaths but rather as health problems mixed with a small number of actual deaths due to varying types of illnesses; the focus here is on finding a relevant proxy measure.

total damages from hazardous waste disposal is about $5.9 billion. To calculate damages per kilogramme, we divide the $5.9 billion in total damages by 125 million metric tonnes of total hazardous waste resulting in a per-kilogramme social damages amount of about 5 cents.

As mentioned above, the firm should not expect to be made responsible for all of the social damages it generates based on the limits of the 'rational regulator' idea. For this example, let us assume that this 'rational regulator' percentage adjustment has already been factored in. This assumption is reasonable in light of the omission of consideration of damages to ecosystems and groundwater caused by disposal of hazardous waste. With the damage rate per kilogramme, we can estimate social damages for each disposal option by simply multiplying the amount of waste generated each year by the 5¢-per-kilogramme social damage estimate. When these social damage amounts are calculated for the ten-year decision span, added to the baseline costs and discounted, the tyre plant faces the total option costs shown in the column labelled 'Social damage approach' in Figure 7.

### Regulatory Future Approach

Next, to illustrate the external environmental costs of the waste management options under the regulatory future approach, the example is split into two sub-options—waste pre-treatment and waste charges.

First, let us suppose that the tyre company's Director of Regulatory Affairs has told all the plant managers of the corporation to expect a requirement for pre-treatment of oily wastes before disposal to land. This law, it is thought, would apply to the Banbury 'oozings' and require the purchase, installation, operation and maintenance of a machine to separate the oils and hazardous materials in the 'oozings' from the rubber components. For this exercise, suppose that the required equipment costs $100,000 in Year 3 plus operating costs of $1,000 per year in Years 4–10.

For the land disposal option, no additional costs are incurred until Year 3 when the pre-treatment machine is purchased. In Years 3–10, 90% of the now separated waste is still sent to the regular landfill, while the 10% containing oils and hazardous materials is sent to a secure landfill in accordance with the new law. Thus the new weighted average waste disposal cost for the land option in Years 3–10 is $11.68 per 1,000 kg. In addition, the $1,000 annual operating cost of the separator is factored into Years 4–10.

For the secure landfill option, all waste, separated or not, is sent to the secure landfill. Costs in Years 1 and 2 are unchanged, then the purchase cost of the equipment is added in Year 3 and the $1,000-per-year operating costs in Years 4–10. Incineration might or might not require the pre-treatment step; this finding is consistent with the uncertainty inherent in trying to predict the regulatory future. Let us suppose that the law makes a specific exception for wastes bound for incinerators (a supposition made slightly more likely by the high energy value of the Banbury 'oozings'). Finally, it is likely that no pre-treatment would be required when the amount of waste generated falls under a certain threshold. This would be the case for the waste minimisation option. The total present value option costs based on this approach are shown in the column labelled 'Regulatory future: pre-treatment' in Figure 7.

Turning to the second option for external cost generation under the regulatory future option, suppose that the general counsel of the tyre company attended a trade association seminar and felt that per-kilogramme charges for wastes were on the immediate horizon. Under this charge scenario, the tyre plant would be required to pay a certain dollar amount per tonne of waste generated. The schedule of charges would probably be created by the rule-making agency as a result of: (a) risk analysis, i.e. higher charges for higher-risk wastes; and (b) lobbying by interest groups, i.e. lower fees for wastes produced by an interest group.[9] In this scenario, **C** is the charge amount and is added to the baseline disposal option costs.

Let us suppose that the tyre plant thinks the schedule of fees will be $0.10 per kg for raw Banbury 'oozings', $0.50 per kg for incineration residuals, and $0.05 per kg for waste oil (this amount being a politically palatable charge for a high-profile waste). These charges are multiplied by the amount of waste generated each year. The annual charge amounts are added to the baseline option costs and are shown as net present values for the ten-year decision life in the column labelled 'Regulatory future: charges' in Figure 7.

### PAC Forecast Approach
Using forecasting tools to predict shifts of external costs requires first that PAC expenditures are isolated from other types of expenditures. However, unlike the paper mill example above, the tyre plant's decision lies completely within the frame of PAC expenditures. The decision being analysed is, itself, a pollution abatement and control decision. As a result, the forecasting methodology is applied to the total value of the waste disposal options because there is no separable PAC portion. The tyre plant's managers have decided to apply the PAC forecasting approach only to the two land disposal options—landfilling and secure landfilling. The managers believe that the incineration and waste minimisation options will remain either beyond or within compliance during the ten-year decision span. Thus it is unlikely that additional PAC costs would be required if either of these two options were selected.

There is a wide range of forecasting tools and data sources available to escalate the land and secure landfill disposal costs. Let us suppose that the tyre plant starts with the growth rates of the data series depicted in Figure 4, focusing in on the current dollar growth rate in PAC expenditures from 1972–75 as the baseline rate. The managers can use this rate, 15.4%, to calculate the escalation amount in land disposal costs for Year 2 by simply adding 15.4% of the first year's disposal cost to Year 2's cost. Similarly, this amount can be calculated for Years 3–10. For the secure landfill option, the managers would probably use a higher rate for the escalation because of higher technology, operation and maintenance costs over regular land disposal and the risk premium charged by secure landfill owners to reflect the hazardous nature of the materials disposed. For the example, suppose that the rate selected is half as high again as the rate for land disposal, 23%. The total option costs under each discount rate are shown in the column labelled 'PAC expenditure forecast' in Figure 7.

### Analysis
If the managers of the tyre plant examine the results of each internalisation approach individually (i.e. each column of Fig. 7), their conclusion would be that land disposal—

---

9. Alternatively, this charge scheme could be viewed, as it would be in the United States, as a 'permit' system where polluters are required to secure permits for the emission of the waste. The analysis is the same; permits are the duals of charges.

the status quo—is preferable. This result is not sensitive to the discount rate nor is it particularly sensitive to changes in the underlying assumptions. However, a key feature of the actual waste disposal decision is deep uncertainty about what the future may hold for the tyre plant. The tyre plant's managers do not know which internalisation option may come to pass or which approach to estimating external costs may be the most accurate proxy. Therefore it would be appropriate for the risk-averse firm to perform some sort of risk analysis. A very conservative approach would be the 'minimax' cost criterion used in the paper mill example outlined above.

The first step in finding the minimax cost decision is finding the worst result produced by each investment option. To do this, we look across each row of Figure 7 to find the maximum value for that disposal option. The right-hand column of Figure 7 shows these row maxima. For all three discount rates, the maximum option costs for land disposal and for secure landfill disposal are found in the 'Regulatory future: pretreatment' column—the external cost approach requiring pre-treatment of the Banbury wastes. The identity of the internalisation approach producing these maxima is not particularly sensitive to changes in assumptions. For example, social damages would have to increase by almost an order of magnitude (to $0.40–$0.45) before that internalisation approach produced the land disposal and secure landfill row maxima. The incineration investment option produces its maximum cost under the 'Regulatory future: charges' scenario. Again, this maximum is not particularly sensitive to changes in assumptions. As above, social damages would have to increase dramatically before that option produced the maximum cost of incineration. The waste minimisation investment option maximum cost is found in the 'Social damage approach' column. This maximum changes only when the per-kilogramme waste charges are doubled across the board and 'Regulatory future: charges' produces the maximum cost of waste minimisation. That three different candidate approaches result in the row maxima for the four investment options illustrates the value of looking at multiple proxies for the future.

It is interesting to note that the PAC forecasting approach (using escalation on the 'No changes' option costs) does not produce a row maximum, even though the mechanics of this method is perhaps the most familiar to many managers. However, the PAC forecasting approach can come into play in this example when managers switch from using an escalation factor to making a more general prediction of future 'lump-sum' PAC expenditures. In 1975, the managers might have sensed that a large shift in PAC expenditures was looming with the passage of the Resource Conservation and Recovery Act (RCRA) of 1976. Also in 1975, the first reports were being released about the problem of abandoned hazardous waste sites; the existence of abandoned disposal sites led to the Comprehensive Environmental Response, Compensation and Liability Act (CERCLA) of 1980 (Superfund). These two laws created cost internalisations that no reasonable growth rate could have captured. For example, if the managers predicted the potential of paying $250,000 in remedial costs—under RCRA or Superfund—in Year 10 to clean up wastes that had been disposed on land, then the PAC forecast method produces the maximum costs for the two land disposal options for the 6% and 10% discount rates; these PAC costs based on a lump-sum forecast are given in the footnote to Figure 7.

The next step in choosing the minimax cost option is to find the lowest of the four maximum option costs under each discount rate. In all three cases, this choice would be land disposal. However, looking at the relative values of the row maxima reveals that the minimax cost decision is very sensitive to the assumptions. For example, for the 6% discount rate, land disposal and waste minimisation option costs are within $4,000 of each other. If the initial investment in the retrofit of the Banbury mixer cost only $4,200 less, waste minimisation would be the minimax cost decision. For the 10% discount rate, land disposal, secure landfilling and waste minimisation are all within $10,700 of each other. If the pre-treatment equipment purchase were moved forward from Year 3 to approximately Year 1½, waste minimisation would become the minimax cost strategy. For the 14% discount rate, land disposal and secure landfilling are within $8,700 of each other, with waste minimisation a slightly more distant third. If the initial waste minimisation investment changes from $100,000 to $80,000 or if the investment in pre-treatment equipment is required in Year 1 rather than in Year 3, then waste minimisation becomes the minimax cost option. Finally, if we make the switch from PAC escalation to a lump sum of future remedial costs as suggested above, the changes in maximum option costs makes waste minimisation the minimax cost strategy choice under the 6% and 10% discount rates. The lump-sum remedial cost would have to increase to $325,000 to make waste minimisation the preferred strategy under the 14% discount rate.

A strict reading of the results of this example suggests that it is not economic to go 'beyond compliance' in the management of hazardous waste from the Banbury mixer if land disposal is allowed. However, the strategy choice is highly sensitive to the assumptions made about the unknown future. It takes only trivial changes in the inputs to the example to produce different decisions about strategy selection. Acceleration in the implementation of waste pre-treatment makes waste minimisation appealing, as does the potential for remedial costs in the final year of the decision's life. This example is a case for spending some time worrying about what might happen next and what those various possibilities might cost the firm—a case for performing external cost accounting as an adjunct to environmental accounting.

## ▮ Conclusion

Milne observes that 'very little research has appeared to date on extending internal information systems to incorporate the wider range of social costs and benefits associated with corporate [environmental] activities' (Milne 1991: 82) This chapter has covered a little theoretical ground and a lot of practical territory in showing that conducting external environmental cost accounting is a valuable adjunct to the environmental accounting process. While the manager may not yet know the final status of the idea of 'beyond-compliance' environmental performance, incorporating external costs in managerial decision-making may be a wise course of action and is most certainly an achievable one.

This chapter has focused on the positive and would not be complete without consideration of the following four objections to external cost accounting. First, Magness points out that 'It is understandable that accountants feel uneasy about external costs. Conventional accounting theory tells us to recognize costs when they are incurred. Externalities do not meet this criterion' (Magness 1997: 17). When costs are shifted across the border of the firm's 'box' (see Fig. 1), then they will be recognised by accountants. Second, according to Milne, 'The lack of research in this area may be attributed to…the philosophical issue of whether private sector organizations should really be concerned with what are effectively public concerns and the province of public decision makers' (Milne 1991: 82-83). Here, the corporation adopts the view that public environmental costs generated by the firm are 'not my problem'. It might even be argued that the corporate taxes paid by the business cover, in whole or in part, these 'public concerns'. Third, according to Owen, 'a major stumbling block is encountered in the accountant's traditional obsession with, and insistence on, objectively verifiable and largely financially based measurement techniques' (Owen 1993: 71). The apparent lack of external cost measurement techniques appropriate for external cost accounting may be due to the lack of external cost data relevant to the *single company's* point of view. Fourth, Rubenstein points out differences between the perception of environmental issues and business issues (1994: 18; 1992: 30). Underlying his analysis is the idea that environmental concerns embody a human dimension while 'traditional' business concerns do not.

These problems boil down to two technicalities and two philosophical differences. The technicalities are the meaning of the term 'incurred' and the source of public cost data; these barriers might be overcome through advances in research in the field. The philosophical differences are the inherently 'public' nature of public environmental costs and the 'human' dimension of environmental concerns; these barriers might be overcome by changes in the culture of corporations or in managerial accounting practice.

Surma captures the essence of the challenge that environmental managers face as they consider how to 'do' external cost accounting as an adjunct to environmental accounting:

> As environmental expenditures become an increasingly substantial cost of doing business, the corporation must…determine how to account for and report the costs of its past, present, and future environmental activities (Surma 1992: 20).

This chapter addresses the latter—future environmental activities—in a systematic way. In the end, dealing with external environmental costs is just one step towards overcoming the fact, as pointed out in *The Economist*, that 'most environmentalists know even less about accounting than most accountants know about the environment' (*The Economist* 1993: 69).

# Part Four

## Implementation

# 21 Implementing Environment-Related Management Accounting

Martin Bennett and Peter James

The preceeding chapters have described and evaluated several different concepts and case studies of initiatives taken in environment-related management accounting in business. To provide practical guidance for those wishing to apply these approaches in their own organisations, the main action points that have been detailed earlier in this book are summarised below.

## ▌ Strategic Actions

- Identify the organisation's significant environmental impacts (which should already have been done by the environmental management function). For each, analyse the potential short-, medium- and long-term cost and benefit implications of possible changes in:
    - Government policies (e.g. on transport or carbon dioxide emissions)
    - Legislation and regulation (e.g. more extensive and severe carbon taxation)
    - Supply conditions (e.g. fewer landfill sites, higher resource costs to reflect the environmental impacts of their production)
    - Market conditions (e.g. export restrictions, changing customer views)
    - Social attitudes (e.g. increased public unacceptability of certain processes or substances)
    - Competitor strategies (e.g. introducing new, more environmentally benign, products or processes)

- If the conclusions are that there could be significant financial implications:
  - Prepare a summary document for communication to senior managers and use this as a background document for capital investment and product development decision processes.
  - Arrange a briefing on the findings.
  - Develop a 'high environmental pressure' scenario to illustrate the risks that the company may be taking by ignoring or underestimating the cost and benefit implications.
  - Designate an environmental 'champion' within the strategic planning or accounting function to ensure that environmental considerations are fully considered, both directly and by involving environmental managers and other experts at appropriate points.
- Consider whether a Baxter-style 'environmental financial statement' (see Chapter 15) could be of value in heightening awareness of environment-related business opportunities and building linkages between the accounting, environmental management and other functions.

## ▌ Processes

- Ensure that environmental staff are at least consulted about, and ideally are involved in, important accounting processes and vice versa.
- Consider establishing a cross-functional green bottom line team to address all or some of our questions and identify opportunities for action.
- Develop checklists and other assistance to allow internal auditors to understand and address key environmental and energy issues.
- Develop environmental managers' accounting and financial management knowledge.
- Develop accountants' environmental understanding and knowledge.
- Nominate an environmental champion among the accounting function and vice versa.

## ▌ Data Collection and Presentation

- Consider what data would be needed to assess fully the green bottom line both now and in the medium to long term (even if this cannot or is too expensive to be collected now, prior consideration makes it easier to set up collection as and when there are changes in circumstances, e.g. introduction of new accounting systems).
- Assess how much of this data is being collected already and whether better use could be made of it for environment-related management accounting purposes.
- Ensure that environmental staff are consulted or involved in all significant changes to data collection activities (e.g. introduction of ABC, business process re-engineering).

- Consider whether general ledger codes could be modified to collect more or better data related to environmental costs and benefits (e.g. different codes for different energy sources).

- Consider whether non-financial data that is important for assessing environment-related costs and benefits could be captured by management accounting or other systems (e.g. data on waste collected at point of generation rather than disposal).

- Assess whether existing data on energy and water consumption is adequate and/or sufficiently reliable to understand and manage their costs effectively. Would it be worth upgrading existing or installing new energy and water meters to obtain better data on how much is consumed, and where?

- Consider whether data presentation could be improved in order to drive environmental improvement more effectively (e.g. by making comparisons between sites, offices, etc. easier to highlight poor performance and generate peer pressure for action).

- Assess the extent to which data collection and usage is being hampered by non-standardised definitions and/or collection procedures. If so, can this be remedied?

## ▌ Costing and Budgeting

- Analyse which process and product cost objects and drivers are environment-related and analyse the potential short-, medium- and long-term implications of possible changes in government policies, legislation and regulation, supply and market conditions, social attitudes and competitor strategies (see above for examples).

- Assess what percentage of total costs is being driven (or could be, in future) to a significant extent by environmental factors and, if significant, communicate the figure within the organisation.

- Analyse whether standard costing models incorporate environment-related costs such as energy and water consumption and waste disposal. If not, revise them. If so, assess whether they are encouraging improved performance over time—both absolutely and with regard to competitors.

- Consider an experimental environmental costing exercise on a product, process or site as a means of building awareness and/or developing a methodology or relevant information for application elsewhere in the organisation.

- Ensure that all significant environment-related costs are allocated to specific budgets.

- Consider whether possible future cost increases or other factors would justify applying an internal tax to activities such as waste disposal.

- Ensure that energy and/or environmental managers receive data about variances in environment-related cost items (e.g. monthly energy consumption).

## ▊ Capital Investment

- Ensure that all business cases are scrutinised/signed off by environmental staff.
- Calculate/consider the end-of-life costs of equipment.
- Check whether all current pollution control and environmental management costs have been included.
- Assess possible increases in the costs of pollution control, energy, water and transport and the effects of these on the business case.
- Check if the approvals involve the use of hazardous or controversial substances. What would be the effect of bans, more stringent regulations and/or higher costs for such substances?
- Consider whether environmental groups or other external stakeholders could have objections to the decisions and whether these might influence the costs, payback periods, etc.
- Assess whether the investment will make use of existing pollution control, take-back or other forms of environmental infrastructure and whether the costs of this could increase in future.
- Consider if the investment could generate long-term liabilities and, if so, whether these have been assessed fully.
- Ask if the investment is achieving a substantial improvement in environmental performance compared to what it is replacing or some other reference point (e.g. comparable investments by other organisations). If not, can this be achieved by revisions or alternative proposals?
- Consider whether the decisions will still be valid, and whether any environment-related risks will be acceptable, in a world where much greater attention is paid to environmental issues.

## ▊ Product Development

- Ensure that there is expert environmental input when business cases are being made.
- Revise product development processes to ensure that environment-related costs and benefits in manufacture, use and end-of-life activities are fully considered (e.g. by involving external recyclers and disassemblers at appropriate points) and, wherever possible, quantified (e.g. by a subheading for end-of-life issues in business case templates).
- For each individual product, identify significant environmental impacts over its life-cycle (which should already have been done). For each impact, analyse the potential short-, medium- and long-term cost and benefit implications of possible changes in government policies, legislation and regulation, supply and market conditions, social attitudes and competitor strategies (see above for examples). Use the results to undertake the following tasks:

- Analyse how far the environmental aspects of both the current product and possible future developments of it will influence revenue streams.
- Calculate the end-of-life costs of the product and consider who will be responsible for them.
- Calculate/consider the environment-related costs in use of the product and how these could change in alternative scenarios. Could this potentially offer a competitive advantage or disadvantage?
- Consider if the product could generate long-term liabilities and, if so, whether these have been assessed fully.
- Analyse whether the decisions will still be valid, and whether any environment-related risks will be acceptable, in a world where much greater attention is paid to environmental issues.
- Consider whether the answers to these questions require modification of the product.

# Bibliography

Accounting Standards Board (ASB) (1997) *Provisions and Contingencies* (Financial Reporting Exposure Draft, 14; Milton Keynes: ASB Publications).

Advisory Council on Business and the Environment (ACBE) (1996) *Environmental Reporting and the Financial Sector: An Approach to Good Practice* (London: Department of the Environment).

Agathen, P.A. (1992) 'Dealing with Environmental Externalities', *Public Utilities Fortnightly* 129.4 (15 February 1992): 23ff.

Akzo Nobel (1996) *ARBO-& Milieujaarverslag 1996* (Hengelo, Netherlands: Akzo Nobel Chemicals).

American Institute for Pollution Prevention (AIPP) with J. Aldrich (1992) *A Primer for Financial Analysis of Pollution Prevention Projects* (Cincinnati: University of Cincinnati).

Anderson, R. (1992) 'Accounting with a Conscience', *CA Magazine* 125.2 (February 1992): 62.

Arora, S., and T.N. Cason (1995) 'An Experiment in Voluntary Environmental Regulation: Participation in EPA's 33/50 Program', *Journal of Environmental Economics and Management* 28: 271-86.

Bailey, P. (1991) 'Full Cost Accounting for Life-Cycle Costs: A Guide for Engineers and Financial Analysts, *Environmental Finance* Spring 1991: 13-29.

Bailey, P., and P. Soyka (1996) 'Environmental Accounting: Making it Work for Your Company', *Total Quality Environmental Management* Summer 1996: 13-30.

Barber, J., F. Daley and M. Sherwood (1997) 'Rating Environmental Risk', *Certified Accountant* March 1997: 42-43.

Barth, M., and M. McNichols (1994) 'Estimation and Market Valuation of Environmental Liabilities Relating to Superfund Sites', *Journal of Accounting Research* 32 (Supplement): 177-209.

Bartolomeo, M., M. Bennett and P. James (1998) *Eco-Management Accounting: A Framework for Analysis and Action* (Wolverhampton, UK: Wolverhampton Business School Environmental Management Accounting Group).

Baxter International (1997) *Environmental, Health & Safety Performance Report* (Deerfield, IL: Baxter International, Inc.)

Bebbington, J., and I. Thompson (1996) *Business Concepts of Sustainability and the Implications for Accountancy* (London: Association of Chartered Certified Accountants).

Belkaoui, A. (1984) *Socio-Economic Accounting* (Westport, CT: Quorum Books).

Bennett, M., and P. James (1998a) *Environment-Related Performance Measurement* (London: Association of Chartered Certified Accountants).

Bennett, M., and P. James (1998b) ECOMAC *Case Study: Baxter Healthcare, Irish Manufacturing Operations* (Working paper WP 004/98; Wolverhampton, UK: Wolverhampton Business School, Environmental Management Accounting Group).

Bierma, T.J., and F.L. Waterstraat (1996) 'P2 Assistance from your Chemical Supplier? How Shared Savings Contracts Work', *Pollution Prevention Review* Autumn 1996: 13-24.

Bierma, T.J., and F.L. Waterstraat (1997) *Innovative Chemical Supply Contracts: A Source of Competitive Advantage* (TR-31; Champaign, IL: Illinois Waste Management and Research Center, September 1997).

Birkin, F., and D. Woodward (1997a) 'Introduction', *Management Accounting* (UK) (Management Accounting for Sustainable Development Series): June 1997: 24-26.

Birkin, F., and D. Woodward (1997b) 'From Economic to Ecological Efficiency', *Management Accounting* (UK) (Management Accounting for Sustainable Development Series): July/August 1997: 42-45.

Birkin, F., and D. Woodward (1997c) 'Stakeholder Analysis', *Management Accounting* (UK) (Management Accounting for Sustainable Development Series): September 1997: 58-60.

Birkin, F., and D. Woodward (1997d) 'The Eco-Balance Account', *Management Accounting* (UK) (Management Accounting for Sustainable Development Series): October 1997: 50-52.

Birkin, F., and D. Woodward (1997e) 'Accounting for Sustainable Development', *Management Accounting* (UK) (Management Accounting for Sustainable Development Series): November 1997: 52-54.

Birkin, F., and D. Woodward (1997f) 'A Zero-Base Approach to Accounting for Sustainable Development', *Management Accounting* (UK) (Management Accounting for Sustainable Development Series): December 1997: 40-42.

Blumberg, J., A. Korsvold and G. Blum (1997) *Environmental Performance and Shareholder Value* (Geneva: World Business Council for Sustainable Development).

Boone, C., and H. Howes (1996) FCA: *Barriers and Opportunities. Ontario Hydro's Experience* (Toronto: Ontario Hydro, March 1996).

Boone, C., H. Howes and B. Reuber (1995) *A Canadian Utility's Experience in Linking Sustainable Development, Full Cost Accounting and Environmental Impact Assessment* (Toronto: Ontario Hydro, June 1995).

Bouma, J. (1995) *Environmental Care in the Royal Dutch Air Force and Industry: A Study of the Integration of Environmental Aspects in Strategic Decision-Making Processes* (PhD dissertation; Rotterdam: Erasmus University).

Bouma, J., and T. Wolters (1998) *Management Accounting and Environmental Management: A Survey Among 84 European Companies* (Zoetermeer, Netherlands: EIM).

Brealey, R., and S. Myers (1991) *Principles of Corporate Finance: Application of Option Pricing Theory* (New York: McGraw–Hill).

Bromwich, M., and A. Bhimani (1994) *Management Accounting: Pathways to Progress* (London: Chartered Institute of Management Accountants).

Brouwers, W., and A. Stevels, (1997) 'A Cost Model for the End-of-Life Stage of Electronic Consumer Goods', *Greener Management International* 17 (Spring 1997): 129-39.

Brown, G., Jr (1992) *Replacement Costs of Birds and Mammals* (Juneau, AK: State of Alaska Attorney General's Office).

Brown, R.M., and L.N. Killough (1988) 'How PCs Can Solve the Cost Allocation Problem', *Management Accounting* (US) November 1988: 34-39.

Burritt, R.L., A.T. Craswell and M.C. Wells (1980) 'The Costs and Benefits of Cost Allocation: A Comment', in M.C. Wells (ed.), *Controversies on the Theory of the Firm, Overhead Allocation and Transfer Pricing* (New York: Arno Press).

Busby, J., and C. Pitts (1997) 'Real Options and Capital Investment Decisions', *Management Accounting* (US) November 1997: 38-39.

*Business and the Environment* (1998) 'New Tools for Getting the Green Message across to the Financial Sector', *Business and the Environment* 9.2: 1-3.

Business Council for Sustainable Development (BCSD) and University of Houston (1997) *Environmental Cost Accounting for Chemical and Oil Companies: A Benchmarking Study* (EPA 742-R-97-004; Washington, DC: EPA).

Business in the Environment and Extel Financial (1994) *City Analysts and the Environment* (London: Business in the Environment).

Butler, D. (1997) 'I'm Green, Buy Me', *Accountancy* March 1997: 36-38.

Cairncross, F. (1991) *Costing the Earth: What Governments Must Do; What Consumers Need to Know; How Business Can Profit* (London: Great Britain Books).

Cairncross, F. (1995) *Green Inc.: A Guide to Business and the Environment* (London: Earthscan).

Canadian Institute of Chartered Accountants (CICA) (1992) *Environmental Accounting and the Role of the Accounting Profession* (Toronto: CICA).

Canadian Institute of Chartered Accountants (CICA) (1994) *Reporting on Environmental Performance* (Toronto: CICA).

Canadian Institute of Chartered Accountants (CICA) (1997) *Full Cost Accounting from an Environmental Perspective* (Toronto: CICA).

Carson, R.T. (1992) *A Contingent Valuation Study of Lost Passive Use Values Resulting from the Exxon Valdez Oil Spill: A Report to the Attorney General of the State of Alaska* (Juneau, AK: State of Alaska Attorney General's Office).

Centraal Bureau voor de Statistiek (CBS), Netherlands (1994) *Milieukosten van bedrijven* (Voorburg/Heerlen, Netherlands: CBS).

Centraal Bureau voor de Statistiek (CBS), Netherlands (1996) *Milieustatistieken voor Nederland 1996* (Voorburg/Heerlen, Netherlands: CBS).

Centre for the Study of Financial Innovation (CSFI) (1995) *An Environmental Risk Rating for Scottish Nuclear* (London: CSFI).

Chambers, R.J. (1966) *Accounting, Evaluation and Economic Behavior* (Houston, TX: Scholars Book Co.).

Charter, M. (ed.) (1992) *Greener Marketing: A Responsible Approach to Business* (Sheffield, UK: Greenleaf Publishing)

Chartered Institute of Management Accountants (CIMA) (UK) (1997) *Environmental Management: The Role of the Management Accountant* (London: CIMA).

CIBA-Geigy (1996) 'Environmental Cost Accounting', presented at the Workshop on Benchmarking for Environmental Cost Accounting for Chemical and Oil Companies, Houston, TX, 27 June 1996.

Clark, K.B., and S.C. Wheelwright (1993) *Managing New Product and Process Development* (New York: The Free Press).

Cohen, M., S. Fenn and J. Naimon (1995) *Environmental and Financial Performance: Are They Related?* (Washington, DC: Investor Responsibility Research Center).

Cohen, W.M., and D.A. Levinthal (1994) 'Fortune Favors the Prepared Firm', *Management Science* 40.2: 227-51.

Colby, S.J., T. Kingsley and B.W. Whitehead (1995) 'The Real Green Issue: Debunking the Myths of Environmental Management', *The McKinsey Quarterly* 2: 132-43.

Cooke, R.M. (1991) *Experts in Uncertainty: Opinion and Subjective Probability in Science* (New York: Oxford University Press).

Cookson, C. (1997) 'World Must Look to Nature's Free Services', *Financial Times* 17 February 1997.

Cooper, R., and R.S. Kaplan (1988) 'Measure Costs Right: Make the Right Decisions', *Harvard Business Review* September/October 1988: 96-103.

Corbett, C.J., and L. Van Wassenhove (1995) 'Environmental Issues and Operation Strategy', in H. Folmer, H.L. Gabel and H. Opschoor (eds.), *Principles of Environmental and Resource Economics* (Aldershot, UK: Edward Elgar).

Cormier, D., M. Magnan and B. Morard (1993) 'The Impact of Corporate Pollution on Market Valuation: Some Empirical Evidence', *Ecological Economics* 8.2: 135-56.

Cormier, D., M. Magnan and B. Morard (1994) *An Empirical Investigation of the Relation between Corporate Pollution and Stock Market Valuation: Some Canadian Evidence* (unpublished manuscript).

Cropper, M.L., and W.E. Oates (1992) 'Environmental Economics: A Survey', *Journal of Economic Literature* 30 (June 1992): 675.

Curvan, M. (ed.) (1995) *Life-Cycle Assessment* (New York: McGraw–Hill).

Datar, S., M.J. Epstein and K. White (1996) *Bristol-Myers Squibb: The Matrix Essentials Product Life Cycle Review* (Stanford, CA: Stanford Business School case).

Davies, A.J. (1997) 'The Environment and Business Today', *Certified Accountant* June: 20-23.

Davy, B. (1996) 'Fairness as Compassion', *Risk: Health, Safety and Environment* 7.2: 99-108.

DeCanio, S. (1993) 'Barriers within Firms to Energy-Efficient Investments', *Energy Policy* 21.9 (September 1993): 906-14.

Deloitte & Touche (1991) *Public Utility Executive Briefs* (London: Deloitte & Touche, 26 April 1991).

Department of the Environment (1996) *Environmental Protection Expenditure by Industry* (London: HMSO).

DeSimone, L., and F. Popoff (1997) *Eco-Efficiency: The Business Path to Sustainable Development* (Cambridge, MA: MIT Press).

Diependaal, M., and F. de Walle (1994) 'A Model for Environmental Costs for Corporations', *Waste Management and Research* 12: 429-39.

Ditz, D., J. Ranganathan and R.D. Banks (eds.) (1995) *Green Ledgers: Case Studies in Corporate Environmental Accounting* (Washington, DC: World Resources Institute).

Dixit, A., and R. Pindyck (1993) *Investment under Uncertainty* (Princeton, NJ: Princeton University Press).

Douglas, M., and A. Wildavsky (1982) *Risk and Culture: An Essay on the Selection of Technical and Environmental Dangers* (Berkeley, CA: University of California Press).

Drury, C., S. Braund, P. Osborne and M. Tayles (1993) *A Survey of Management Accounting Practice in UK Manufacturing Companies* (London: Chartered Association of Certified Accountants).

Drury, C., and M. Tayles (1998) 'Cost System Design for Enhancing Profitability', *Management Accounting* (UK): January: 40-42.

Du Pont (1993) *Corporate Environmentalism: Progress Report* (Wilmington, DE: Du Pont).

Earl, G. (1996a) 'Total Value Analysis of Environmental Investments: A Model', *Greener Management International* 15: 87-102.

Earl, G. (1996b) 'An Objective Model for Economic Evaluation of Environmental Investment Projects', *The Environmentalist* 16: 327-30.

*The Economist* (1993) 'A Green Account: A Management Focus', *The Economist* 4 September 1993: 69.

Elkington, J. (1997) *Cannibals with Forks* (Oxford: Capstone Publishing).

*Environmental Accounting and Auditing Reporter* (EAAR) (1997) 'Are Financial and Environmental Performance Interlinked?', *EAAR* May 1997: 4-8

Environmental Law Institute (1993) *A Framework for Understanding the Relationship between Environmental Liability and Managerial Decisions Affecting Pollution Prevention* (prepared for the US Environmental Protection Agency; Washington, DC: Environmental Law Institute, September 1993).

EPON (1991) *Environmental Report* (Dutch language; Zwolle, Netherlands: EPON).

Epstein, M.J. (1996a) 'Accounting for Product Take-Back', *Management Accounting* August 1996: 29-33.

Epstein, M.J. (1996b) 'Improving Environmental Management with Full Environmental Cost Accounting', *Environmental Quality Management* Autumn 1996: 11-22.

Epstein, M.J. (1996c) *Measuring Corporate Environmental Performance: Best Practices for Costing and Managing an Effective Environmental Strategy* (Chicago: Irwin/Institute of Management Accountants).

Epstein, M.J. (1996d) *Tools and Techniques of Environmental Accounting for Business Decisions* (Management Accounting Guideline No. 40; Hamilton, Ontario: Society of Management Accountants of Canada).

Epstein, M.J. (1996e) 'You've Got a Great Environmental Strategy: Now What?', *Business Horizons* September/October 1996: 53-59.

Ethridge, J., and V. Rogers (1997) 'Transactions that may Prompt Environmental Reporting Problems', *Management Accounting* (US) July 1997: 57-58.

European Commission (EC) (1995) *JOULE ExternE: Externalities of Energy* (6 vols.; DG XII; Luxembourg: European Commission).

European Environment Agency (EEA) (1996) *Environmental Taxes: Implementation and Environmental Effectiveness* (Copenhagen: EEA).

Fédération des experts comptables européens (FEE) (1996) *Expert Statements on Environmental Reports* (Research paper; Brussels: FEE).

Fiskel, J., and K. Wapman (1994) *How to Design for Environment and Minimize Life Cycle Cost* (Mountain View, CA: Decision Focus Inc.).

Freedman, M., and B. Jaggi (1992) 'An Investigation of the Long-Run Relationship between Pollution Performance and Economic Performance: The Case of Pulp and Paper Firms', *Critical Perspectives in Accounting* 3: 315-36.

Freeman, A.M., III, D. Butraw, W. Harrington and A.J. Krupnick (1992) 'Accounting for Environmental Costs in Electric Utility Resource Supply Planning' (Resources for the Future Discussion Paper QE92-14).

Freeman, H.M. (ed.) (1995) *Industrial Pollution Prevention Handbook* (New York: McGraw–Hill).

Fussler, C., with P. James (1996) *Driving Eco-Innovation: A Breakthrough Discipline for Innovation and Sustainability* (London: Pitman Publishing).

General Electric and ICF Incorporated (1987) *Financial Analysis of Waste Management Alternatives* (Fairfield, CT: General Electric Corporate Environmental Programme).

George Beetle Company (1989) *PRECOSIS* (Philadelphia, PA: George Beetle Company).

Gitjenbeek, M., J. Piet and A. White (1995) 'The Greening of Corporate Accounting', in P. Groenwegen, K. Fischer, E. Jenkins and J. Schot (eds.), *The Greening of Industry: Resource Guide and Bibliography* (Washington, DC: Earth Island).

Global Environmental Management Initiative (GEMI) (1994a) *Finding Cost-Effective Pollution Prevention Initiatives: Incorporating Environmental Costs into Business Decision Making. A Primer* (Washington, DC: GEMI, September 1994).

Global Environmental Management Initiative (GEMI) (1994b) 'Business Environmental Cost Accounting Survey', in *Global Environmental Management Initiative '94 Conference Proceedings*, Arlington, VA, 16–17 March 1994: 243.

Golany, B. (1993) 'A Multi-Criteria Evaluation of Methods for Obtaining Weights from Ratio-Scale Matrices', *European Journal of Operational Research* 69: 210-22.

Graff, R.G., E.D. Reiskin, A.L. White and K. Bidwell (forthcoming) *Snapshots of Environmental Cost Accounting* (prepared by Tellus Institute for the US Environmental Protection Agency; Boston, MA: Tellus Institute).

Gray, R.H. (1990) *The Greening of Accountancy: The Profession after Pearce* (London: Certified Accountants Publications).

Gray, R.H., J. Bebbington and D. Walters (1993) *Accounting for the Environment* (London: Paul Chapman and Association of Chartered Certified Accountants).

Gray, R.H., D. Owen and C. Adams (1996) *Accounting and Accountability: Changes and Challenges in Corporate Social and Environmental Reporting* (Hemel Hempstead, UK: Prentice–Hall).

Grayson, L., H. Woolston and I. Tanega (1993) *Business and Environmental Accountability: An Overview and Guide to the Literature* (London: Technical Communications).

Green, K., A. McMeekin and A. Irwin (1994) 'Technological Trajectories and R&D for Environmental Innovation in UK Firms', *Futures* 26.10: 1047-59.

*The Green Business Letter* (1997) 'Less is More: How Du Pont is Turning Sustainability into an Engine for Profitability', *The Green Business Letter* October 1997: 1ff.

Grimstead, B., *et al.* (1994) 'A Multimedia Assessment Scheme to Evaluate Chemical Effects on the Environment and Human Health', *Pollution Prevention Review* Summer 1994: 259-68.

Haas, M. (1996) *Milieu-Classificatie-model Bouw* (Bussum, Netherlands: Nederlands Instituut voor Bouwbiologie en Ecologie bv).

Haasis, H.D. (1996) *Betriebliche Umweltökonomie* (Berlin: Springer Verlag).

Hahn, R.W. (1994) 'United States Environmental Policy: Past, Present And Future', *National Resources Journal* 34.1 (Winter 1994): 305-48.

Hallay, H., and R. Pfriem (1992) *Öko-Controlling. Umweltschutz in mittelständischen Unternehmen* (Frankfurt: Campus).

Hamlen, S.S., W.A. Hamlen and J.T. Tschirhart (1977) 'The Use of Core Theory in Evaluating Joint Cost Allocation Schemes', *The Accounting Review* 52: 616-27.

Hamner, B., and C. Stinson (1993) *Managerial Accounting and Compliance Costs* (University of Washington discussion paper; Washington, DC: University of Washington; reprinted in *Journal of Cost Management* Summer 1995: 4-10).

Harker, P.T. (1989) 'The Art and Science of Decision Making', in B.L. Golden, E.A. Wasil and P.T. Harker (eds.), *The Analytic Hierarchy Process Applications and Studies* (Berlin: Springer Verlag).

Harr, J. (1995) *A Civil Action* (New York: Random House).

Hayes, R.H., S.C. Wheelwright and K.B. Clark (1988) *Dynamic Manufacturing: Creating the Learning Organization* (New York: The Free Press).

Hiromoto, T. (1988) 'Another Hidden Edge: Japanese Management Accounting', *Harvard Business Review* July/August 1988: 23-26.

Hongisto, M. (1997) 'Assessment of External Costs of Power Production', paper presented to *IIIEE and VTT Seminar on Total Cost Assessment*, Nagu, June 1997 (Helsinki: IVO).

Hopfenbeck, W., and C. Jasch (1993) *Öko-Controlling: Umdenken zahlt sich aus! Audits, Umweltberichte und Ökobilanzen als betriebliche Führungsinstrumente* (Eco-Controlling: New Philosophy Pays! Audits, Environmental Reports and LCA as Management Tools) (Landsberg: Verlag Moderne Industrie).

Horngren, C.T., G. Foster and S.M. Datar (1994) *Cost Accounting: A Managerial Emphasis* (Englewood Cliffs, NJ: Prentice–Hall).

Huber, J. (1991) 'Ecologische modernisering: Weg van schaarste, soberheid en bureaucratie?' ('Ecological Modernisation: Abandoning Society, Sobriety and Bureaucracy?'), in A.P.J. Mol (ed.), *Technologie en Milieubeheerr* (Den Haag: Staats Drukkereruen Ultseveru).

Hughes, S., and D. Willis (1995) 'How Quality Control Concepts can Reduce Environmental Expenditures', *Journal of Cost Management* Summer 1995: 15-19.

Humphreys, K. (ed.) (1984) *Project and Cost Engineers Handbook* (American Association of Cost Engineers; New York: Marcel Dekker).

Hunt, C.B., and E.R. Auster (1990) 'Proactive Environmental Management: Avoiding the Toxic Trap', *Sloan Management Review* Winter 1990: 7-18.

ICI (1994) *Environmental Report 1994* (London: ICI).

INFORM (1985) *Cutting Chemical Wastes* (New York: INFORM).

INFORM (1992) *Environmental Dividends: Cutting More Chemical Wastes* (New York: INFORM).

Institute of Chartered Accountants in England and Wales (ICAEW) (1996) *Environmental Issues in Financial Reporting* (London: ICAEW).

Institute of Management Accountants (IMA) (US) (1981) *Definition of Management Accounting* (Statement on Management Accounting, 1A; Montvale, NJ: IMA, 19 March 1981).

Institute of Management Accountants (IMA) (US) (1990) *Management Accounting Glossary* (Statement on Management Accounting, 2A; Montvale, NJ: IMA).

Institute of Management Accountants (IMA) (US) (1993) *Practices and Techniques: Implementing Activity-Based Costing* (Statement on Management Accounting, 4T; Montvale, NJ: IMA, 30 September 1993; reprint of the Society of Management Accountants of Canada Management Accounting Guideline, 17, *Implementing Activity-Based Costing*).

Institute of Management Accountants (IMA) (US) (1995) *Practices and Techniques: Implementing Corporate Environmental Strategies* (Montvale, NJ: IMA).

International Accounting Standards Committee (IASC) (1997) *Provisions, Contingent Liabilities and Contingent Assets* (Exposure Draft, 59; London: International Accounting Standards Committee).

International Institute for Industrial Environmental Economics (IIIEP) and VTT Non-Waste Technology (1997) *Challenges and Approaches to Incorporating the Environment into Business Decisions: International Expert Seminar* (Lund, Sweden: IIIEP; Helsinki: VTT).

Islei, G., and A.G. Lockett (1988) 'Judgmental Modelling Based on Geometric Least Squares', *European Journal of Operational Research* 36: 27-35.

Ito, Y. (1995) 'Strategic Goals of Quality Costing in Japanese Companies', *Management Accounting Research* 6: 383-97.

Jaffe, A.B., S.R. Peterson, P.R. Portney and R.N. Stavins (1995) 'Environmental Regulation and the Competitiveness of US Manufacturing: What Does the Evidence Tell Us?', *Journal of Economic Literature* 33 (March 1995): 132-63.

James, P., M. Prehn and U. Steger (1997) *Corporate Environmental Management in Britain and Germany* (London: Anglo-German Foundation).

Johnson, H.T. (1992) 'It's Time to Stop Overselling Activity-Based Concepts', *Management Accounting* (US) September 1992: 26-35.

Johnson, H.T. (1994) 'Relevance Regained: Total Quality Management and the Role of Management Accounting', *Critical Perspectives on Accounting* 5: 259-67.

Johnson, H.T., and R.S. Kaplan (1987) *Relevance Lost: The Rise and Fall of Management Accounting* (Boston, MA: Harvard Business School Press).

Johnston, N. (1994) *Waste Minimisation: A Route to Profit and Cleaner Production* (London: Centre for Exploitation of Science and Technology).

Jurewitz, J.L. (1978) *The Internalization of Environmental Costs in the Private Electric Utility Industry* (Thesis; Madison, WI: University of Wisconsin at Madison).

Kainz, R., M. Prokopyshen and S. Yester (1996) 'Life Cycle Management at Chrysler', *Pollution Prevention Review* Spring 1996: 71-83.

Kaplan, R., and D. Norton (1992) 'The Balanced Scorecard: Measures that Drive Performance', *Harvard Business Review* January/February 1992: 71-79 (reprint 92105).

Kaplan, R., and D. Norton (1993) 'Putting the Balanced Scorecard to Work', *Harvard Business Review* September/October 1993: 134-42 (reprint 93505).

Kaplan, R., and D. Norton (1996a) 'Using the Balanced Scorecard as a Strategic Management System', *Harvard Business Review* January/February 1996: 75-85 (reprint 96107).

Kaplan, R., and D. Norton (1996b) *The Balanced Scorecard* (Cambridge, MA: Harvard Business School Press).

Keeney, R.L. (1976) *Decisions with Multiple Objectives: Preferences and Value Trade-Offs* (New York: John Wiley).

Kirschner, E. (1994) 'Full Cost Accounting for the Environment: Gold Mine or Minefield?', *Chemical Week* 9 March 1994: 25-26.

Kite, D. (1995) 'Capital Budgeting: Integrating Environmental Impact', *Journal of Cost Management* Summer 1995: 11-19.

Klassen, R.D., and D.C. Whybark (1995) *Plant-Level Choices of Environmental Technologies: The Influence of Environmental Management Strategy* (Working Paper Series, No. 96-01).

Koechlin, D., and K. Müller (1992) 'Environmental Management and Investment Decisions', in: D. Koechlin and K. Müller (eds.), *Green Business Opportunities: The Profit Potential* (London: Pitman).

KPMG, *Survey of Environmental Reporting 1997* (London: KPMG).

Kreuze, J.G., and G.E. Newell (1994) 'ABC and Life-Cycle Costing for Environmental Expenditures', *Management Accounting* (US) February 1994: 38-42.

Kreuze, J.G., G.E. Newell and S.J. Newell (1996) 'Environmental Disclosures: What Companies are Reporting', *Management Accounting* (US) July 1996: 37-43

Kula, E. (1992) *Economics of Natural Resources* (London: Chapman & Hall).

Kunert AG (1995) *Environmental Report* (Immenstadt, Germany: Kunert AG).

Lascelles, D. (1993) *Rating Environmental Risk* (London: Centre for the Study of Financial Innovation).

Lawrence, J.E., and D. Cerf (1995) 'Management and Reporting of Environmental Liabilities', *Management Accounting* (US) August 1995: 48-54.

Lefebvre, L.A., E. Lefebvre and M.-J. Roy (1995) 'Integrating Environmental Issues into Corporate Strategy: A Catalyst for Radical Organizational Innovation', *Creativity and Innovation Management* 4.4: 209-22.

Leggett, J. (1995) *Die Klimaveränderung und der Finanzsektor* (*Climate Change and the Financial Sector*) (Hamburg: Greenpeace).

Leonard-Barton, D. (1992) 'Core Capabilities and Core Rigidities: A Paradox in Managing New Product Development', *Strategic Management Journal* 13 (special issue): 111-25.

Leonard-Barton, D. (1995) *Wellsprings of Knowledge: Building and Sustaining the Sources of Innovation* (Boston, MA: Harvard Business School Press).

Linnerooth-Bayer, J., and K.B. Fitzgerald (1996) 'Conflicting Views on Fair Siting Processes', *Risk: Health, Safety and Environment* 7.2: 119-34.

Lober, D., D. Bynum, E. Campbell and M. Jacques (1997) 'The 100 Plus Corporate Environmental Report Study', *Business Strategy and the Environment* May 1997: 1-12.

Long, G.E. (1997) 'Justice, Culture and NIMBYism', in C. Ashton and M. Nicholas (eds.), *Engineering Doctorate Annual Conference 1997* (Guildford, UK: University of Surrey, Centre for Environmental Strategy).

Magness, V. (1997) 'Environmental Accounting in Canada: New Challenges to Old Theory', *CMA: The Management Accounting Magazine* 71.1 (February 1997): 15-18.

Mansley, M. (1995) *Long Term Financial Risks to the Carbon Fuel from the Climate Change* (London: The Delphi Industry Group).

Maunders, K.T., and R.L. Burritt (1991) 'Accounting and Ecological Crisis', *Accounting, Auditing and Accountability Journal* 4: 9-26.

McAuley, L., G. Russell and J. Sims (1997) 'How Do Financial Directors Make Decisions?', *Management Accounting* November 1997: 32-34.

McLaughlin, S., and H. Elwood (1996) 'Environmental Accounting and EMSs', *Pollution Prevention Review* Spring 1996: 13-21.

McLean, R., and J. Shopley (1996) 'Green Light Shows for Corporate Gains', *Financial Times*, 3 July 1996.

Mercer, D. (1995) 'Simpler Scenarios', *Management Decision* 33.4: 32-40.

Milne, M.J. (1991) 'Accounting, Environmental Resource Values, and Non-Market Valuation Techniques for Environmental Resources: A Review', *Accounting, Auditing, and Accountability Journal* 4.3: 81-109.

Ministerie van Volkshuisvesting, Ruimtelijke Ordening en Milieubeheer (VROM), Netherlands (1994) *Methodiek Milieukosten* (publikatiereeks Milieubeheer, 1; Zoetermeer, Netherlands: VROM).

Ministerie van Volkshuisvesting, Ruimtelijke Ordening en Milieubeheer (VROM), Netherlands (1995) *Paasbrief 1995* (Rijksgebouwendients, Inspectie Milieuhygiene voor de Rijkshuisvesting; Zoetermeer, Netherlands: VROM).

Moilanen, T., and C. Martin (1996) *Financial Evaluation of Environmental Investments* (Rugby, UK: Institute of Chemical Engineers [IChemE]).

Monty, R.L. (1991) 'Beyond Environmental Compliance: Business Strategies for Competitive Advantage', *Environmental Finance* 1.1 (Spring 1991): 3-11.

Morgan, M., and P. Weerakoon (1993) 'Japanese Management Accounting: Its Contribution to the Japanese Economic Miracle', in J. Ratnatunga, J. Miller, N. Mudalige and A. Sohal (eds.), *Issues in Strategic Management Accounting* (Sydney: Harcourt Brace Jovanovich).

Müller, K., J. de Frutos, K. Schüssler and H. Haarbosch (1994) *Environmental Reporting and Disclosures: The Financial Analyst's View* (London: Working Group on Environmental Issues of the Accounting Commission of the European Federation of Financial Analysts' Societies).

Müller, K., J. de Frutos, K. Schüssler, H. Haarbosch and M. Randel (1996) *Eco-Efficiency and Financial Analysis* (Basel: European Federation of Financial Analysts' Societies).

Nagle, G. (1994) 'Business Environmental Cost Accounting Survey', presented at *the Conference on Environmental Management in a Global Economy*, Arlington, VA: 243-48.

National Research Council (NRC) (1991) *Environmental Epidemiology: Public Health and Hazardous Wastes* (Washington, DC: National Academy Press).

New Jersey Department of Environmental Protection and Energy (NJ DEPE) Office of Pollution Prevention (1995) *Preliminary Finding* (Trenton, NJ: NJ DEPE, February).

Northeast Waste Management Officials Association (NEWMOA) (1994) *Improving Your Competitive Position: Strategic and Financial Assessment of Pollution Prevention Projects* (Boston, MA: NEWMOA).

Oak Ridge National Laboratory and Resources for the Future (1992–96) *External Costs and Benefits of Fuel Cycles* (8 vols.; Washington, DC: US Department of Energy).

Office of Technology Assessment (1994) *Studies of the Environmental Costs of Electricity* (Washington, DC: US Government Printing Office).

Ontario Hydro (1993a) *Full Cost Accounting Workshop: Proceedings* (Toronto: Ontario Hydro, June 1993).

Ontario Hydro (1993b) *A Strategy for Sustainable Energy Development for Ontario Hydro* (Toronto: Ontario Hydro, December 1993).

Ontario Hydro (1993c) *Full Cost Accounting for Decision Making* (Toronto: Ontario Hydro, December 1993).

Ontario Hydro (1994) *Sustainable Development/Environmental Performance Report* (Toronto: Ontario Hydro).

Ontario Hydro (1995a) *Sustainable Energy Development: Policy and Principles* (Toronto: Ontario Hydro, April 1995).

Ontario Hydro (1995b) *Ontario Hydro's Corporate Guidelines for Full Cost Accounting* (Toronto: Ontario Hydro, September 1995).

Organisation for Economic Co-operation and Development (OECD) (1994) *Managing the Environment: The Role of Economic Instruments* (Paris: OECD).

Organisation for Economic Co-operation and Development (OECD) (1997) *Environmental Taxes and Green Tax Reform* (Paris: OECD).

Owen, D. (ed.) (1992) *Green Reporting: Accountancy and the Challenge of the Nineties* (London: Chapman & Hall).

Owen, D., R. Gray and R. Adams (1997) *Corporate Environmental Disclosure: Encouraging Trends* (London: Association of Chartered Certified Accountants).

Owen, D. (1993) 'The Emerging Green Agenda: A Role for Accounting?', in D. Smith (ed.) *Business and the Environment: Implications of the New Environmentalism* (New York: St Martin's Press): 55-74.

Owen, J.V. (1995) 'Environmental Compliance: Managing the Mandates', *Manufacturing Engineering* March 1995.

Parker, J.N. (1995) 'Profits and Ethics in Environmental Investments', *Management Accounting* (US) October 1995: 52-53.

Parker, J.N. (1996) 'The Importance of Environmental Cost Accounting', *Management Accounting* (US) (December 1996): 63.

Popoff, F., and D. Buzzelli (1993) Full Cost Accounting, *Chemical Engineering News* 11 January 1993: 8-10.

Porter, M.E. (1985) *Competitive Advantage: Creating and Sustaining Superior Performance* (New York: The Free Press).

Porter, M.E. (1991) 'America's Green Strategy', *Scientific American* April 1991: 168.

Porter, M.E., and C. van der Linde (1995a) 'Toward a New Conception of the Environment–Competitiveness Relationship', *Journal of Economic Perspectives* 9.4 (Fall 1995): 97-118.

Porter, M.E., and C. van der Linde (1995b) 'Green and Competitive: Ending the Stalemate', *Harvard Business Review* September/October 1995: 120-34.

Post, J.E., and B.W. Altman (1994) 'Managing the Environmental Change Process: Barriers and Opportunities', *Journal of Organizational Change Management* 7.4: 64-81.

President's Council on Sustainable Development (1996) *Sustainable America: A New Consensus* (Washington, DC: US Government Printing Office).

Rappaport, A. (1986) *Creating Shareholder Value: The New Standards for Business Performance* (New York: Free Press).

Rice, V.R. (1994) 'Regulating Reasonably', *The Environmental Forum* May/June 1994: 16-23.

Rijksgebouwendienst (Rgd), Netherlands (1996) *Milieukosten Rijkshuisvesting: Belastinggebouw Enschede* (Arnhem, Netherlands: Rgd).

Roberts, J.A. (1996) 'Green Consumers in the 1990s: Profile and Implications for Advertising', *Journal of Business Research* 36: 217-31.

Rogers, E.M. (1983) *Diffusion of Innovations* (New York: The Free Press).

Roome, N.J. (1994) 'Business Strategy, R&D Management and Environmental Imperatives', *R&D Management* 24.1: 65-82.

Ross, S.M. (1985) *Introduction to Probability Models* (New York: Harcourt Brace Jovanovich, 3rd edn).

Roth, H.P., and C.E. Keller (1997) 'Quality, Profits and the Environment: Diverse Goals or Common Objectives?', *Management Accounting* (US) July 1997: 50-55.

Roth, U. (1992) *Umweltkostenrechnung: Grundlagen und Konzeptionen aus betriebswirtschäftlicher Sicht* (Wiesbaden: Deutscher Universitätsverlag).

Royal Society of Arts (RSA) (1994) *Tomorrow's Company* (London: RSA).

Rubenstein, D.B. (1992) 'Bridging the Gap between Green Accounting and Black Ink', *Accounting, Organizations and Society* 17.5: 501-508.

Rubenstein, D.B. (1994) *Environmental Accounting for the Sustainable Corporation* (Westport, CT: Quorum Books).

Rückle, D. (1989) 'Investitionskalküle für Umweltschutzinvestitionen' ('Investment Appraisal for Environmental Investments'), *Betriebswirtschaftliche Forschung und Praxis (BFuP)* 41.1/89: 51-65.

Russell, C.S. (1996a) *Textbook: What does Economics have to Say about the Environment?* (unpublished manuscript).

Russell, C.S. (1996b) *Direct Methods of Benefit Estimation: Thoughts on the Past and Future from a US Point of View* (Working paper AKF ver. 2; Nashville, TN: Vanderbilt University, October 1996.

Russell, C.S., and P.T. Powell (1994) *Report to the Office of Technology Assessment on the Efficiency and Fairness of Candidate Approaches to Environmental Pollution Management* (Nashville, TN: Vanderbilt University, 31 May 1994).

Russell, W., S. Skalak and G. Miller (1995) 'Environmental Cost Accounting: The Bottom Line for Environmental Quality Management', in J. Willig (ed.), *Auditing for Environmental Quality Leadership* (Chichester, UK: John Wiley).

Saaty, T.L. (1980) *The Analytic Hierarchy Process* (New York: McGraw–Hill).

Savage, D., and A. White (1994) 'New Applications of Total Cost Assessment: Exploring the P2-Production Interface', *Pollution Prevention Review* Winter 1994–95: 7-15.

Schaeffer, R., and J. McClave (1990) *Probability and Statistics for Engineers* (Boston, MA: PWG-Kent, 3rd edn).

Schaltegger, S., and F. Figge (1997) *Environmental Shareholder Value* (Basel: Centre of Economics and Business Administration, University of Basel).

Schaltegger, S., with K. Müller and H. Hindrichsen (1996) *Corporate Environmental Accounting* (Chichester, UK: John Wiley).

Schmidheiny, S., and F. Zorraquin (1996) *Financing Change: The Financial Community, Eco-Efficiency and Sustainable Development* (Cambridge, MA: MIT Press with the World Business Council for Sustainable Development).

Schoemaker, P.J.H. (1991) 'When and How to Use Scenario Planning: A Heuristic Approach with Illustration', *Journal of Forecasting* 10: 549-64.

Schoemaker, P.J.H. (1995) 'Scenario Planning: A Tool for Strategic Thinking', *Sloan Management Review* 36 (Winter 1995): 25-40.

Schoemaker, P.J.H., and J.A. Schoemaker (1995) 'Estimating Environmental Liability: Quantifying the Unknown', *California Management Review* 37.3 (Spring 1995): 29-61.

Schulz, E., and W. Schulz (1994) *Umweltcontrolling in der Praxis (Environmental Controlling in Practice)* (Munich: Vahlen).

Shane, P.B. (1982) *Internalization of Social Costs and the Value of the Firm: A Descriptive Empirical Study* (PhD thesis; University of Oregon).

Shell UK (1996) *Brent Spar: Next Stage in New Way Forward* (*Shell UK Bulletin*; London: Shell UK, July 1996).

Shrivastava, P. (1995) 'Environmental Technologies and Competitive Advantage', *Strategic Management Journal* 16: 183-200.

Simmonds, K. (1991) 'Strategic Management Accounting', *Management Accounting* (UK): April 1991: 26-29.

Skea, J. (1995) 'Environmental Technology', in H. Folmer, H.L. Gabel and H. Opschoor (eds.), *Principles of Environmental and Resource Economics* (Aldershot, UK: Edward Elgar).

Spitzer, M. (1992) 'Calculating the Benefits of Pollution Prevention', *Pollution Engineering* 1 September 1992: 33-38.

Statistical Abstract of the United States (1996) (Washington, DC: US Census Bureau).

Statistical Abstract of the United States (1997) (Washington, DC: US Census Bureau).

Sterling, R.R. (1979) *Toward a Science of Accounting* (Houston, TX: Scholars Book Co.).

Stewart, G.B. (1991) *The Quest for Value* (New York: Harper Business).

Stewart, G.B. (1994) 'EVA: Fact and Fantasy', *Journal of Applied Corporate Finance* Summer: 71-84.

Stickney, C.P., Weil and Davidson (1991) *Financial Accounting: An Introduction to Concepts, Methods, and Uses* (New York: Harcourt Brace Jovanovich, 6th edn).

Surma, J.P. (1992) 'Tackling Corporate America's Environmental Challenge', *Price Waterhouse Review* 36.2: 10.

SustainAbility/United Nations Environment Programme (UNEP) (1997) *Engaging Stakeholders: The Third International Progress Report on Company Environmental Reporting* (London: SustainAbility; Paris: UNEP).

Tellus Institute (1991) *Alternative Approaches to the Financial Evaluation of Industrial Pollution Prevention Instruments* (Boston, MA: Tellus Institute).

Tellus Institute (1992) *Total Cost Assessment: Accelerating Industrial Pollution Prevention through Innovative Project Financial Analysis. With Applications to the Pulp and Paper Industry* (EPA-741-R-92-002; Washington, DC: EPA, May).

Tellus Institute (1995) *Environmental Cost Accounting for Capital Budgeting: A Benchmark Study of Management Accountants* (EPA 742-R-95-005; Washington, DC: EPA, September 1995).

Tellus Institute (1997) *Strengthening Corporate Commitment to Pollution Prevention in Illinois: Concepts and Case Studies of Total Cost Assessment* (Prepared for the Illinois Waste Management and Research Center; Boston, MA: Tellus Institute).

Thomas, A.L. (1969) *The Allocation Problem in Financial Accounting Theory* (Evanston, IL: American Accounting Association).

Thomas, A.L. (1974) *The Allocation Problem: Part Two* (Evanston, IL: American Accounting Association).

Thomas, A.L. (1982) 'Goals for Joint-Cost Allocation: An Incompatibility in the Literature', *Abacus* 18: 166-74.

Thomas, S.T., V. Weber, S.A. Berger and I.L. Klawiter (1994) *Estimate the Environmental Cost of New Processes in R&D* (Prepared for American Institute of Chemical Engineers [AIChE] Spring National Meeting; April 1994).

Thompson, M., R. Ellis and A. Wildavsky (1990) *Cultural Theory* (Oxford, UK: Westview Press).

Todd, R. (1992) 'Accounting for the Environment: Zero-Loss Environmental Accounting Systems', Presented at the *National Academy of Engineering, Industrial Ecology/Design for Engineering Workshop*, 13–17 July 1992.

Tuppen, C. (ed.) (1996) *Environmental Accounting in Industry: A Practical Review* (London: British Telecom).

United Nations Development Programme (UNDP) (1997) *Valuing the Environment: How Fortune 500 CFOs and Analysts Measure Corporate Performance* (ODS Working Paper; New York: United Nations Office of Development Studies).

US Environmental Protection Agency (EPA) (1975) *Assessment of Industrial Hazardous Waste Management Practices, Rubber and Plastics Industry* (EPA 530-SW-163c.1, 163c.2, 163c.3 and 163c.4; Washington, DC: EPA/Florham Park, NJ: Foster D. Snell, Inc.).

US Environmental Protection Agency (EPA) (1980) *Hazardous Waste Generation and Commercial Hazardous Waste Management Capacity: An Assessment* (EPA SW-894; Washington, DC: EPA).

US Environmental Protection Agency (EPA) (1984) *National Survey of Hazardous Waste Generators and Treatment, Storage and Disposal Facilities Regulated under RCRA in 1981* (EPA 530/SW-84-005; Washington, DC: EPA).

US Environmental Protection Agency (EPA) (1989) *Pollution Prevention Benefits Manual* (Washington, DC: EPA).

US Environmental Protection Agency (EPA) (1993a) *Life Cycle Assessment: Inventory Guidelines and Principles* (EPA-600-R-92-245; Washington, DC: EPA).

US Environmental Protection Agency (EPA) (1993b) *Life Cycle Design Guidance Manual: Environmental Requirements and the Product System* (EPA-600-R-92-226; Washington, DC: EPA).

US Environmental Protection Agency (EPA) (1994) *Stakeholders' Action Agenda: A Report of the Workshop on Accounting and Capital Budgeting for Environmental Costs, 5–7 December 1993* (EPA 742-R-94-003; Washington, DC: EPA, May 1994).

US Environmental Protection Agency (EPA) (1995a) *An Introduction to Environmental Accounting as a Business Management Tool: Key Concepts and Terms* (EPA 742-R-95-001; Washington, DC: EPA).

US Environmental Protection Agency (EPA) (1995b) *Environmental Accounting Case Studies: Green Accounting at AT&T* (EPA 742-R-95-003; Washington DC: EPA).

US Environmental Protection Agency (EPA) (1996a) *Incorporating Environmental Costs and Considerations into Decisionmaking: Review of Available Tools and Software* (EPA 742-R-95-006; Washington, DC: EPA).

US Environmental Protection Agency (EPA) (1996b) *Valuing Potential Environmental Liabilities for Managerial Decision-Making: A Review of Available Techniques* (EPA 742-R-96-003; Washington, DC: EPA).

US Environmental Protection Agency (EPA) (1996c) *Full Cost Accounting for Decision-Making at Ontario Hydro* (EPA 742-R-95-004; Washington, DC: EPA).

US General Accounting Office (GAO), Superfund (1996) *Outlook for and Experience with Natural Resource Damage Settlements* (RCED-96-71).

US Office of Management and Budget (1976) *Circular A-109* (Washington, DC: Office of Management and Budget, 5 April 1976).

Verschoor, C.C. (1997) 'Principles Build Profits', *Management Accounting* (US) October 1997: 42-46.

Vogan, C.R. (1996) 'Pollution Abatement and Control Expenditures, 1974–1994', *Survey of Current Business* 76.9 (September 1996): 48ff.

Von Weizsäcker, E.U., A. Lovins and L.H. Lovins (1997) *Factor Four: Doubling Wealth, Halving Resource Use* (London: Earthscan).

Wagner, B. (1995) *Arbeitsmaterialien Umweltmanagement* (*Working Materials: Environmental Management*) (Augsburg: Scriptum).

Wagner, G.R., and H. Janzen (1993) 'Ökologisches Controlling: Mehr als ein Schlagwort?', *Controlling* 3: 120-29.

Walley, N., and B. Whitehead (1994) 'It's Not Easy Being Green', *Harvard Business Review* May/June 1994: 46-52. (See also follow-up letters: 'The Challenge of Going Green', *Harvard Business Review*, July/August 1994: 37-50.)

Washington State Department of Ecology (1993) *Pollution Prevention Planning in Washington State Businesses* (Olympia, WA: Washington State Department of Ecology, May 1993).

Waste Reduction Institute for Training and Applications Research (WRITAR) (1997) *Applying Environmental Accounting to Electroplating Operations* (EPA 742-R-97-003; Washington, DC: EPA).

Wells, M.C. (1980) *Controversies on the Theory of the Firm, Overhead Allocation and Transfer Pricing* (New York: Arno Press).

Western Mining Corporation (WMC) (1995) *Environmental Progress Report 1994–95* (Melbourne: WMC).

White, A., and M. Becker (1992) 'Total Cost Assessment: Catalysing Corporate Self Interest in Pollution Prevention', *New Solutions* Winter 1992: 34.

White, A., and D. Savage (1995) 'Budgeting for Environmental Projects: A Survey', *Management Accounting* (US) October 1995: 48-54.

White, A., M. Becker and J. Goldstein (1991) *Total Cost Assessment: Accelerating Industrial Pollution Prevention through Innovative Project Financial Analysis* (Sponsored by the US Environmental Protection Agency; Boston, MA: Tellus Institute).

White, A., M. Becker and D. Savage (1993) 'Environmentally Smart Accounting: Using Total Cost Assessment to Advance Pollution Prevention', *Pollution Prevention Review* Summer 1993: 247-59.

White, A., D. Savage, J. Brody, D. Cavanader and L. Lach (1995) *Environmental Cost Accounting for Capital Budgeting: A Benchmark Survey of Management Accountants* (Sponsored by the US Environmental Protection Agency; Boston, MA: Tellus Institute).

Williams, T.A., S.S. Brewer, P.N. Mishra and O.W. Underwood (1995) 'Pollution Prevention at General Motors', in H.M. Freeman (ed.), *Industrial Pollution Prevention Handbook* (New York: McGraw–Hill).

Wilson, R. (ed.) (1997) *Strategic Cost Management* (Aldershot, UK: Ashgate).

Winsemius, P. (1986) *Gast in eigen huis* (Alphen aan den Rijn, Netherlands: Sansom H.D. Tjiink Willink).

Winsemius, P., and W. Hahn (1992) 'Environmental Option Assessment', *The Columbia Journal of World Business* Fall/Winter 1992: 249-66.

Winterfeld, D., and W. Edwards (1987) 'Public Risk Values in Risk Debates', *Risk Analysis* 7.2: 141-58.

Wolters, T., and M. Bouman (eds.) (1995) *Milieu-investeringen in Bedrijfseconomisch Perspectief* (Alphen aan den Rijn/Zaventem, Netherlands: Sansom Bedrijsinformatie).

Wood, J. (1998) 'Environmental Life Cycle Costing' (Mimeo available from Life Cycle Dimensions, 30 Cambridge Road, Bedminster, NJ 07921, USA).

World Commission on Environment and Development (WCED) (1987) *Our Common Future* (The 'Brundtland Report'; New York: Oxford University Press).

Zadek, S., P. Pruzan and R. Evans (eds.) (1997) *Building Corporate Accountability: The Emerging Practices in Social and Ethical Accounting, Auditing and Reporting* (London: Earthscan).

Zeller, T.L., and D.M. Gillis (1995) 'Achieving Market Excellence through Quality: The Case of Ford Motor Company', *Business Horizons* May/June 1995: 13-24.

Zimmerman, J.L. (1979) 'The Costs and Benefits of Cost Allocations', *The Accounting Review* 54: 504-21.

# Biographies

**R. Darryl Banks** is a Senior Fellow for technology and environmental issues at the World Resources Institute (WRI). He directs and conducts research activities in the following areas: environmental technology policy; technology innovation and environmental regulatory policy reform; strategic corporate environmental management including ground-breaking work on environmental accounting and performance measurement; and international technology transfer.

A scientist and expert on environmental technology, Dr Banks has extensive experience in environmental policy management. Prior to joining WRI in 1992, he served as Deputy Commissioner for Environmental Quality at the New York State Department of Environmental Conservation for eight years. Before his appointment as Deputy Commissioner in New York in 1983, he was on the research staff at the Rand Corporation, and was executive assistant to the Assistant Administrator for Research and Development at the US Environmental Protection Agency. He serves on numerous boards of directors and advisory groups for a range of organisations. He is a Rhodes Scholar, and holds degrees in biochemistry and chemistry from Oxford University and Coe College.

---

**Matteo Bartolomeo** has a Degree in Economics, and a Masters Degree in Environmental Management. He is a PhD student of Environmental Benchmarking at Erasmus University, Rotterdam. As a senior researcher at Fondazione Eni Enrico Mattei, Milan, Italy, he is currently working in the field of business and environmental issues. He is also the co-founder of an innovative research institute dealing with environmental management issues from a multidisciplinary perspective. He is the author of several publications, a member of the Expert Working Group on Environmental Accounting for UNCTAD-ISAR, and is also a member of the Editorial Board for the *Environmental Accounting and Auditing Reporter* and for the *Social and Environmental Accounting Newsletter*.

---

**Beth R. Beloff** has been Director of the Institute for Corporate Environmental Management at the University of Houston, Texas, since 1992 when she founded ICEM. Her research efforts have been focused in the area of environmental accounting and interdisciplinary education. She is also the co-founder of BRIDGES, a non-profit-making organisation whose aim is to foster global sustainable development through innovative partnerships, leadership and education. Beth Beloff received an MBA from the University of Houston, an MArch degree from UCLA and a BA from UC Berkeley. Her publications include her contribution to the WRI Publication, *Green Ledgers: Case Studies in Corporate Environmental Accounting*, and she serves on a number of committees and boards, including the Business Council for Sustainable Development, Gulf of Mexico, where she chairs the Education and Training Task Force.

---

**Martin Bennett** is Principal Lecturer in Financial Management at the Business School of the University of Wolverhampton, and leads the Environmental Management Accounting Group in the School's Management Research Centre. He was previously in the accountancy profession with KPMG and BDO Binder Hamlyn, in commerce with Great Universal Stores, and in education with Nottingham Trent University and Ashridge Management College. He researches into environmental management accounting and performance measurement., and publications (with Peter James) include *The Green Bottom Line: Environmental Accounting for Management, Current Practice and Future Trends* (Greenleaf, forthcoming); *Environment under the Spotlight: Current Practice and Future Trends in Environment-Related Performance Measurement for Business* (Association of Chartered Certified Accountants Research Report); and a substantial contribution to *Environmental Accounting in Industry: A Practical Approach* (British Telecom, ed. C Tuppen). Martin runs seminars and courses on environmental accounting and performance measurement at Brunel, Carnegie Mellon and Ghent Universities, and is a member of the editorial board of the journal '*Greener Management International*'.

---

**Thomas J. Bierma** is Professor and Acting Director of the Environmental Health Programme at Illinois State University, USA. He holds a PhD in Public Health and a Master of Business Administration degree from the University of Illinois. His work in the field of business and the environment includes a series of course modules for graduate schools of business, a study of cleaner production decision-making among small businesses, the application of marketing principles to the diffusion of cleaner production technologies, and the evaluation of innovative chemical supply strategies. Dr Bierma's current research focuses on assessing the total cost of ownership for chemicals in manufacturing.

---

**Jan Jaap Bouma** is a senior researcher at the Erasmus Centre for Environmental Studies (ECES) at the Erasmus University, Netherlands. He is a business economist and holds a PhD in environmental management. ECES is involved in research projects that focus on environmental management and its relationship with management accounting, to gain more knowledge about the shortcomings of current management accounting systems and develop techniques to achieve an adequate environmental management. Within ECES, a group of (business) economists work on adapting existing management accounting and designing new accounting systems and techniques.

---

**Roger Burritt** is a Senior Lecturer in Commerce at The Australian National University and Co-ordinator of the Asia Pacific Centre for Environmental Accountability (APCEA). His research interests include environmental accountability and management accounting.

---

Professor **Roland Clift** is Professor of Environmental Technology and Founding Director of the Centre for Environmental Strategy (CES) at the University of Surrey, UK. CES was set up in 1992 as a multidisciplinary research centre concerned with long-term environmental problems. In addition to its research activities, CES runs MSc, PhD and EngD programmes, the last of these being an innovative Doctor of Engineering programme in Environmental Technology. He is a member of the Royal Commission on Environmental Pollution and the UK Eco-labelling Board. He is a Fellow of the Royal Academy of Engineering and of the Institution of Chemical Engineers.

---

**Daryl Ditz** is Director of Environmental Management Programmes at the Environmental Law Institute, Washington, DC. He is currently managing a major project on environmental law and industrial management in India, overseeing a set of dialogues on sustainable development in the US South, and a series of programmes in corporate environmental management. Before joining ELI, he led several projects at the World Resources Institute, including innovative work with the business community such as *Green Ledgers: Case Studies in Corporate Environmental Accounting*. His interest in environmental accounting led to work on environmental performance metrics, as reflected in the 1997 WRI report with Janet Ranganathan, *Measuring Up: Toward a Common Framework for Tracking Environmental Progress*. He has conducted research on hazardous waste management in India, western Europe and the United States. Dr Ditz has lectured and published widely on risk management, information disclosure and other aspects of environmental policy. He holds a BSc in Chemical Engineering from the University of Wisconsin and a PhD in Engineering and Public Policy from Carnegie Mellon University.

---

**Graham Earl** is a Research Engineer with the Engineering Doctorate (EngD) programme and is sponsored by the Centre for Environmental Strategy at the University of Surrey, UK, and Paras Ltd, Isle of Wight, UK. His doctoral research project which is currently in its final year has involved extensive liaison and test case applications with leading multinational companies and has resulted in the development of the Stakeholder Value Analysis Toolkit. This is a hybrid model which links together a collection of decision-support tools and aims to support decision-makers in identifying, measuring and linking the stakeholder values driving environmental investment decisions. The results of his research have been widely presented in leading journals and conferences. Prior to this research, Mr Earl—a Chartered Engineer—worked as a consultant to the petrochemical industry, providing expertise on safety, quantitative risk and reliability of complex systems, including offshore oil installations, oil and gas pipelines and onshore chemical plants.

---

**Marc J. Epstein** is presently Price Waterhouse Visiting Professor of Accounting and Control at INSEAD. Formerly a professor at both Harvard and Stanford Business Schools, he is also the editor-in-chief of *Advances in Management Accounting* and author of over fifty articles and twelve books on accounting and environmental topics.

---

**Terry Foecke** is the President of WRITAR (Waste Reduction Institute for Training and Applications Research, Inc.), an independent non-profit organisation devoted to the development and dissemination of information which will facilitate the implementation of waste reduction. Before joining WRITAR, Terry was employed as a waste reduction specialist by the Minnesota Technical Assistance Program (MnTAP). Terry has a degree in Technical Communications from the University of Minnesota, and thirteen years' experience in the electroplating industry. His responsibilities have included education, research and evaluation in the areas of waste reduction and management, with a special focus on the metal-finishing industries.

---

**Timothy T. Greene** received his BA in Philosophy at Vanderbilt University, USA, and went on to found a contracting company specialising in the design and construction of unusual residential projects. At the Ingram Barge Company, he developed a PC-based transportation scheduling system and established an environmental, health and safety risk analysis database and reporting system for the company's fleet. He is currently completing applied projects to facilitate environmental management as the Bridgestone/Firestone, Inc. Fellow in Environmental Management at the Vanderbilt Center for Environmental Management Studies. His current academic research explores the complexities of managerial decisions and involves the search for new ways to encourage managers to consider environmental costs in these decisions.

---

**Mark Haveman** is a former Vice President and Director of Manufacturing Programmes for the Waste Reduction Institute. In this capacity he led the Institute's research programmes in the area of manufacturing management and environmental excellence, and worked with manufacturing extension programmes across the US to develop and deliver integrated assistance services.

Mr Haveman is currently as an independent consultant working to build expertise within financial service firms on environmental issues as they pertain to equity research, industry analysis, and socially responsible investing.

---

**Miriam Heller**, Assistant Professor of Industrial Engineering, University of Houston, Texas, researches tools for sustainable development. Recent and ongoing research includes knowledge-based conceptual design for preventing metal-finishing pollution; neural network prediction of potable water demand; life-cycle costing and system dynamics of infrastructure renewal technologies; environmental cost accounting for pollution prevention and remediation management. She was a key investigator on the Amoco and Du Pont case studies reported in World Resources Institute's *Green Ledgers: Case studies in Corporate Environmental Accounting* and EPA's *Environmental Cost Accounting for Chemical and Oil Companies: A Benchmarking Study*. Heller serves on the American Water Works Association (AWWA) Research Foundation Project Advisory Committees for the Utility Business Architecture, AWWA Computer Research Committee, Yale's Environmental Reform: Next Generation Project and the Houston Environmental Foresight Science Subpanel. She is a member of the International Society of Logistics Engineers, the American Institute of Industrial Engineers, IEEE, and other societies. Miriam received her PhD in Environmental Engineering and System Analysis form The John Hopkins University; served as Congressional Science and Engineering Fellow at the Office of Technology Assessment; managed fraud as Assistant Vice President at Citibank Credit Services and was a software consultant and design engineer at Digital Equipment Corporation.

---

**Peter James** is director of the Sustainable Business Centre, which conducts and disseminates research on how organisations can integrate sustainable development into their activities. It has particular expertise in the areas of environmental change management, environmental performance measurement, environmental benchmarking, and environmental accounting. Peter is also a Visiting Professor at Wolverhampton Business School and an associate of Ashridge Management College, where he was employed prior to founding the Sustainable Business Centre. Before joining Ashridge, Peter held positions at Stirling, Warwick and Limerick business schools, latterly as Professor of Management. He has published a number of articles on environmental benchmarking, environmental performance measurement and quality approaches to environment and was a co-author with Claude Fussler of *Driving Eco-Innovation* (FT Pitman, 1996).

---

**David A. Miller** is a Research Analyst at the Tellus Institute, Boston, USA, where he studies environmental accounting and environmental policy issues for the Risk Analysis group. He has investigated public-sector sustainable development accounting practices for the Government of Canada, the effect of waste minimisation on hazardous waste combustion pricing for the US EPA, and capital budgeting practices in the chemicals and printing industries for the state of Illinois, among other projects. He holds a bachelor's degree in Environmental Science and Public Policy from Harvard University.

---

With an unusual background combining corporate finance, economics and business administration, **Tuula Moilanen** has a unique insight into the financial side of environmental decision-making. She was educated in the US, Finland and Switzerland, gaining Masters both in Corporate Finance and Business Administration, the latter from the prestigious International Management Institute (IMI) in Geneva. She initially worked in merchant banking and corporate treasury management, then becoming Finance Director of the Valmet Automation Group and the Nordic Finance Manager of ICI. She is the main author of the IChemE publications, *Financial Evaluations of Environmental Investments* and *Environmental Investment Appraisal*, and the main developer of the Paras model. Her present consultancy work concentrates on internal marketing of environmental issues in large organisations.

**Kaspar Müller** is the founder and partner of Ellipson Ltd, a management consulting company with a special focus on sustainable strategies (financial and environmental issues). Until 1989, he was financial analyst and head of the corporate finance department at a Swiss private bank, and is currently Co-Chairman of the Commission on Accounting of EFFAS (European Federation of Financial Analysts' Societies).

**Joyce A. Ostrosky** is Associate Professor of Accounting at Illinois State University, USA. She holds a PhD in Accountancy from the University of Mississippi and a Master of Science degree in Accounting from Illinois State University. Her work in the field of business and the environment includes an evaluation of innovative chemical supply strategies. Dr Ostrosky's current environmental research focuses on assessing the total cost of ownership for chemicals in manufacturing.

**Janet Ranganathan** is an Associate in the Management Institute for Environment and Business at the World Resources Institute in Washington, DC, where she works on the Sustainable Enterprise Initiative—a major collaborative effort to engage business and others in the transformation toward business Sustainability. Janet has also worked on corporate environmental accounting, business environmental performance measurement, and the role of emerging environmental monitoring technologies for better management and accountability. She is co-author and editor of a WRI publication, *Green Ledgers: Case Studies in Corporate Environmental Accounting*, with Daryl Ditz and Darryl Banks. Her most recent WRI publication is *Measuring Up: Toward a Common Framework for Tracking Corporate Environmental Performance*, which she co-authored with Daryl Ditz. Janet is a member of the President's Council for Sustainable Development's Environmental Management Task Force. Prior to joining WRI, she worked on business and environmental issues in the UK both as a Senior Lecturer at the University of Hertfordshire and in a regulatory capacity with the Department of Environment and Hertfordshire Waste Regulatory Authority. Janet Ranganathan received a BSc (Hons) from Imperial College of Science, Technology and Medicine, London, in 1983, and an MSc with distinction in Environmental Technology from Imperial College in 1990.

**Edward D. Reiskin** is an Associate Scientist at the Tellus Institute, Boston, USA. His work is focused on environmental accounting, pollution prevention and sustainable communities. Prior to joining Tellus, Mr Reiskin spent seven years with the United Technologies Corporation as a Mechanical Engineer and later as a Field Manager. Following his work with UTC, he completed a research engagement with the Global Environment Programme at New York University. He holds a BSc in mechanical engineering from the Massachusetts Institute of Technology and a MBA degree with Distinction from NYU's Stern School of Business.

**Marie-Josée Roy** is a PhD candidate in the Management of Technology programme at École Polytechnique de Montréal. She received her MBA degree from Université Laval in Québec.

**Deborah E. Savage**, a PhD Chemical Engineer, is a Senior Scientist at the Tellus Institute, Boston, USA, where she focuses on work in the areas of pollution prevention, environmental accounting and environmental performance indicators. She has worked with firms in a number of industry sectors including chemicals, automotive, printing, telecommunications and others, and she has conducted numerous training workshops for industry environmental managers, accountants and state technical assistance staff. In addition to her work at Tellus, Dr Savage also teaches a course for Chemical Engineering Seniors at MIT.

**Stefan Schaltegger** is Assistant Professor at the Wirtschaftswissenschaftliches Zentrum WWZ (Centre of Economics and Business Administration) at the University of Basel, Switzerland, specialising in environmental management, business economics and public policy. He was head of the Priority Programme on the Environment (PPE), a co-ordinated project on LCA and eco-controlling by the Swiss

National Science Foundation (1992–95), and is Scientific Advisor to the 'Sustainable Development Fund EcoSar' at the Sarasin Bank. In addition, he is a board member of the Environmental Management Programme at the Herning School of Business Administration and Technology (HIH), Herning, Denmark.

---

**Georg Schroeder** is Director for Quality and Environment for the Sulzer Hydro product division. He is responsible for the promotion of total quality and the introduction and maintenance of quality and environmental management systems at all sites.

---

**David Shields** is Professor of World Business at Thunderbird, the American Graduate School of International Management. He received his PhD from the University of Michigan in 1980, after receiving the BA and MBA degrees from the Ohio State University. He is a CPA, and was affiliated with Coopers & Lybrand in Columbus, Ohio. Previous teaching positions include the University of Florida, the University of Michigan, Rice University, Texas A&M University and the University of Houston. He is on the editorial boards of *Accounting Horizons* and the *Journal of Accounting and Public Policy*. His publications have appeared in numerous journals, including *Accounting, Organizations and Society*; *Auditing: A Journal of Practice & Theory*; the CPA journal, *Journal of Accounting and Public Policy*; *Journal of Business Finance and Accounting*; *Journal of Economic Literature*; *Journal of Marketing*; *Managerial and Decision Economics*; and others, as well as in numerous books and monographs. He is Principal Investigator on several studies on Environment Cost Accounting in the chemical and refining industries, including studies funded by the National Science Foundation, the Gulf Coast Hazardous Substance Research Foundation, and the US Environmental Protection Agency.

---

**Frank L. Waterstraat** is Assistant Professor and Director of the Health Information Management programme at Illinois State University, USA. He holds a Master of Business Administration degree from San Diego State University. His work in the field of business and the environment includes a series of course modules for graduate schools of business, a study of cleaner production decision-making among small businesses, the application of marketing principles to the diffusion of cleaner production technologies, and the evaluation of innovative chemical supply strategies. He is currently completing his doctoral dissertation on the total cost of ownership for computers in education.

---

**Matthias Winter** is Research Associate at the International Institute for Management Development (IMD), working in the field of environmental management. In his research he focuses on environmental accounting and environmental performance indicators, and is running pilot projects with several blue-chip companies.

# Index